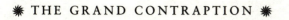 THE GRAND CONTRAPTION

THE GRAND CONTRAPTION
✳ ✳ ✳

The World as Myth, Number and Chance

David Park

PRINCETON UNIVERSITY PRESS

PRINCETON AND OXFORD

Published by Princeton University Press, 41 William Street, Princeton, New Jersey 08540
In the United Kingdom: Princeton University Press, 3 Market Place, Woodstock,
Oxfordshire OX20 1SY

ISBN 0-691-12133-8

This book has been composed in Sabon

Printed in the United States of America

✳ CONTENTS ✳

List of Illustrations   ix

Preface   xiii

ONE
Voices from the Sands   1

  1.1 The Biblical Universe   1
  1.2 Tales from Sumer and Egypt   6
  1.3 Two More Worlds   15
  1.4 Deluge   18
  1.5 The Twisted Axle   22

TWO
Managing the World   26

  2.1 Dramatis Personae   27
  2.2 The Lower Tier   31
  2.3 The Shape of the World   35
  2.4 Fortune-Telling   40
  2.5 The Stars Move Westward   44
  2.6 Guiding Hands   47

THREE
Guesswork   54

  3.1 A Mass of Rock   54
  3.2 Ionians   57
  3.3 Earth, Sun, Moon, and Law   60
  3.4 A World Made of Numbers   62
  3.5 Change and Eternity   68
  3.6 Theories of Matter   72
  3.7 Atoms and the Pursuit of Happiness   78

FOUR
Earth and Heaven   83

    4.1 Law and Nature   83
    4.2 Measuring Months and Years   87
    4.3 Plato's Fantasy   88
    4.4 Aristotle's Optimism   96

FIVE
Beginnings and Endings   107

    5.1 Time and Space   107
    5.2 Creation   111
    5.3 The Universe Recycled   116
    5.4 The End of Everything   120

SIX
Philosophy Continued   126

    6.1 The Stars in Motion   126
    6.2 Stars, Earth, and Numbers   132
    6.3 Omens and Demons   136
    6.4 Remembrance of Things Past   142
    6.5 Motes of Dust   148
    6.6 The Great Design   153

INTERLUDE
The World Map   161

    I.1 Earth and Cosmos   162
    I.2 Explorers and Traders   164
    I.3 The Christian Earth   168
    I.4 Travelers' Tales   172
    I.5 The Age of Exploration   180

SEVEN
Toward a New Astronomy   190

    7.1 The Sun Stands Still   191
    7.2 The Mathematical Plan   197

7.3 The World Observed   204
7.4 A World Invented   210
7.5 Isaac Newton   217

EIGHT
What Is the World Made Of?   *225*

8.1 Atoms Reborn   226
8.2 Transformations   228
8.3 A Theory of Matter   231
8.4 Atoms and Numbers   233
8.5 Ether and the Nature of Light   239

NINE
The Universe Measured   *245*

9.1 Surveyors at Work   245
9.2 The Age of the Earth   251
9.3 The Long Descent of Man   255

TEN
The Exploding Universe   *269*

10.1 The Cosmos in Motion   269
10.2 The Big Bang   272
10.3 What's Out There?   279

ELEVEN
The View from Here   *283*

11.1 Is There Anyone Else?   283
11.2 The Best of All Possible Worlds?   286
11.3 Will It Ever End?   289
11.4 Reflections   292

References and Further Reading   *297*

Bibliography   *311*

Index   *327*

FIGURES

1.1 Moses receives the tablets of the Ten Commandments   4

1.2 A nest of monsters, possibly representing those born
by Tiamat   7

1.3 Mesopotamian fish-god carved on a wall in Nimrud   9

1.4 Mesopotamian world map centered on Babylon   11

1.5 The ziggurat at Ur   20

1.6 Circle of the zodiac with Earth in the center   24

2.1 Satan at the bottom of the pit of Hell   30

2.2 Vase painting of three keres   33

2.3 *Defixio* intended to fix a chariot race   50

3.1 The geometry of numbers   63

3.2 Pythagorean ratios in the plan of the Parthenon   67

3.3 Plan and vaulting of St. Wolfgang's church   68

4.1 The five Platonic solids   94

4.2 Eudoxus's model of the Sun   101

4.3 Artillery theory in the sixteenth century   104

5.1 The heavens will be rolled up like a scroll   122

6.1 Claudius Ptolemaeus (Ptolemy)   129

6.2 The medieval universe as a system of concentric
spheres   131

6.3 Berthold the Black   141

6.4 Ptolemy's five climatic zones on a flat Earth   147

6.5 Whale and calves   155

6.6 Zodiacal man   158

I.1 If the Earth were round   170

I.2 Cosmas's diagram of the civilized world as a
    vaulted box   170

I.3 Cross section of the box   171

I.4 Situation of the box   171

I.5 A seal and a hippopotamus   172

I.6 World map, 1483   175

I.7 Illustrations from tales of Sir John Mandeville   179

I.8 Part of a world map, 1508   184

I.9 Map of America, 1540   185

I.10 From Mercator's world map, 1587   187

7.1 An angel turns the Earth   193

7.2 Copernicus's drawing of the solar system   194

7.3 Copernicus's planetary spheres inside the
    Platonic solids   199

7.4 Drawing an ellipse   202

7.5 Galileo's drawings of the half-moon   206

7.6 Galileo's notes on the appearance of three
    little stars   207

7.7 René Descartes   211

7.8 Stars surrounded by Cartesian vortices   214

8.1 Symbolic representation of the Philosopher's Stone   230

8.2 Lavoisier's table of the elements   235

9.1 Measurement of the Moon's distance   246

9.2 Thomas Wright's model of the Galaxy   247

9.3 William Herschel's map of the Galaxy   250

9.4 Darwin compares the beaks of finches   262

9.5 Last page of the Book of the World   268

10.1 The expanding scale of the universe   280

10.2 The history of the universe in a single diagram   281

TABLES

6.1 The Zodiacal Man   157

6.2 Hours and Planets   157

THIS IS A BOOK about ways in which people in the Western World and a few nearby places have imagined the universe and their place in it. The universe? Anything we can talk about pertains to some part of it, but I want to consider how people saw the whole of it: What is it? Where is it? How did it get here? What lives in it? Is what we see all there is? But it's useless to list questions; there are too many of them. The universe is experience and conjecture, not answers to questions.

This is a book about exploration. It turns out that since the beginning people have imagined our world embedded in a larger, supernatural reality. The eternal procession of the stars in their mysterious groupings have long been taken as signs of this larger world. It lies beyond our senses and can be explored only by inference and invention, but what purpose could stars have if not to communicate some message? Perhaps the sky contains the answers to great questions of good and evil, of power, of the origin of everything. Do our comfort and safety depend on our relations with this unknown world? Weather lurches from extreme to extreme while the motion of the stars (all except five of them) is the same every year. How can that be—is this motion being controlled somewhere, does it control itself, or is the world constituted so that they cannot move in any other way? The world throws us hints: eclipses and other portents that seem to depart from the natural order, but are they perhaps required by it? That has turned out to be an answerable question through it has taken careful measurement and thought.

By now, planning and experience have helped provide most of us with reasonably reliable food, water, and shelter, but in other aspects we live close to some of knowledge's ragged edges: we are warned against polluting the atmosphere but not told just what the penalty will be, and even after all this time, no one can accurately predict an earthquake or an epidemic.

I have tried to write the history of some of what we know, or think we know, or believe about our world. The discussion is chronological and geographical and makes no pretense of completeness. It skips over names that would appear in a proper historical account in order to focus on people chosen because they represent a historical period or a kind of thinking; in this way it resembles three earlier books I have written,[1] but each of those focused on a single topic while here there are many. The

---

[1] Noted in the section References and Further Reading in the back of the book.

idea of the natural world as some kind of mechanism was not the only possible organizing theme for the pages to follow, but it provides boundaries for a work that otherwise would swell to gigantic proportions. Because every model of the world originates in some thinker's historical present, I include some potted history when it seems relevant. Also, the text is perhaps overloaded with quotations from many authors. They are here because they illustrate people's thought processes better than any secondhand account can do: their assumptions, what they considered to be evidence, and, just as important, their personal styles. The story line ends at the present, with current ideas concerning how the universe started and grew. Throughout, the material was chosen because it interested me. I thought you might find it interesting, too.

Thanks to Ilse Browner, Lenore Congdon, Marek Demiansky, Chris Holmes, and David Smith for advice; to Christine Ménard for life-saving help with publication; to Alice Calaprice, who believes that everything that can be said can be said clearly; and to Katharine Park, who lovingly pointed out many errors of judgment but is not responsible for what I did to mend them.

✳ THE GRAND CONTRAPTION ✳

ONE

Voices from the Sands

When my hair was half done
I remembered I love you
I forgot my hair
I ran to find you
Now let me finish
I'll only be a minute.

—Egyptian, before 1200 BCE

IN ANCIENT MESOPOTAMIA or in the valley of the Nile, you could look around at a landscape interrupted only by a farmhouse or a town or a temple and imagine that the world continued like that. What would happen if you just walked in a straight line, day after day? If there is an edge, would it be a wall or more like the edge of a table? If a wall, what lies behind it? If a table, what would you see if you looked down? And if you could fly straight up, would you hit something? And finally, in those days everybody knew that the world is full of supernatural beings that you never saw. Is there a domain somewhere on Earth or perhaps up in the sky where these beings spend their time? Unless they are invisible, they have to be somewhere. This chapter is about how people have imagined the layout of the Earth and the regions around it, how the Earth started, and some things that have happened to change it. It is just a beginning; these questions will be looked at from many angles as the story develops.

1.1 THE BIBLICAL UNIVERSE

In the beginning God created the Heaven and the earth. And the earth was without form, and void; and darkness was upon the face of the deep. And the Spirit of God moved upon the face of the waters. And God said, Let there be light: and there was light. And God saw the light, that it was good: and God divided the light from the darkness. And God called the light Day, and the darkness he called Night. And the evening and the morning were the first day.

And God said, Let there be a firmament in the midst of the waters, and let it divide the waters from the waters. And God made the firmament, and divided the waters which were under the firmament from the waters which were above the firmament: and it was so. And God called the firmament Heaven. And the evening and the morning were the second day.[1]

On the third day God separated land from water beneath the firmament and created the vegetable kingdom; on the fourth: Sun, Moon, and stars; on the fifth: birds and fish; on the sixth: animals, a man, and a woman, and on the seventh day he rested.

The mind cries out for more. Why was all this done, and how? The Bible doesn't say. Was the watery waste there to begin with? Some Judaic commentators imagine a wild chaos that had to be tamed by an act of will before the work of creation could begin. Others say no: first there was nothing, then water, then light. We shall return to the question in chapter 5. What is clear is that the watery waste came first, and next came light and the imposition of a plan where there was no plan. Where do we look if we want to see the plan? Look at a forest or a pond. Each is a collection of plants and animals, insects and creatures too small to see. No appearance of order there, but think how they combine to make a living environment. One creature eats another; later it nourishes a third. A bee on the way to its hive leaves a grain of pollen on a flower whose seed will nourish other creatures. Flowers bloom at different times so that bees will be kept busy. The order of the natural world is more apparent in the way it functions than in the way it looks.

When the Bible begins all we see is water, but underneath is a layer of earth. So that this can appear, God creates a great sky-vault known as the firmament. One might have supposed that its purpose, in that barren land, was to raise the celestial waters above the earth, but the Bible says it was to separate water from water. Even in a dry countryside, every inhabited place is near water, flowing on the surface or a few feet down in a well. You may remember that a few generations later, when the Flood came, water spurted out of the ground to augment the rain. The authors of Genesis saw humanity living in a bubble with water above and below. A *midrash,* or comment, says: "Why did he separate them? Because the upper water is a male, whilst the nether water is female, and when they desired to unite they threatened to destroy the world. The water roared up mountains and hurtled down hills, the male in hot pursuit of the female, until the Holy One, blessed be He, rebuked them. . . . Between the

---

[1] For most of this book, after much reflection, I am using the text of the Oxford Study Bible, entered in the Bibliography as Oxford 1992. Though I miss the cadence of the King James version, its echo is there and the translation is often more exact. But for those who were brought up with them, the stately words in which the King James version brings the universe into being speak with the voice of God, and I use them here and in a few other places.

upper and nether worlds are but three finger breadths, and the vault of the firmament interposes to keep them apart." We shall see that the bubble has a long history.

The vault is *raqia* in Hebrew; the word often refers to a pot hammered out of copper. Later, when Job's young neighbor Elihu reproaches him for his protest against God's injustice, he contrasts God's greatness with that of any mortal: "Can you as he [did] beat out the vault of the skies, hard as a mirror of cast metal?" Beneath the vault moved the Sun, stars, and angels. How high was it? When Moses took Aaron and seventy elders of Israel to meet with the Lord, they walked up Mount Sinai "and they saw the God of Israel. Under his feet there was, as it were, a pavement of sapphire, clear blue as the very heavens." There is an inconsistency here, for later the Lord told Moses, "No man shall see me and live." The conventional explanation is that the seventy elders saw the vault from below. Figure 1.1, from the Regensberg *Pentateuch,* c. 1300, shows Moses receiving the Ten Commandments and handing them to the Elders below. The Lord stands on a vault painted blue. Moses stands on what is perhaps a tree stump and is careful not to look behind him.

Here, then, is a vision of a world that functions with the aid of divine powers. In modern terms it resembles a submarine with windows. From time to time the windows are opened to let water come in and nourish the rivers and soil, and for a few years manna dropped down from Heaven to relieve the Lord's people as they wandered in the desert. Above the vault was water; perhaps on its shore was the City of God where Ezekiel, seated on a sapphire throne, saw a figure resembling a man who spoke to him and told him to prophesy.[2]

That is about all we learn about the geography of Heaven from the canonical scriptures, but later writers filled what they must have perceived as a vacuum. The most complete account, and the source of many conventional ideas of Heaven, is the apocryphal Apocalypse of Paul, a Greek text that probably originated in Egypt about the middle of the third century. It takes off from a passage in 2 Corinthians 12 in which Saint Paul says that he once felt he was "caught up into paradise, and heard unspeakable words, which it is not lawful for a man to utter." Later, his guardian angel shows him the mysteries of Heaven and Hell. After Paul has seen how the souls of the recently dead are sorted out according to their deserts, he is raised to the third and highest heaven. The angel leads him through a gold door above which are inscribed the names of the just—not only their names but their pictures, so that every angel will know them. Paul is greeted by Enoch and Elijah and is shown the premises, but he is forbidden to tell anyone what he has seen or heard.

---

[2] Psalm 104 reminds God, "You have . . . laid the beams of your dwelling on the waters."

Figure 1.1   Moses receives the tablets of the Ten Commandments and hands them to the elders below. From the Regensburg Pentateuch, c. 1300. (Photo © The Israel Museum, Jerusalem.)

Then the angel carries Paul down to a place where they stand on top of the firmament. This is paradise, the second heaven. As one might expect, there is a river of milk and honey, and countless trees bear a variety of fruits. A grape arbor contains ten thousand vines, each one supporting ten thousand thousand bunches and in each of these a thousand single grapes. Paul and the angel walk to the Acherusian Lake, whiter than milk, on which is a golden ship, "and about three thousand angels were singing a hymn before me till I arrived at the City of Christ, all of gold and encircled with twelve walls. . . . And there were twelve gates in the circuit of the city, of great beauty, and four rivers that encircle it." The river of honey is called Pison, that of milk is Euphrates, that of oil is Gion, and the river of wine is Tigris. On their banks he is greeted by several Patriarchs. Except for Tigris, the names of these rivers are the same as those in Genesis 2:10 given to the four rivers that flow from (and not around) the Garden of Eden. In the center of the city, next to a great altar, stands David, holding a psaltery and harp. He sings "Alleluia!" in a voice that fills the city, and the people respond with an alleluia that shakes its foundations.

The story skips over the first heaven, which I suppose is Eden (we will look for it in the Intermission); then Paul is shown the torments of Hell. This much will do, and we can go back to the scriptural account.

• • •

Below the ground and its surface waters, far down, lay Sheol, where the dead pass their silent existence. Classical Judaism is concerned with the fate of the community of Israel more than with that of individuals, but in about the fourth century BCE the author of the Book of Ecclesiastes tells what the dead may expect: "One and the same fate comes to all, just and unjust alike, good and bad. . . . True, the living know that they will die; but the dead know nothing. There is no more reward for them; all memory of them is forgotten." But Ecclesiastes always takes a gloomy view, and later books of the Bible suggest an afterlife. Perhaps two hundred years after Ecclesiastes, the prophet Daniel foretells the end of the world when the Jews will at last be delivered, and "many of those who sleep in the dust of the earth will awake, some to everlasting life and some to the reproach of eternal abhorrence. The wise leaders will shine like the bright vault of Heaven, and those who have guided the people in the true path will be like the stars for ever and ever."

It is easy enough to draw diagrams of the cosmos as the Bible describes it, and many have done so, but as one reads the text it is clear that the writers were not thinking in diagrams. The visions are fragmentary, but no one tells us what lies below Sheol or how a city can be poised above the firmament. These are idle questions; they have nothing to do with the story and are not thought about.[3] In fact, in the *Mishnah,* a collection of

---

[3] For miscellaneous but well-documented speculations concerning details the Bible leaves out, see Graves and Patai 1964, chapter 3.

teachings of early rabbis, one of the rabbis declares: "Whosoever reflects on four things, it were better for him if he had not come into the world— what is above; what is beneath; what is before, and what is after."

Compare this sketch of the cosmos with the actuality of Palestine's stony landscape, and see how much imagination has added to it, all around, above, below. That is the biblical model, but what a model looks like is only part of the story. Much more interesting is how it functions. But before we go further with the miraculous bubble that Genesis describes, we had better look inside some other bubbles that had already formed nearby.

1.2 Tales from Sumer and Egypt

History starts in Sumer and Egypt; before that we had spearpoints, pots, and silence. History is defined as written, and since about 3300 BCE Sumerian and Egyptian texts have survived: Sumerian on clay tablets marked with a stick and then baked, Egyptian scratched into stone. Sumerian was spoken in what is now southern Iraq, where the Tigris and Euphrates flow into the Persian Gulf, and its speakers were the dominant power there for the next thousand years. The oldest Sumerian writings are receipts and tax records, but after a few centuries came libraries and literature. The Sumerians' language is unrelated to any other that is known. It was deciphered because when the Akkadians, a new population, arrived, they produced bilingual inscriptions as well as handbooks for translating Sumerian documents into their own Semitic tongue which can be read. Spoken Sumerian died out, but just as ancient Greek survives among us, the richness of Sumerian literature kept the written language alive for another two millennia.

One broken Sumerian tablet, inscribed about 2100 BCE—a thousand years before the earliest parts of the Bible were written—introduces an epic poem with a preface that tells how the world began:

> After Heaven had been moved away from Earth,
> After Earth had been separated from Heaven,
> After the name of man had been fixed . . .

From the Akkadians a couple of centuries later, we learn how the separation took place. In this version, known as the *Enuma elish,* Apsu and Tiamat are lovers; Apsu is the fresh water under the earth, and his consort Tiamat is the stormy and untamed sea. (The watery waste in Genesis is called *t^ehōm,* related to *Tiamat.*) The story begins:

> When skies above were not yet named
> Nor Earth below pronounced by name,

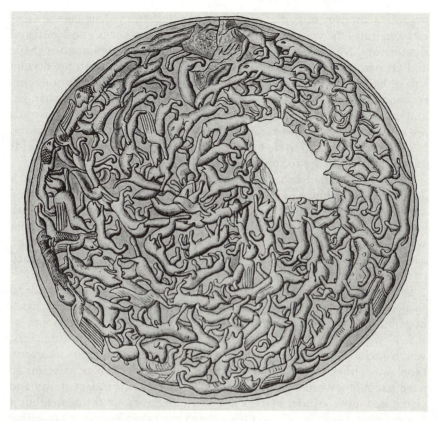

Figure 1.2   A nest of monsters, possibly representing those born by Tiamat (see Dalley 1989, 237), on a bronze dish found in Nineveh. (From Ball [1899], plate 2.)

> Apsu, the first one, their begetter
> And maker Tiamat, who bore them all,
> Had mixed their waters together,
> But had not formed pastures, nor discovered reed-beds;
> When yet no gods were manifest,
> Nor names pronounced, nor destinies decreed,
> Then gods were born within them.

Apsu and Tiamat are fresh and salt water, but they are bodies also, and the sons born of their union are imprisoned inside Tiamat or between the loving parents. After a while, the sons begin to make so much noise that Apsu can't sleep. With the aid of an evil counselor he plots to kill them, but they learn of the plan. There are struggles; Tiamat brings forth an army of dragons and poisonous snakes, perhaps as depicted in figure 1.2. The fighting goes on, and finally Marduk, one of Tiamat's descendants, organizes an army of gods, defeats Tiamat, and, in the language of Genesis, separates the waters from the waters. The story turns into blood and thunder and Marduk kills his mother. "He divided her monstrous shape and created marvels from it. He sliced her in half like a fish for drying: half of her he put up to roof the sky"; then he puts up constellations corresponding to the great gods, makes the Moon and decrees its changes, and creates various geographical features out of her entrails; the details are not pretty. Finally, he executes the evil counselor, and from his blood he creates humankind so that the gods will no longer have to toil in fields and irrigation ditches to support themselves.

Those first human beings were useful for labor but they were rough and barbarous, almost like animals. Then out of the Persian Gulf crawled a strange creature. It had the body of a fish, but attached to it underneath were the head and feet of a man (fig. 1.3). It announced its name as Oannes and began teaching humans the arts of civilization: writing, mathematics, agriculture, how to build a city, how to make laws. Each night it returned to the water, and after it had finished its mission it was seen no more.

Is it strange to portray the sky as a creature's body, or half of one? I suspect nobody, if asked, would have said he or she thought that Tiamat's huge bulk was actually up there. The history of language gives some insight. There is no gender in Sumerian, but in the old Semitic languages, which include Akkadian and the ancestors of modern Hebrew, everything was either masculine or feminine. Proto-Indo-European seems to have had a few words with neuter gender, and its descendants Latin and Greek had more. Modern Greek and German have kept the neuter, but it has dropped from French and Italian and the other Romance languages descended from Latin. English speakers encounter gender as a ridiculous and unnecessary bother, but at the time the ancestral tongues were developing it seems that their speakers regarded everything around them as having some qualities of life and every process as more or less a living process. If this is so, then for them distinctions of gender must have been as essential to talking about a thing as they are for us when we talk about a person. Collectors of the world's myths find that in many of them, as in the Akkadian story, Earth and Sky are portrayed as living creatures.

A thousand years later, in the eighth century BCE, the Greek poet Hesiod tells a similar tale. It begins with the same ordering of the cosmos: "First Chaos came to be, but next wide-bosomed Earth." Mother Earth, Gaia, gives birth to many offspring, including Ouranos the Sky, by whom she later has a trio of violent children, each with fifty heads and a hundred arms. Ouranos hates them and hides them inside their mother, where they cause discomfort that can only be imagined, but with Gaia's help they escape after one of them, Kronos, manages to castrate Ouranos. This sets off a string of complications that can be read about elsewhere. Three cen-

Figure 1.3   Mesopotamian fish-god carved on a wall in Nimrud. (From Layard 1856, 301.)

turies after Hesiod, a fragment from Euripides refers to another version of the story. In a play, *Melanippe* (now lost except for this fragment), a character says that "Sky and Earth were once one and when they were separated they gave birth to all things and brought them forth: trees, birds, beasts, creatures that live in the sea, and the race of men."

Almost within human memory the same story was told in Polynesia. Here the original pair are named Rangi and Papa. Their embrace produced children who stayed trapped between their motionless bodies and unable to move. "There was darkness from the first division of time to the tenth, to the hundredth, to the thousandth." Finally the sons resolved to escape. One by one they tried and failed until the last one, Tane-mahuta, succeeded in prying his parents apart. "It was the fierce thrusting of Tane which tore the Heaven from the Earth, so that they were rent apart, and darkness was made manifest, and so was the light." When this was done the brothers found a multitude of hidden siblings who then spread out to settle the watery domain that had been revealed. Then, as in the other stories, begin the endless quarrels and fighting which seem to be the way all heroes of mythical antiquity spent their time.

I repeat this myth in different versions because their resemblance seems to me astonishing, and there are still other versions. Can it date back to a time when the human population was are still small, before it spread over the world? If it lasted so long, I suspect it told more than just how the world started; it must have explained something about how it *is,* but that revelation is lost to us now.

• • •

Let's pack up the remains of Tiamat and bury them kindly. They have done their job, which was to explain the Beginning. Now, what actually is the world? Babylonian prayers and incantations have allowed archaeologists to reconstruct it. There are different forms. Here is one.

Imagine a great stone disk in the sky, perhaps six or seven hundred miles across. On it, in a palace, lives Anu, the chief Babylonian god. There are smaller accommodations for three hundred of the lesser gods known collectively as *igigi.* One tablet says that during the Flood all the gods took refuge up there. From this level, a stairway leads down to a second disk, which is made of a blue stone that may be lapis lazuli. Here live the rest of the *igigi.* The disk below this one is made of a semitransparent stone through which the blue of the upper one can be seen. On its underside are engraved the stars, and there are tracks on which the planets move. I suppose this disk rotates once a day around the point that today is marked by the North Star but was not then. Around the edge of the disk was a wall with gates through which planets enter and leave. The common term for any one of these three disks, or all three together, is *šamû,* a plural form that is an ancestor of the Hebrew *šamayim,* the heavens. (This multiplicity of levels, or perhaps the multiplicity of heavenly spheres that we shall find in chapter 6, may explain what puzzles every child: Why do people say heavens when there is only one?)

Next in order comes the disk on which we live, and by great good luck the Babylonians have left us a map of some of it. It dates from about 900 BCE, when some of the Genesis account had already been written, and it lives in the British Museum (fig. 1.4). It is centered on the region of the Two Rivers, but imagination has expanded it into a circle. It is bounded by a circle of water labeled *marratu,* ocean; later the Greeks called it

Figure 1.4   Mesopotamian world map centered on Babylon, c. 900 BCE, British Museum no. 92687. The circular band is Oceanus. (© Copyright The British Museum, London.)

Oceanus. The horizontal rectangle above the middle is labeled Babylon, and this makes it likely that the two vertical lines represent the banks of the Euphrates, which used to run though the city. The region into which the lines run is labeled "marsh," which some of it still is today. There seem to have been eight regions beyond *marratu,* indicated here as triangles.

Below the disk we live on are two more, making six in all. The fifth is Apsu, Tiamat's watery husband, and it now belongs to a powerful god named Ea. The sixth and final one is the underworld, home of six hundred gods. It is dry and dusty, and the dead go there. Finally, the six levels are connected by the World Tree, *mesu,* "whose roots reach a hundred leagues to the depths of the underworld, whose crown, in the heavens, leans on the Heaven of [Anu? The name is uncertain]."

I mentioned that there are lands beyond *marratu.* Inscriptions on the map suggest that they are inhabited by strange animals and special people, and one of these places—we don't know which—was visited by Gilgamesh. I digress to say something about him.

The story is told in eleven clay tablets, dating from about 1900 BCE, which identify Gilgamesh as king of Uruk (now Warka, in central Iraq). His name occurs in early king-lists and he may actually have existed, but by the time history catches up with him he has grown into a huge and boisterous man from whose violence no man and from whose lust no woman is safe. In desperation, his subjects beg the gods to create another being equal to Gilgamesh who can restrain him, and from a bit of clay a goddess forms a creature named Enkidu. He has the shape of a man but she puts him down to live with the animals. To bring him into the city the citizens tempt him with a harlot; after he has united with her his animal friends forsake him and he follows her to Uruk. There Enkidu meets Gilgamesh and they wrestle. Gilgamesh finally wins, but they become loving friends and soon go off on an adventure to kill a murderous giant named Humbaba who lives among the cedars of Lebanon. They succeed, but later they antagonize a great goddess, Ishtar. She sends the Bull of Heaven against them, they kill it, and the gods decree that one of them must die. Enkidu dies, but not as heroes die. An injury from the fight with Humbaba slowly drags him down, and Gilgamesh is desolate: not only has he lost his friend, but for the first time he faces his own mortality. One day there will be no more adventures or glory, only age, failing powers, sickness, and finally descent among the strengthless dead.

The Akkadians had a flood legend like the one told in the Bible. Its hero, Utnapishtim, who corresponds to Noah, saved humanity and was rewarded with eternal life in one of the lands beyond the *marratu.* Gilgamesh, desperate, resolves to ask him for the secret of immortality. After a hard journey he reaches an inn at the edge of the water. The innkeeper, a woman named Siduri, tells him that no one has ever crossed over the water, but with her help and that of a supernatural boatman he does so and steps ashore. He finds Utnapishtim and asks his question. Utnapishtim tells him the quest is hopeless:

One day Death comes
For Gilgamesh as for a little man—

No one sees Death
No one sees his face or hears his voice
But cruel Death harvests all mankind.

Nevertheless, as a parting gift he tells Gilgamesh of a plant called Man Becomes Young in Old Age that grows deep under the water, and Gilgamesh manages to gather a bit of it before he starts homeward across the river. As he swims along, a water snake steals the leaves, and then he understands: time will not be cheated, and he must walk the same road as everyone else.

A thousand years later, Homer composed the *Iliad* as the story of ten crucial days toward the end of the Trojan War, and the *Odyssey* is told in a similarly compressed time frame. Aristotle called them both tragedies: they deal with the acts of great men and they inspire pity and fear. The *Gilgamesh* epic is the world's first tragedy.

There are striking parallels between *Gilgamesh* and the Homeric epics. In the Iliad, Patroclus is the bosom companion of Achilles as Enkidu is to Gilgamesh. Both Patroclus and Enkidu die, and each tells his friend of the miserable half-life in the land below. In the *Odyssey*, book 11 tells how Odysseus, like Gilgamesh, crosses Oceanus to consult the soul of a wise man, Tiresias, whose intelligence was exempt from death. He is sent there by Circe, the enchantress whose palace serves as an inn. Like Siduri, the innkeeper at the water's edge, she steps out of legends even more remote to help the traveler. Neither woman is explained; hearers are supposed to recognize them. Gilgamesh persuades a boatman to take him across the river. Odysseus has his own boat, but the Greeks also told of a boatman named Charon, to whom the newly dead give a coin to take them across. Homer wove episodes from ancient epics in another language into his poems.

• • •

The Gilgamesh story takes place somewhere on the map (fig. 1.4), which represents, in whole or in part, the fourth disk from the top of the Babylonian cosmos. Egyptian texts sometimes refer to the Mediterranean Sea as the "Great Encircler," which shows how little they knew about it, and perhaps Babylonians also thought of it that way. The image of the world as inhabited land encircled by sea has a long history, and it persisted even after the time of Plato, when people had begun to talk seriously about a sphere.

Homer's Earth is essentially that of the Babylonian map but centered in Greece. In the *Iliad*, he describes a shield made for Achilles by the smith Hephaestus. It is embossed with scenes of war and peace, life as it is and as it ought to be. The description is long, and last comes Oceanus, "the mighty Ocean River," which runs around the shield's edge.

Hesiod, writing at about the same time, describes a simplified version of the Babylonian cosmos. In his *Theogony* he tells how the Olympians defeated the Titans who had risen against them. They were bound in chains and thrown down to misty Tartarus,

> as far beneath the earth as heaven is above earth; for so far is it from Earth to Tartarus. For a brazen anvil falling down from heaven nine nights and days would reach the earth upon the tenth: and again, a brazen anvil falling from Earth nine nights and days would reach Tartarus upon the tenth. Round it runs a fence of bronze, and night spreads in triple line all about it . . . while above it grow the roots of the Earth and unfruitful sea.[4]

Thus: a cosmos of three disks (I assume they are disks), far apart and wide enough for the middle one to hold the known world, with separate domains for gods, mankind (living and dead), and immortals who had disturbed the peace. Nothing could be simpler.

The Egyptians had no fixed cosmology or story of creation. Anyone composing a poem or a prayer was free to explain the motion of the Sun or Moon as desired, but the world was generally imagined as a flat region with mountains at each side and bisected by a river; at the four corners, high peaks supported a flat canopy. Each creation story starts with water, personified as a god named Nun. The story as told in Hermopolis tells how a mound of wet earth rose out of this water, perhaps suggested by what happens as the Nile's flood abates each year. In the water around the mound float deities named after boundlessness, mystery, chaos, darkness, and infinity. The mound is deified as a god whose name varies, and life begins with him. Another story, celebrated at Heliopolis in the third millennium BCE, starts with a solitary god Atun who, sitting on the mound, spills his own semen. From this arise a boy, Shu, and a girl, Tefnut. Tefnut gives birth to Geb and Nut, a loving brother and sister who couple in motionless embrace in the water (here the story becomes the same as the one told in Babylon) until their father, Shu, separates them by force, then raises Nut to form the sky with open air below. In every direction is a watery chaos which, if Shu allowed, would pour in and put an end to the world, but he never lets it happen. Egyptians may have felt as the inhabitants of the polder country in Holland do today: the ocean level is far above their heads. They know that the seawalls that protect them have broken in the past and may break again.

• • •

---

[4] Hesiod 1914, 131.

Every human society has myths that tend to bring its members together. Most Americans believe that all men are created equal. Even though it is hard to say what the words mean, and though what they mean to us today is probably not what they meant to Jefferson when he wrote them, most of us feel that this is a good thing to believe and we are better because we believe it. Many who do not take the myths of religion literally nevertheless draw peace and strength from a service of worship. Myth is a way of talking that brings people together by transcending differences of opinion; it is a language, not a story, and terms such as "true" and "false," which may apply to a story, do not apply to a language. The story of Adam and Eve, whether or not one takes it literally, has implications that concern men and women, knowledge and ignorance, obedience and disobedience, but its overarching message is that humanity is one. A myth that situated the Egyptian people in a bubble of air entirely surrounded by water held back by the kindness of a single god may have helped hold a diverse society together. Statements of fact end and fall silent, but myth echoes in the memory of generations. The languages of Egypt and Mesopotamia are related, and their populations are probably related, too. We should not be surprised that the founding myths of these two ancient peoples told similar stories of creation.

None of these stories tells how or why the world was created. In all of them—except perhaps the biblical one—something preexists: the watery chaos or a divine pair that represents it. The true origin of everything is lost in darkness.

1.3 TWO MORE WORLDS

For those who wondered what is actually up there, the Bible gives an answer that could be taken literally: above the flat Earth, the Sun and stars; above them, the firmament; above that, fresh water, and somewhere, far above, the City of God. The Egyptians had water above as well as below, while the Babylonians and early Greeks had their disks. A book as kaleidoscopic as this one cannot be ordered in strict chronology, so here I will mention two more models of the world, probably of later origin and still current in our own time. The first, from Tibetan Buddhism, inspires devotion and meditation today; the other is Native American, one version chosen from among many because we know quite a lot about it.

Traditional Hindu cosmology in the *Brahmana Vyasa* is similar to the Buddhist version but much more elaborate, and it varies greatly from one account to another. This Buddhist version at least has the merit of being possible to summarize. It is familiar, in one form or another, to a substantial fraction of the world's people.

The story begins with particles of matter floating in space, the wreckage of whatever catastrophe destroyed a previous world, for time in Buddhism has neither beginning nor end. Particles drift together to form water, and rain pours down. Wind drives water into a circular flow that creates a great golden disk; then other materials pile up on it to build a world. The disk itself is deep under water. At its center is Mount Meru, the world's axis. It is 160,000 leagues high with half of it above water.[5] The mountain's four sides are of gold, blue beryl, crystal, and ruby, and it is surrounded by four square terraces on which live the Great Kings. On top of the mountain, and extending above it, is the Heaven of the Thirty-three Gods with palaces, parks, playing fields—every amenity for the good life. Below the water level is an elaborate arrangement of hells where evil is punished.

Surrounding Mount Meru are seven mountain ranges, each in the form of a square, with seven seas in between. The names of the seven ranges derive from the way certain of the mountains look: Yoke, Plow, Acacia Forest, Pleasing-to-the-Eye, Horse's Ear, Bent; the outermost one, of solid iron, is called the Rim. The seas between them are of milk, wine, *ghee* (clarified butter), molasses, and other liquids, and in the outermost and largest sea, which is salt water, contains several islands and four continents. We live the farthest south on these continents, called Jambu after a tree that grows there. The other continents are inhabited too, but Jambu is the only one in which the balance of suffering and ease encourages the meditation that finally leads to release from the cycle of birth and rebirth. It is on Jambu that the Buddha achieved nirvana.

The geography of Jambu suggests India. In the north are snowy mountains enclosing a cold lake from which flow four great rivers. There is indeed such a lake in the Tibetan Himalaya. It is called Manasarovar, and pilgrims journey for months to bathe in its icy waters. Nearby is Mount Kailash, considered to be Jambu's embodiment of Mount Meru, and it takes a pilgrim several days to walk around it. Four of the great rivers of Asia—the Ganges, Indus, Sutlej, and Brahmaputra—do in fact originate nearby. As to the scale of this entire world-picture, given that one of four continents in one of the seven concentric seas is the size of India, we can make our own estimates, but nobody seems to worry about the scale.

Western commentators have reported this contraption with much merriment as a version invented by simple people who did not know the facts. These critics seem to forget the firmament that figures in their own scriptures; they take allegory as a geography lesson. The map of this

---

[5] The league is variously and always vaguely defined, something like four or eight earshots, an earshot being defined as the distance beyond which one has the impression of being alone—perhaps about a mile.

world directs the travel of the mind, not the feet, for Buddhists have always considered the world to be the product of mind. Mind would not have created the lumpy and odd-shaped land masses that our atlas shows us. The worldview in which most of us were tacitly brought up holds that we are spectators and actors in a universe that would be pretty much the same without us. In the East, that possibility is almost unimaginable. The Buddhist universe (which is certainly not the Buddha's invention) is a mental creation that embodies order and justice in a physical world. Along with suffering and hard work it contains good food and amusement, leisure and joy. It encourages a comparatively low opinion of what we see as we look around us. To this day, there are many men in India who, as they get old, abandon the life of a householder and take to road or forest or monastery in search of the eternal.

• • •

The Babylonian-Greek model survives in smaller versions for smaller societies. The Chumash were a Native American nation of about 15,000 who lived in several groups in the region of present Santa Barbara, California. They no longer exist as a nation, but the last speaker of their language, who died only in 1965, helped anthropologists to record Chumash customs and beliefs. The people lived on the coast and on nearby islands in tribal areas, some of which were small, only a few miles wide. Each was a world of its own. Its cosmos, no larger than itself, consisted of three (occasionally five) disks, suspended one above another with spaces in between. The top level was the world of celestial beings, the bottom one held dangerous spirits, and the Chumash lived on the middle one, except for a region in the center called *'antap* that was full of spirits. In the upper world, in a crystal house with two wives named Morning and Evening and two daughters, lived Sun. Each day he walked across the sky on a tightrope, carrying his torch and noticing what everyone did. Occasionally, at the end of the day, he snatched a Chumash off the ground to be carried home for dinner.

If we live north of the equator, the Sun shines longest in June; after June 21, if we have a clear view of the point where the Sun rises or sets, we can see it moving farther south every day as the days get shorter. On about December 21 the southward movement is supposed to come to a stop and turn back toward the north. What if it doesn't? What if the Sun keeps going farther south and finally disappears altogether? Today this isn't a problem for most people, but some traditional societies still do the best they can to make sure it won't happen. Ceremonies with this mission have been studied among the Chumash as well as the Hopi, the Zuni, and other Pueblo Indian groups that until recently were not much affected by the opinions of the world beyond their fields.[6]

Sun's tightrope proclaimed the instability of the universe. Every change in the natural world—the alternations of wet and dry and of hot and cold, the northward and southward drift of the Sun and Moon—was understood not as part of a cycle but as a reminder that the world balances on the edge of disaster. It barely survives from year to year, and shamans and other cult leaders had to work hard to keep it going. I remarked that most of us live in a universe that would be pretty much the same if we did not exist. For the Chumash this was not so; humanity was enmeshed in every part of it.

In contrast with this precarious existence, the island of Jammu will suffer troubles but apparently no major disasters until, at the end of a very long time, it falls apart. The Chumash lived in constant danger of boiling or freezing or being eaten. The older civilizations, also, lived in danger, for they lived surrounded by powers stronger than themselves and were usually the ones that suffered when something went wrong.

1.4 Deluge

In the 1850s there lived in London an engraver of banknotes named George Smith, a man with an immense beard who spent his lunch hours in the British Museum because he was interested in cuneiform tablets from Mesopotamia. Scholars were beginning to read them; he picked up the language and deciphered a few himself. A staff member noticed him at work and offered him a small job in the museum. While sorting some of the thousands of tablets that had been carried back from the library of King Ashurbanipal in Nineveh, George Smith saw what seemed to be part of a story of the Flood, but the tablet was broken and incomplete. He announced his discovery in 1862. This was a time when Darwin's ideas were in the air and geological discoveries were putting scriptural chronology under severe strain. The public was much interested in Smith's claim, and the *Daily Express* offered to send him to Nineveh to see if he could find the rest of the tablet. This he actually did, and it turned out to be Tablet XI of the Akkadian Gilgamesh epic which, as we have seen, dates from about 1900 BCE, a thousand years before the oldest bits of Genesis were written down. He returned to Mesopotamia to dig again but died of dysentery in Aleppo at the age of thirty-six.

---

[6] The Chinese, who until the eighteenth century lived on a square flat earth covered by a round rotating dome, told of an ancient emperor Yao who sent out four magicians twice a year for the same purpose.

In Tablet XI, after Gilgamesh has crossed *marratu* and found Utnapishtim, the old king tells him how he became immortal. One day a minor god had whispered to him through the reed wall of his house that his chief, Enlil, was angry with humanity, that Utnapishtim must tear down his house and give up his possessions and build a great square barge an acre in size. He built it and loaded it with his family, all kinds of craftsmen, his possessions, and his livestock. Wind and rain started, land was washed away, even the gods were afraid and cowered like dogs. On the seventh day the storm abated, the Sun came out, and the barge fetched up on top of a mountain. After a week Utnapishtim released a dove, but she found no place to rest and returned. Next day he sent out a swallow, but she too came back. Next, a raven, which did not return. Then he set out an offering of sweet-smelling herbs for the gods, and the story goes on its way.

What was the reason for such a disaster? The Gilgamesh tablet doesn't say, but a fragment of an Old Babylonian version explains: humans were making too much noise. Consider that the gods did not live in the sky or on the tops of mountains because there was no food or drink up there. At first they had worked in the fields, but later they had created mankind, "the black-headed ones,"[7] to work for them. When cities were built, each contained a ziggurat, a home for the gods in the form of a huge temple with a tower in the middle. Figure 1.5 shows the partially restored ziggurat in Ur. The walls are about fifty feet high. In such a ziggurat, Enlil and his friends had settled down, but their black-headed employees gave such wild parties that Enlil couldn't sleep. He tried to quiet them with a drought, then with a pestilence, but the noise went on, so over the objections of the other gods, who valued cheap labor, he decided to drown them all. He would have succeeded if someone hadn't warned Utnapishtim and Utnapishtim hadn't done as he was told. In gratitude, the other gods made him immortal.

Later, on the other side of what is now Syria, the story was told again, but now the savior's name is Noah. The flood begins and ends much as in the Gilgamesh tablet, and not only were the windows of Heaven opened but also "all the springs of the great deep burst out." And there is even a connection with the Old Babylonian fragment: the only specific charge that Noah's God brings against humanity is its violence.

There is also a flood story told in Greece, though it is hard to see how it could have occurred in that dry and mountainous landscape. A collector of myths named Apollodorus tells how Zeus once took the form of a man and walked around to inspect his property. One evening a local

---

[7] Prof. Eric Henry tells me that the same phrase occurs in old Chinese texts. Is it possible that Chinese and Mesopotamian gods, like some of the Olympians, were fair-haired?

Figure 1.5   The ziggurat at Ur. (By permission of Hirmer Verlag München.)

tyrant named Lycaon invited him in and served him human flesh for dinner. Naturally, Zeus struck the man dead, but on further reflection he decided that was not nearly enough; he would wipe out the rest of humanity with a flood. As always there was an exception. Zeus chose a blameless old man named Deucalion and told him to build and provision a boat so that he and his wife Pyrrha could save themselves. They did, and then, by a procedure involving stones, they repopulated the Earth. Ovid, in *Metamorphoses,* tells the same story with the added detail that water poured up out of the earth to increase the deluge.

I mention these groundwaters because they refer to humanity's perilous existence on land completely surrounded by water, Apsu below and Tiamat above. Even as late as our own era, some time about the year 50, the Roman senator and dramatist Lucius Annaeus Seneca (c. 4 BCE–65 CE) wrote that we live in a bubble and, apparently, that its destruction is scheduled. "There are immense lakes hidden underground, a vast quantity of hidden sea. . . . For a long time they have been restrained, but they will conquer." In four chapters of his *Natural Questions,* he describes the horrors of universal destruction borrowing heavily from Ovid. Rivers and seas rise, solid earth turns into water, houses totter and fall, forests float away, cities are dragged to their ruin. "Nature has put water everywhere so she can attack us from all sides when she chooses." He describes all this with such zest that he surely thought he had a long time to live, but he was mistaken. Learning that Emperor Nero had ordered him extinguished, and having no other choice, he opened his own veins.

• • •

In 1929 an Anglo-American expedition explored a cemetery outside an old city at the lower end of the Euphrates that the Bible calls Ur of the Chaldees. As they dug they found a deposit, many feet deep, of pure, clean, water-laid mud. Above and below it were samples of the pottery and stone tools of a neolithic culture known as al' Ubaid. Evidently at some time not very far from 5000 BCE there was a severe flood in the area of Ur. It did not wipe out the population nor did it leave significant traces elsewhere, but perhaps, just perhaps, archaeologists thought, this was the origin of the ancient story.

More recently, another catastrophe has been studied. Geological evidence suggests that where the Black Sea now is, there was once a deep valley with a fresh-water lake in the bottom. At a time datable by radiocarbon at 5650 BCE, its fresh-water shellfish were abruptly covered over by salt-water forms. Study of the geology of the Bosphorus Strait, which now connects the Black Sea with the Mediterranean, suggests that the strait was originally only an inlet, and that as one of the minor ice ages ended, melting glaciers raised the level of the Mediterranean until it began to pour into the valley below. In a year or so, a large area of what had been lakeside territory would have been flooded with salt water.[8] If the area was inhabited, people must have fled in every direction, and some would have turned up in Mesopotamia. The flood is probable; the people are conjectural, and underwater archaeology has found no sign of human occupation.

But was there a historical deluge at all? Some mythical events—the Trojan War, for example—are probably rooted in history. Others, like the story of Jason and Medea, probably are not. If we look just at the biblical flood the question is hard to decide, but cultures everywhere have flood myths and in many, perhaps most places, people have known great floods. Anyone who has experienced one can imagine a deluge. Do we need a real one?

• • •

Imagine what it was like to live before people looked for the physical and biological causes of weather and climate and illness. Rejoicing in rich harvests and the pleasures of life and love, or beset by droughts, floods,

---

[8] The Roman encyclopedist Pliny the Elder, writing in the first century CE, deduced from the local geography that this must have happened. At the beginning of book 6 of his *Natural History*, he writes: "That this event occurred against the will of the Earth is shown by the number of narrows, and by the smallness of the gaps left by nature's resistance."

lightning, and disease, they lived surrounded by powers from which they could not defend themselves, powers that without warning reached into every household to carry off a spouse, a parent, or a child. To turn them aside, there had to be some way to protest, to bargain and flatter and persuade. Thus the operations of nature received names. They were addressed like people and some became gods. But there had to be more to it than that. What if one addressed a god and he simply said no? A god cannot be threatened; there must be an inducement, something he needs or wants that could be withheld. One thing he needs is nourishment. Suppose a god would grow thin without the labor of the black-headed ones or the smoke that rises from a sacrificial altar. Now a sort of contract exists. In the Babylonian story Enlil broke it when he sent the flood, but the other gods were appalled and warned Utnapishtim in time to save what had to be saved. Afterward, Utnapishtim's offering of herbs showed that for him the contract still held.

What happened in the Bible is much the same. There also, when God saw the ark's people climb down the ladder and begin to look around, he lit a rainbow over the sodden landscape as a promise that he would not do it again. By the time the Greeks came along, their gods had their own sources of food and were economically independent of earthlings, but an hors d'oeuvre of sacrificial smoke was pleasant to their senses and flattering to their self-esteem, and so the contract, though weaker in its terms and often broken, continued in effect.

In this way humans gained a degree of influence over the immortals; this meant that as long as nothing went seriously wrong they exerted some control over their own destinies. Chapter 2 will recount some of the ways this was done. Occasionally, though, even gods lost control.

1.5 The Twisted Axle

After several generations, Mesopotamia and its black-headed folk recovered from the Flood and life went on as before, but another time the results of a disaster were permanent. The story of Phaethon's ride is well known. I summarize it here as Plato tells it in *Timaeus;* it is told much more fully in Ovid's *Metamorphoses.*

The wise Solon, traveling in Egypt and studying to become still wiser, met an old priest who was not impressed by what passed for wisdom in Greece:

> I mean to say, he replied, that in mind you are all young; there is no old opinion handed down among you by ancient tradition, nor any science which is hoary with age. And I will tell you why. There have been, and will be again, many destructions of mankind arising out of many causes. . . . There is a story which

even you have preserved, that once upon a time Phaethon, the son of Helios, having yoked the steeds in his father's chariot, because he was not able to drive them in the path of his father, burned up all that was upon the earth, and was himself destroyed by a thunderbolt. Now this has the form of a myth, but really signifies a declination of the bodies moving in the heavens around the earth and a great conflagration of things upon the earth which recurs after long intervals.

What is this declination? In Egypt, almost a thousand years after Plato, Nonnus of Panopolis assembled a collection of myths in a work called the *Dionysiaca*. At that late time he no longer takes them seriously, but occasionally he mentions a detail that has otherwise dropped from memory. At the climax of Phaethon's ride, he writes:

There was tumult in the sky shaking the joints of the immovable universe: the very axle bent which runs through the middle of the revolving heavens. Lybian Atlas himself could hardly support the rolling firmament of stars, as he rested on his knees with bowed back under this greater burden.

Plato and Nonnos knew the Earth is round, so even though the people who first told the myth probably didn't know it, it is easiest to imagine ourselves sitting somewhere on a motionless spherical Earth looking out at the Sun and, at night, the stars. In figure 1.6 the diagram is oriented so that the north and south poles are straight up and down. Note that these poles, labeled by stars, are in the sky, not on the Earth. In the old cosmologies and pretty much up to the seventeenth century, it was the framework of the sky that rotated so that the sky had poles and not the Earth. The same is true of the equator, not labeled, and as late as the eighteenth century, navigators spoke of sailing *under* the equator. The imaginary plane that contains the equator is called the *equatorial*. When we hold the framework still for a moment and let the Earth rotate, what we call the Earth's poles are the only points on its surface that do not move.

As they travel through the sky, the Sun, Moon, and planets follow paths that go through or near the twelve constellations of the zodiac. In the figure, these paths lie in an inclined circular band marked with signs that denote the constellations. An imaginary plane through this band is called the *ecliptic*. The two planes, equatorial and ecliptic, are about 23 degrees out of alignment. If we hold the Earth still, the Sun moves steadily around the ecliptic, a complete circuit in 365¼ days, while the planets move at uneven speeds, sometimes even reversing their motion for a while. Where they will be at a given date can only be predicted by calculation.

The whole arrangement would have been neater, and more as a god might have designed it, if the ecliptic were not tilted. Then day and night would be equally long and there would be no seasons. As it is, when the Sun is in Cancer it shines longer on the Northern Hemisphere and brings

Figure 1.6   Earth is in the center, the poles of the celestial sphere are above and below, and the circle of the zodiac is inclined with respect to the equator. (From Regiomontanus 1496; courtesy of the Chapin Library, Williams College.)

summer while the Southern Hemisphere has winter; when the Sun is in Capricorn, the situation is reversed. Day and night are equally long at the equinoxes, the two points where the ecliptic crosses the equatorial. Our arrangement of leap years is designed to keep March 21 as close as possible to the vernal equinox.

Not only is the ecliptic tilted with respect to the poles; it also wobbles, like a top getting ready to settle down but much more slowly: a complete wobble takes almost 26,000 years. This motion is called the *precession of the equinoxes*. Horoscopes published in newspapers and magazines tell us that at the vernal equinox the Sun is just entering Aries. Actually, for about two thousand years it has been in Pices. Precession has carried the zodiac backward through almost one full sign since the Aries convention was established, but since this error makes no difference to anything it need not be corrected. In a few more centuries the equinox will cross over into Aquarius, inaugurating the New Age of that name.

In Greek mythology, the tilt of the equinox is the result of Phaethon's disaster. In Book 10 of *Paradise Lost* John Milton gives a different reason, the Fall of Man, but the result is the same. As punishment for the original sin, God sentences humankind to live in a less comfortable world. Instead of Eden's warm and steady climate there will now be seasons that oscillate between hot and cold, wet and dry, and to accomplish this:

> Some say he bid his angels turn ascanse
> The Poles of Earth, twice ten degrees and more
> From the sun's Axle; they with labour push'd
> Oblique the Centric Globe.

Phaethon's error or Adam's sin, something went terribly wrong, so now we have summer and winter.

• • •

This was a meager sketch of some old versions of the Grand Contraption. Next we want to know, how does it work? The mind craves explanation. How did it come to exist, how did animals and mankind get here, why does the world attack us with wind and water, drought and disease? Prayers and hymns and rituals embody answers to these questions. They represent men and women as involved with characters who are like them but more than human, who can do what men and women do but also what they wish they could do: be invisible, fly through the air, change their forms, control the forces of nature. They never get older, they know everything we do; they have to be dealt with. The next chapter tells who some of them were, what they did, how they spoke to humanity, and finally, how humanity used them for its own benefit.

TWO

Managing the World

> It's better to go along with the stories about gods than give in to
> what the natural philosophers call Fate. If there are gods there
> is some hope of appeasing them with a little worship; if not, we
> are ruled by something that no one can appease.
>
> —Epicurus (c. 300 BCE)

IN 1890 the literary critic Goldsworthy Lowes Dickinson, a member of
London's Bloomsbury coterie, published a book called *The Greek View
of Life*. As background for what he planned to say about Greek religion,
he imagined the predicament of "primitive man" faced with the power of
nature:

> Naked, homeless, weaponless, he is at the mercy, every hour, of this im-
> mense and incalculable Something so alien and so hostile to himself. As fire
> it burns, as water it drowns, as tempest it harries and destroys; benignant
> it may be at times, in warm sunshine and calm, but the kindness is brief
> and treacherous. Anyhow, whatever its mood, it has to be met and dealt
> with. . . . What is it then, this persistent, obscure, unnameable Thing? What
> is it? The question haunts the mind; it will not be put aside; and the
> Greek at last, like other men under similar conditions, only with a lucidity
> and precision peculiar to himself, makes the reply, "It is something like
> myself."

Dickinson's education set the Greeks in the center of the stage and swept
everyone else into the wings, but long before Greek was spoken in Hellas,
other people all over the world had arrived at the same conclusion, multiplied by thousands. Beneath their feet, on the mountains above them,
thick in the air about them, were living beings. They had their own problems and agendas, but they were something like us. Dickinson is trying to
touch the mind of primitive man; thousands of years later, Epicurus's
remark higher on this page explains why people assigned supernatural
causes to natural events, and in our own time newspaper polls remind us
how little has changed since then.

2.1 DRAMATIS PERSONAE

To distinguish even a few of the supernatural beings who have crowded into the universe, let us start with some names and habitats, noting a few of the things these beings have done and perhaps still do.

Gods and Goddesses

The chief Babylonian god was known as Enlil by the Sumerians and Marduk by the Akkadians. Around him was a court of gods and goddesses with various powers, looking down upon the swarm of *igigi*. They in turn looked down on the black-headed ones, who appealed to members of the divine court when they needed help. Having created humanity to be their slaves, the gods were responsible for their welfare, and since they lived in temple compounds built for them in the middle of the city, they knew everything that was going on. Helping, punishing, or ignoring, they were a continual presence.

Since Egyptian gods are often represented with animal heads they seem to have sprung from a form of nature worship, but in a culture in which most people were illiterate the heads also helped to identify them. Chepre (a dung beetle) created himself and was afterward involved in creating the world; Horus (a hawk) was a Sun-god; Hathor (often a cow) was the goddess of love and beauty; and there were many others. Each had a place in the funerals of important people and in ceremonies which, every day, were necessary to keep the world going.

The principal Greek gods arrived with Greek-speaking invaders from the north in the second millennium BCE. Their leader was Zeus, an Indo-European sky god known in India as Dyaus Pater, Sky-Father, and in the Roman empire as Dies Pater, which became Jupiter. They settled on Mount Olympus where they lived in the style of Norse legends: drinking, fighting, and seducing the women of the land below. Zeus was the strongest, keeping order among the members of his court, while among men he enforced justice and the rules of hospitality. Homer and Hesiod make clear, however, that his authority was not absolute—in the war over Troy, for example, he favored the Trojan side, but the Greeks won.

In ancient Greece, which had dozens of gods, in modern India where there are still thousands, thoughtful people sense that the world has a unity that such multitudes do not express. Greeks spoke of *ho Theos*, the God; Hindus speak of *Brahman*. In the Land of Israel, the Lord said, "You must have no other god besides me." This doesn't mean that there were no other gods, for the Bible mentions several of them. The emphasis was on *you*. One has the impression that the early Jews lived surrounded by strange gods who pulled at their sleeves and demanded worship. The Book of Judges tells how as the Israelites spread out to occupy the newly conquered land of Canaan, "they forsook the Lord, their fathers' God who had brought them out of Egypt, and went after other gods, the gods of the peoples among whom they lived. . . . In his anger the Lord made them the prey of bands of raiders and plunderers." In Psalm 82, "God takes his place in the court of Heaven to pronounce judgment among the gods." Monotheism came later, after the days of Moses, with the revelation of a single God who first created the world and all it contains, and continued as the ruler and judge of mankind.

Angels

Angels (the word comes from Greek *angelos,* messenger) are the agents through whom divine power carries out its plan and guides the world toward its appointed conclusion. An example that shows the role of God as their overseer occurs in the Second Book of Samuel when he orders a pestilence upon Israel. An angel engineers the pestilence and gets it going; but at the moment when it is about to strike Jerusalem, the Lord tells him to stop. Enough is enough. Later, a restraining order does not come and another angel kills 185,000 Assyrian invaders overnight.

The Bible refers to the angels as Sabaoth, the Heavenly Host, and as Sons of God. They appear as men in several episodes of the Old Testament, and one wrestled all night with Jacob (though Jacob was sure it was God who was his opponent). Their leader is the Archangel Michael; Gabriel is the chief messenger, and two more archangels are mentioned in the noncanonical scriptures: Raphael, God's helper; and Uriel, the watcher. The Sons of God were present at the Creation—in fact they themselves had just been created—and they shouted for joy. Christians say that later they announced the birth of Jesus. How big an army were they? In Daniel's vision (7:10), "thousands upon thousands served him, and myriads upon myriads were in attendance." A myriad is ten thousand. The vaguely feminine figures of angels that ornament trees and cards at Christmas are the visible remains of that army of tough, competent, and hard-working helpers.

But all did not go smoothly among the Host, for it seems that some of them defied their Creator. The Bible is vague on the subject. In chapter 6 of Genesis we read: "When men began to increase on Earth and daughters were born to them, the sons of God saw how beautiful the daughters of men were and took wives from among those that pleased them. . . . It was then, and later too, that the Nephilim appeared on earth—when the divine beings cohabited with the daughters of man, who bore them offspring. They were the heroes of old, the men of renown." They are called Nephilim, and later we learn what happened to them. When Moses led his people out of Egypt to occupy the promised land of Canaan, he sent scouts to survey it. They returned to say that Nephilim were there and could not possibly be defeated, but it turned out that only three were left in the region and the Israelites prevailed. But this was not the end of the Nephilim, for a few were said to survive in Gaza, Gath, and Ashdod. Goliath came from Gath.

The Genesis account continues: "The Lord saw how great was man's wickedness on earth, and how every plan devised by his mind was nothing but evil all the time. And the Lord regretted that he had made man on earth, and his heart was saddened." But what is the connection between the descent of angels and man's wickedness? Some time around the year one, an apocalyptic treatise called the Book of Enoch was compiled. There are several versions and I mention only one. It says that angels, "sons of the gods," were sent to watch over humans but were not to mix with them. As happens so often in history, sex took control. About two hundred angels came down and moved in with the human colonists. An angel named Azaz'el taught them spells and astrology and magical cures. He taught them warfare and cosmetics and sexual perversions; all of these were things they were not supposed to know. Adam and Eve had sinned and been punished for their disobedience, but they were not evil. It was Azaz'el who introduced evil into a good world.

The Nephilim born of these unions were immense creatures, and when they had eaten up the available food they ate people. The survivors cried to Heaven for help. Archangels Michael and Gabriel heard them and told the Most High what was happening, and on learning of this situation God decided to put an end to his experiment. He told Raphael, "Bind Azaz'el hand and foot and throw him into the darkness." He decreed other punishments and prepared the Flood.

There is also a slightly later work called the Book of Giants, but only scraps of it are legible. They repeat and embellish the themes of Enoch, and they say, two thousand years after the Sumerian tablets, that one of the Nephilim was named Gilgamesh and another was Humbaba.[1] There is no reason to think that *Enoch* and *Giants* represent material as old as the Old Testament, but it is hard to read the sparse Genesis account of early days without wanting to know a little more. History abhors a vacuum, and many stories must have been told to fill it.

---

[1] This diffusion of the Babylonian legend is not mysterious. In the seventh century, Assyrians invaded and conquered Palestine and the Nile Valley (they already had Syria), creating a web of trade, travel, and intermarriage. In Palestine, archaeologists have found tablets containing bits of the Gilgamesh epic.

Figure 2.1   Satan at the bottom of the pit of Hell, from Gustave Doré's illustration of Canto 34 of Dante's *Divine Comedy.* Dante and Virgil are in the upper right. (From Doré's illustrations for the *Inferno.*)

By New Testament times, the fall of the angels was common knowledge. Saint Jude refers to "those angels who were not content to maintain the dominion assigned to them, but abandoned their dwelling-place; God is holding them, bound in darkness with everlasting chains, for judgment on the great day." In chapter 20 of the Book of Revelations, "the old serpent, which is the Devil and Satan," has been bound with chains and thrown into the abyss (fig. 2.1), though for only a thousand years. "When the thousand years are ended Satan will be let loose from his prison, and he will come to seduce the nations in the four quarters of the earth." He will raise a great army but will finally be "defeated and flung into a lake of fire and sulfur . . . to be tormented day and night forever." It is not clear when the Devil was first imprisoned and, therefore, when the thousand years ended or will end. People at many times in history have felt that the end times were close upon them. As the year 1000 approached, both anxiety and hopeful expectation abounded, and another millennium has recently released the same emotions. More of this in section 5.4.

How is it that so little is made of the Devil in the Old Testament and so much in the new one? It seems that in the intertestamentary period, about

100 BCE to 100 CE, people began to feel a philosophical necessity to imagine God as just and perfect—but then, how could they explain evil and suffering? The authors of the Old Testament had been content to give God motives of jealousy, anger, cruelty, and revenge, but by now Judea was an eastern province of the Roman Empire. Wealthy Greeks bought large tracts of the Holy Land, Greek towns and cities sprang up, and many middle-class Jews acquired a passion for Greek civilization and philosophical clarity. Greek games were celebrated. Once again the athletes competed nude, and young Jews underwent painful plastic surgery to replace the foreskin. This generation was no longer content with a God who seemed arbitrary and inconsistent. They concluded from the world's vast supply of misery that evil must be more than simply the absence of good and must be concentrated in some malignant power. There were plenty of candidates.

2.2 THE LOWER TIER

The word "demon" began honorably as the Greek *daimon*. This was a very general term, and basically it meant god. Homer refers to "Zeus and the other *daimones*," but many of them were not what we usually think of as gods, for they were everywhere. Hesiod says they are the departed spirits of the first generation of men, "kindly, delivering from harm, and guardians of mortal men; for they roam everywhere over the earth, clothed in mist, and keep watch on judgments and cruel deeds." Every person had a daimon, a sort of higher self; for the Romans this was a *genius,* and for us today it is a guardian angel. A rock, a spring, an old tree had its daimon; more than that, it *was* its daimon. In works by Hesiod and other early writers, any geographical feature—an ocean, a range of hills—is born and in turn gives birth. They are all persons, and some are gods. Hades isn't just the king of the underworld; he *is* the underworld, and powerful river gods abound in the old legends. In Book 21 of the Iliad, for example, Achilles fights a fierce battle with the river Scamander which runs below Troy. The river speaks like a god but fights like a river, rising in a flood "seething with foam and blood and dead men" that almost drowns Achilles before Hera steps in to rescue him.

But something is missing from this sketch of gods and daimones, a question left unanswered. How can anything be a daimon and a tree at the same time? It is because they belong to different categories of existence. For Roman Catholics, as for many others, there is a moment in the Mass when a bit of bread becomes the Body of Christ. How can that be? Because at that moment the bread has entered another world, the world of the sacred. In his great study *The Elementary Forms of Religious Life,*

Emile Durkheim situates objects of contemplation in two realms: the sacred and the secular. There are sacred trees, houses, sets of words. To the believer, a sacred tree looks like any other tree but there the resemblance stops. Religion does not need a creator or a supreme god—Buddhism has neither—but it exists when the division is recognized. It is different, says Durkheim, from any other division we may make because it is absolute. "In the history of human thought there is no other example of two categories of things as profoundly differentiated or as radically opposed to one another." We smile when we read about the medieval king who caused a man's body to be weighed immediately before and after death in order to detect the soul's departure. Whether or not we happen to believe in souls, most of us detect a confusion of categories. Of course, different individuals or groups place the dividing line in order to extend or restrict the domain of the sacred. There are people who acknowledge only one domain—every year a few of them risk their lives in the snows of Mount Ararat looking for Noah's Ark—but for most of us the distinction exists.

Demons

Angels crowded out the good daimones in the course of time; evil demons, though, were already plentiful. Babylonia was infested with them. Some were dead people returning to settle old scores, some personified flood or drought or disease. An accident, a sudden fit of anger, was the work of a demon. For each part of the body, a demon was waiting to strike. One named Labartu attacked women in childbirth and tried to steal the baby. Demons lived in the lower world and bore names like Pestilence, Destroyer, and Storm and they roamed in groups of seven, for troubles never come singly. A sad old poem begins:

> There are seven
> There are twice seven
> Evil ones
> Born in the lake
> They wander
> No breasts, no balls
> They drift
> No wife, no child
> No mercy.

The situation was different in Egypt but no better. There the world was in a tense and unstable equilibrium between two supernatural armies, one trying to destroy everything inside the bubble and the other trying to preserve it. The goddess Nut, for example, was responsible for the Sun. During the day, the Sun travels across the sky in its Day Boat, accompa-

Figure 2.2   Vase painting of three keres
flying out of a *pythos* or grave-jar
opened by Hermes Psychopompos, one
going in. (From Harrison 1903, 43.)

nied by an army of deities to protect it from a demonic force of about
equal size. At nightfall Nut swallows it again. Another version held that
after sunset the Sun travels back through the dangerous underworld in
a Night Boat, sometimes called the Boat of the Millions because of the
army of protectors it carried.

The old Greeks lived surrounded by destructive spirits called *keres*. A
myth explains their origin. After Prometheus stole the secret of fire and
passed it on to men (there were no women then), Zeus, in revenge, ordered his craftsman Hephaestus to create a lovely young girl named Pandora, whom he sent down to Earth together with a sealed jar that the
gods had filled with the keres of old age, sickness, death—every kind of
suffering and bad luck. Innocently, Pandora peeked inside, and out they
flew. Keres were everywhere, usually connected with death. Figure 2.2
shows three of them flying out of a grave jar and one going in. In the
Iliad, Sarpedon is not exaggerating when he reminds a friend of the nearness of death, "Keres by the thousands stand around us and no man can
escape them." Plutarch quotes an unidentified poet as saying that "all the
air is so crowded with them that there is not one empty chink into which
you could push the spike of a blade of corn." They were the bacteria of
the Greek world, lurking everywhere, and nobody was immune. Every
family knew death intimately as one member after another fell sick for no

apparent reason and dropped away. It was not unreasonable to imagine beings, invisible and voiceless, who devoted themselves to the ruin of humankind.

If the Jews were not already aware of the importance of demons in Babylon, they surely learned about them in the sixth and fifth centuries BCE when large numbers of the community were enslaved there. In those years they were also in contact with followers of Zoroaster, for whom the world was the scene of an unending battle between the army of angels led by Ahura Mazda and Angra Mainu's legion of demons. The Jews already had a Heavenly Host; now they imagined Satan as the commander of a demonic host whose aim was to lead people away from the Lord.

Satan

The Bible occasionally refers to Satan as the ruler of the world. This should not be taken literally, but as we look around us and read the papers, it is hard not to feel the presence of some huge obstacle that stands in the way of decent and sensible hopes and aspirations. Faced with any kind of destructive force, humans tend to give it a name and a personality; hence Satan. If my discussion here is short, it is because excellent books have been written in recent years that treat Satan as he deserves.

In Hebrew the word *satan* means adversary. This is the person we encounter in the Book of Zechariah, written in about the sixth century BCE. In this account, the Jews have returned to their own country after the Babylonian exile and an angel shows Zachariah a vision in which the Lord reveals that he will live among the Jews. It seems that the high priest Joshua has been slow to rebuild the Temple in Jerusalem after its destruction. Now he is in trouble and stands before the angel with the satan (there are no capital letters in Hebrew) standing beside him to accuse him. The *satan* seems to be an official, something like a district attorney. His title and function may derive from the Persian secret police, and his responsibility seems to have been to seek out any who denied the dominion of God. Apparently the angel thinks the satan has gone too far and rebukes him, "The Lord silence you, satan! May the Lord, who has chosen Jerusalem, silence you!"

The other early appearance of the name, or word, occurs in the prologue to the Book of Job: "The day came when the members of the court of Heaven took their places in the presence of the Lord," and among them was the satan. The Lord asks him where he has been. "Ranging over the earth," he says, "from end to end." God then authorizes him to test the piety of Job. The story is painful and any summary would diminish it.

After the Book of Job, the First Book of Chronicles portrays Satan as setting himself against Israel, and then and forever afterward he appears as a personal enemy. He becomes the serpent in the Garden and the leader of the fallen angels; he is called Beëlzebub, Lord of the Flies; he works nothing but evil. In New Testament times he assumes the role of a demon who opposes Christ's work of redemption and tries to buy him off with a promise of riches. Thus Satan, who began as part of God's administration, draws apart from him and ends as his opponent. Azaz'el's introduction of evil practices among humankind represents one solution of the problem of evil. Satan is a second attempt. A third? In an old comic strip, a possum named Pogo says, "We have met the enemy and he is us."

2.3 The Shape of the World

Come outside on a summer evening and look at the sky with a simple mind. Light fades, a point of light appears in the west. It wasn't there a moment ago. It brightens, and soon there are other lights. We need a name for them—all right, stars, but let the word signify a point of light that appears when the Sun leaves, nothing more. Other stars turn on, and familiar patterns emerge. Later, if the sky is very clear, we see an irregular swath of pale light extending across it. Someone suggests it looks like a path through a field; for someone else it looks like milk spilled on a table. It acquires a name: the Milky Way. Darkness deepens and the evening star dips toward the horizon. Other stars begin to rise in the east and set in the west, just as the Sun does, but not all stars do that. Those in the northern part of the sky never set but only revolve around a particular star, the one that never seems to move. It is Polaris, the Pole Star. In ancient times, a constellation of bright stars was named the Seven Oxen because the celestial objects seemed endlessly to plow the same field, going around and around. In England it is called the Plow; for Americans it is the Big Dipper, and for the Greeks and Romans it was a Bear.

Someone remembers: a few months ago the evening star wasn't there at all. Someone else recalls that before that you could see a bright star in the morning before dawn. It turns out there can be a bright light in the evening *or* in the morning but never both. If you observe it and mark its position for a few weeks, you will find that is has traveled a little way across the starry background. It turns out that of all the stars in the sky, exactly five of them move visibly. In Greek, *planetes* means wanderer; so let's call them planets. One more strange thing: I mentioned Orion, and people say it looks like a hunter with his hands up and legs spread apart, a belt around his middle with a sword hanging from it. But now, on a summer evening, where is it? Nowhere. In the autumn it will show up but it isn't there now. If you watch night after night at the same hour, you will find that the immense show of constellations drifts slowly across the sky, from east to west, where they disappear. But later they reappear in the east and finally they return exactly to where they were twelve months ago.

I tell about the sky in such detail only to suggest how it changes and how it might raise questions in a curious mind. Why are there lights up there? Why do five of them move with respect to the others? What do these motions mean? Is there some way of guessing what they mean? Life four thousand years ago was more uncertain than it is now. Flood, famine, and pestilence struck without warning; a woman hearing a noise outside opens her door and discovers an invading army. Is it possible that lights have been put into the sky to send us messages, to encourage us or warn us? Most of them just drift across the sky, night after night; they tell us nothing except, if we know how to read them, what time it is. But suppose the planets, like the hills and trees around us, are gods and their motions are voluntary. Their places in the sky, like the courses of our lives, are always changing. Is there a connection? Since we don't know their names, we name them after other gods: Mars, Venus, and so on. If we think of Venus we can look up in the morning or evening and actually see her. In the second century CE, Emperor Marcus Aurelius asks himself how he knows the gods exist. "For one thing," he says, "they are perfectly visible to the eye," which helps us understand a remark made by Aristotle: "The Pythagoreans marveled greatly at anyone who said he had never seen a divine being."

The Sun and the Moon affect our lives more directly. The Sun lights our footsteps, ripens our crops, and warms our bodies. I have shown how important it is that the Sun reverses its southward drift in December and its northward drift in June. The ceremonies that helped these reversals had to be planned in advance. Several ancient peoples have left monuments that mark the solstices and provide a few days' advance notice; Stonehenge is an example. Building and rebuilding these monuments, neolithic British society seems to have devoted more of its GNP to its space program than modern nations do. This effort would not have been made if it had not served important purposes. Perhaps the Chumash Indians have shown us one of them.

The Moon's influence on the course of human life is not as obvious as the Sun's, but consider the following: Tides rise and fall according to where the Moon is, and the menstrual cycle of women and the agitation of insane people during the full Moon (which is why they were called lunatics for *luna*, Moon) left no room for doubt that the Moon affects humanity. Basil of Caesarea, a Father of the Church, writing his study of the Six Days of Creation in the fourth century CE, tells how the Moon's influence changes living things: "When she wanes they lose their density and become void. When she waxes and is approaching her fullness they appear to fill themselves with her, thanks to an imperceptible moisture that she emits mixed with heat, which penetrates everywhere." This belief, much older than Saint Basil and still current all over Europe as late as the seventeenth century, led to Moon-governed rules for planting and harvesting that some farmers and gardeners follow even today.

A long catalog of influences was also attributed to the other planets. The Alexandrian astronomer Ptolemy (first century CE; more about him later) mentions some of them: "It is Saturn's quality to cool and, moderately, to dry. . . . The nature of Mars is chiefly to dry and to burn. . . . Jupiter has a temperate acting force. . . . Venus . . . warms moderately because of her nearness to the Sun, but chiefly humidifies. . . . Mercury in general is found at certain times alike to be drying and absorptive of moisture, . . . and again humidifying, because he is next above the sphere of the Moon." I mention these influences only to point out how easy it was (and for many, still is) to conclude that the gap between Heaven and Earth is not as wide as at first it seems.

There are two reasons, not entirely distinct, for trying to read the planets. It may be that those divine intelligences know the future and are telling us about it, or perhaps they play a more active role and control what will happen. In either case, if men want to live wisely, they must know the planets' language.

• • •

A basic principle exists in every ancient culture I know about: the world is One. Dreams reach us from other times and places. Someone sticks pins into a doll in order to injure a peson in a distant town. Every thing, every thought in the world is connected to every other by a thin, fine web of causality. This instinct—I won't call it a belief—is illustrated by little things people do, a golfer's "body English," for example. It is hard to see how a rabbit's foot carried in one's back pocket and stroked a few times can, for example, cause a policeman not to notice a double-parked car, but someone must buy them in those little shops; the buyer senses an order underlying everyday events that modern science does not know. It is also hard to see why being born when the Sun is thought to be in a particular constellation can have any predictable effect on one's future life, but to exploit this connection astrologers have toiled for thousands of years. In the United States there are ten of them for every astronomer.

Astrology celebrates a marvelous unity that runs or is felt to run through the world. Some of the linkages I have been describing began to loosen in about the year 1, but they are not gone, and there are always new ones. The size of next year's wheat harvest may not be connected with the shape of the liver of a sheep selected at random, as the Akkadians thought, but it may be connected with the number and size of sunspots. In the last century much has been learned about the basic nature of different kinds of matter and radiation, and it is almost certain that this knowledge applies without change everywhere in the universe. It also applies to plants and living creatures, and to the material inside our heads. It doesn't explain the wonders of life and consciousness, but understanding it seems to be necessary if they are ever to be explained.

• • •

The Babylonians did not try to visualize or explain the celestial display in terms of objects moving around up there; for them it was more like a light show or a musical performance. But whether the planets are gods or things or points of light (or all three), diviners need to talk about where they are from night to night, and for this purpose it is lucky that the fixed stars are there, for they serve as a background for locating the planets. There is no sense in naming all the stars; there are too many, so the Sumerians divided them into constellations to which they gave names that connect them with cities and gods and characters from myth.

Some of these names (Lion, Bull, Scorpion) have long endured. For Sumerians, the constellation we know as Taurus was the Bull of Heaven, *Gu.ud an.na,* that Gilgamesh and Enkidu killed. For Akkadians, it was the same bull under a different name. In Greek, Latin, Sanskrit, Persian, and Arabic, it is just a bull, though it doesn't look like a bull or anything else. There are, however, constellations whose names are not so easily explained. Orion is known all over Europe and in Asiatic regions to the north and east as a hunter, and for some Native American peoples he is a hunter of bears. What most Americans call the Big Dipper was known in classical times as *Arktos Megalē,* the Great Bear, even though with its long tail it doesn't look at all like a bear, and the final *e* indicates that it is a she-bear. Anthropologists report that it is also a bear over the whole length of Siberia and across the Bering Strait over to Indian tribes in North America. If the name was not brought by European visitors (and it seems unlikely that it was), the Bear may have received its name long before the last Ice Age.

• • •

The planets move from night to night through the constellations of the zodiac, but they do not move at a steady rate. If you keep track of the motion of Mars, for example, against the background stars, you will find that the planet generally drifts from east to west but sometimes slows, stops, reverses its motion for a couple of months, then reverses again and goes on as before. It took even more record keeping and study to notice that there is regularity in that hesitant motion. It required someone—a sequence of people—for whom the matter was important. For centuries, Babylonian astronomers measured and recorded planetary positions, night after night, and contemplated those lists of numbers. They were looking for signs of some mathematical order that would enable a skilled person to predict where the planets would be at a given time in the future.

In this work the Babylonians had a great advantage over other nations, for they wrote their numbers in a positional notation, much like the decimal system we use today. Imagine studying long tables of Roman numerals, calculating with them, checking the calculations. Try dividing CCVII by XXVII, expressing the result exactly. The Egyptian system was essentially like the Roman one, and Greek numbers were hardly more convenient. The Babylonian system was based on patterns of 60 rather than 10. We count from 0 to 9 and then go on combining the same set of digits;[2] they counted from 0 to 59 and then made analogous combinations. We write $2 \times 10 \times 10$ as 200. They wrote the number 7200, using modern symbols, as $2 \times 60 \times 60 = 2;0;0$ (the semicolons indicate that the base is 60); 74 would be 1;14. When we express an angle in degrees, minutes, and seconds, we are using Babylonian notation. Where we write $43°31'55"$ they used just the symbols for three digits: 43;31;55, and the same for hours, minutes, and seconds. Because a 60×60 multiplication table contains 1770 nontrivial entries as compared with the 45 of our own, they used tables more than we do, but with that difference their arithmetic was much like what we learned in school, or would have learned except, nowadays, for pocket calculators.

It was not until the fourth century BCE, after centuries of observation, that Babylonian astronomers discovered the regularities of planetary motion and began to forecast them. When they finally learned how to do so with reasonable accuracy, it was not because they had formed a mental image of the planets as objects moving through space; it was only because they had noticed certain patterns in the numbers. Before that time, eclipses and planetary reversals were inscrutable signs of celestial wars and upheavals, but now that they could be predicted they needed a new interpretation. It was not hard to find. Gods intervene in human affairs and often send omens that warn and advise. These are generally unpredictable and must be dealt with individually, but planets follow a schedule, and if the world is truly one, their motions are also connected with human affairs. With this knowledge one could begin to predict the course of those affairs, and that is astrology. But gods had many ways of telegraphing the future, more ways than one might at first expect.

---

[2] Of course, the overwhelming majority of all calculations made today use only the digits 0 and 1.

2.4 FORTUNE-TELLING

The gods of ancient Mesopotamia knew the future, but humans could know it too. If the world's structural unity allowed gods to read human minds, knowledge traveled the other way too, and if you looked carefully there were omens everywhere to tell you what the gods were thinking. About seven thousand omens are listed in a series of tablets called *Enuma Anu Enlil*. They are dated c. 700 BCE but their contents are probably much older, and there are other collections as well. Consider a tablet in the British Museum that provides ready-made interpretations more specific than what the stars normally say:

> When a yellow dog enters a palace there will be destruction within its gates. When a piebald dog enters that palace the king will make peace with his enemies. When a dog enters and someone kills it, the palace will enjoy an abundance of peace. When a dog enters and lies on a bed, no man will capture the palace.

And so on and on and on. There is a whole book on the meanings of monstrous births, disturbances of the natural order so great that they could threaten a kingdom:

> If a *suppu* sheep gives birth to a goat there will be destruction in the land. The reign of the king will end and the son of a widow will seize the throne. Pestilence will follow. If a mare bears a colt and a filly and they each have one eye, an enemy will attack and overthrow the land of Akkad.

The next one is less likely, though of course one should be prepared for anything that might happen: "If a woman gives birth to an elephant the land will be laid waste." We may be surprised that in these examples small events could announce such big consequences. But it was the coming catastrophe that caused the event, not the other way around. In spite of the way the omens are described, the catastrophe didn't have to happen. The omen was more like a warning, transmitted by the gods in a special language that only a few people could understand. Something was wrong, and steps should be taken to correct or prevent it. If an omen predicted that the king would be assassinated in Nineveh, he stayed away from Nineveh. If pestilence or general destruction was a threat, everyone understood that the gods could make them happen if they wanted to, but priests had rituals called *namburtu* that almost always dissuaded them. These matters were very important. The surviving scraps of correspondence between temple secretaries and palace officials—only a small fraction of what must have been a continual flood of messages—show that a bump on a sheep's liver could change the policy of a kingdom.

On the other hand, suppose something strange happened in your own neighborhood: a goat dropped a kid with extra legs, a chicken behaved strangely, lightning hit a cow. Experience of human nature tells me that the last thing sensible people did was phone City Hall. They buried the evidence, nothing happened, all was well.

• • •

Gods could communicate with people by sending dogs, chickens, or elephants if they wished. They could also change the weather and send eclipses, and as long as people believed that gods could and did control the movements of the planets, no one tried to forecast these events. It was different with the Moon, which kept to a regular schedule and was therefore used, as it is for Passover and Easter today, to set the dates for festivals and assign an auspicious time for building a bridge or starting a war. In ancient Babylon as in modern Islamic countries, the lunar month began when the new Moon was first seen. Starting then, and on each day until the next new Moon, there were omens to be looked for, acts to be performed, acts to be avoided. Consider a tenth-century Mesopotamian forecast for 1 Nisan, the day of a certain new Moon:

> God Enlil: sinister; difficult for the sick; a physician may not lay his hand upon the sick, a prophet may utter no word; it is not suitable for doing anything desirable. The king and lord may speak boldly. Lucky. Fish and lovage [an herb] may be eaten. The king shall clean his garment. King must make offering to Enlil, Ninlil, Shamash, and Nusku.

There will be serious consequences, including the displeasure of four gods, if the priest misses the new Moon. A new Moon is faint, seen when it is still in the sky after the Sun has set. If the sky was cloudy, astronomers had to be able to calculate when the Moon would be seen if the weather were clear. This is a delicate matter, for its visibility depends on the length of twilight, and that, in turn, because of the tilt of the Earth's axis, depends on the time of year. If the Sun descends along an inclined path, twilight lasts longer than if it goes straight down. Taking this and other effects into account, Babylonian astronomers slowly perfected a mathematical procedure that predicted each new Moon.

As for eclipses, it was early noticed that solar eclipses happen shortly before a new Moon, while the Moon is eclipsed when it is full but not every time. Why not? By the early fifth century BCE, the Babylonians had noticed that if the Moon was eclipsed on a certain night, it would again be eclipsed nineteen years later, about five hours earlier on the same calendar date. (There might be other eclipses in between.) In this way, many lunar eclipses could be predicted. Predicting eclipses of the Sun is much harder. Whenever the Moon is eclipsed, everyone who can see it at all sees the eclipse. But the Moon does not cast a very big shadow on the Earth, and so if there was an eclipse at Babylon there might be no sign of it at Nineveh, a few hundred miles to the north. By the third century, astronomers could predict every lunar eclipse, but for the Sun they never learned to do more than identify days on which it might go dark.

All these calculations were made without any diagrams to illustrate the actual layout of the objects being studied. There is not even any indication that Babylonians knew the Earth is round. There were no pictures in their heads, only rhythms—of night and day, of the seasons, of lunar and planetary motions. I recall that the historian of science Giorgio de Santillana said that nothing in the modern world resembles this ancient cosmology more closely than Bach's *Art of the Fugue*.

•••

The book of Genesis doesn't say much about the sky. On the Fourth Day, "God said, let there be lights in the vault of Heaven to separate day from night, and let them serve for signs both for festivals and for seasons and years." With these words, he created the Sun, Moon, and stars. The Sun regulates our daily and seasonal activities; the Moon sets Passover, Easter, and the Muslim calendar; and stars announce the seasons. Planets, precisely because their motions do not follow a simple pattern, are necessary to mark years. If an event is known to have taken place during the year in which Mars and Saturn were in conjunction in Aquarius, that marks the date because any such combination rarely recurs.

The Bible does not suggest that the celestial lights are meant to serve for signs of events to come, and Isaiah, calling down God's judgment on Babylon, cries, "Let your astrologers, who contemplate the heavens and calculate the new Moons to tell you what is to come, stand up and save you. But they are like stubble, and fire burns them up."

And what, exactly, is wrong with stargazing? Much later, in the non-canonical Book of Jubilees, c. 120 BCE, Abraham sits up all night watching the stars to find out the prospects for rain in the coming year. "And a word came into his mind, saying, 'The signs of the stars and the Sun and the Moon are all in the hands of the Lord. Why do I search them out?'" I read this to mean that the Lord will send rain or withhold it as he pleases, and that anyone who even thinks these acts are predictable is assuming that the Lord's freedom to act is restricted by his own creation.

In the following century, as enthusiasm for reading the stars spread around the eastern Mediterranean, the author of the Book of Enoch was more open to astrology. As we have already seen, he wrote that the secrets of astrology were part of the forbidden knowledge that Azaz'el revealed to humanity when he led his companions down from Heaven. Enoch knows that the stars must go only where God intends them to go, for on a celestial journey he sees a special prison house where stars that have transgressed God's commandments by not arriving on time get roasted. Thus, even if the knowledge imparted by Azaz'el was forbidden it was nonetheless knowledge, and as the news of astrology spread westward an increasing number of people began to agree.

• • •

It was ancient doctrine in Mesopotamia that astronomers could read in stars and clouds whether the grain harvest would be rich or poor, or whether this was a good time to start a war. Such predictions involved the country as a whole; they assumed that the gods had a plan for it or for its king. To assume the gods cared enough about each citizen to have a plan for him or her involved a great leap of imagination, but in the late fifth century BCE tailor-made nativities of individual people began to appear in Babylonia. As the years went on, astrological judgments became more specific thanks to long observation of the planets and better ways of calculating their future positions. At the same time, theoretical doctrines covered a widening range of outcomes.

Astrology spread to Greece and Egypt. In about 290 BCE a Greek-speaking Chaldean (Babylonian priests were known as Chaldeans) named Berossus settled on the Aegean island of Cos and opened a school where he taught divination as well as Babylonian history and astronomy. He is the source of the story of the sea monster Oannes. Especially as commerce expanded after Alexander's conquest of Babylonia in 334 BCE, agents and bankers and shippers moved into one another's countries and brought ideas with them. The Roman encyclopedist Pliny (of whom more later) says that Berossus's predictions were so remarkable that the Athenians set up a statue of him with a golden tongue. Pliny does not mention any of the predictions that came true, but some of them must have. Berossus was highly educated, and if he brought his library with him he must have given the Greek astronomers a chance to learn what the Chaldeans had been doing. Until archaeologists began their digs during the mid-1800s, most of what the West knew of old Mesopotamia came from Greek accounts of Berossus's teachings.

Theophrastus, the scholar who followed Aristotle as director of the Lyceum in Athens, is reported to have said that in his time "the Chaldeans' knowledge [of astrology] was stupefying, since they could predict among other things how individuals would live and die, and not just general events like storms and good weather." He gives no examples. Strangely, the Babylonians, the first to cast horoscopes, seem also to have been the first to become disillusioned with them. At the turn of the era, the Greek geographer Strabo visited Babylon. Though most of the city lay in ruins it was still inhabited, with astrologers living in a special quarter of it. Strabo says that the few who cast horoscopes were held in low opinion by the rest. But what had become a fringe occupation in Babylon was off to a good start in the west.

2.5 THE STARS MOVE WESTWARD

In 459 BCE a series of armed conflicts began between Athens and Sparta for supremacy in the Greek world. The Peloponnesian War went on intermittently until 404, but an event that may have decided its outcome occurred August 27, 413, in Sicily. It was an omen.

The whole Athenian fleet and much of the army were fighting to capture the Sicilian city of Syracuse (the logic connecting this operation with the war against Sparta is too complex for this simple discussion). The action had turned against the Athenians, and that night the commander, Nicias, decided he had better retreat, using the full Moon to light the army's way. I will let Plutarch (in Sir Thomas North's sixteenth-century translation) tell what happened:

> When they had put al things in readinesse for their departure, without any knowledge of the enemie, or suspition thereof, the moone began to eclipse in the night, and suddenly to lose her light, to the great feare of Nicias and divers others, who through ignorance and superstition quaked at such sights. For, touching the eclipse & darknesse of the sunne, . . . every common person then knew the cause to be the darknesse of the body of the moone betwixt the sunne and our sight. But the eclipse of the moone itselfe, to know what doth darken it in that sort, and how being at the full it doth suddainely lose her light and change unto so many kinds of colour: that was above their knowledge, and therefore they thought it very strange, perswading themselves that it was a signe of some great mischief the gods did threaten unto men.

Nicias, able and experienced, was terrified, and the army's soothsayers persuaded him to wait another month before leaving. Remember that at this time very few people thought the Moon's light comes from the Sun. For most, the Moon was its own source of light, and someone, something had turned it off. The Syracusans used the month to regroup their forces and strengthen their defenses. In the end, the Athenians lost everything, and most of the soldiers as well as their commanders were thrown into a quarry outside the city where they perished of hunger and thirst. Thucydides sums it up: "They were, as the saying goes, everywhere and at all points defeated, and of many who set forth, few returned home."

• • •

The war ended with the defeat of Athens, and the following century was a bad time for the city, militarily and politically. Its fall from its status as the Greek world's admired and respected center of culture and strength shook serene assumptions of superiority and forced the Athenians to take a new look at the world around them and their place in it.

In 404, with its walls leveled and its fleet destroyed, Athens was starved into surrender. It lost its empire and navy but not all of its authority, for Athens was still the great center of art and learning, and the Spartan conquerors, remembering how Athens had led the war against Persia, resisted calls for its destruction. They set up an administration that degenerated into chaos and anarchy, and in 338 Athens was conquered by Philip of Macedon who added it to his empire. Alexander came and went; after him came more chaos, which in 295 culminated in a long siege during which many people died. Ten years later Athens, though much reduced in power and influence, was once more free, and it stayed that way until, in 146, the Romans came.

During the bad years a new religion, Stoicism, was born in Athens. It was founded by Zeno of Citium (c. 336–263) and brought its followers peace of mind by teaching that whatever happens is part of the rational plan of a god who is purely good. God is physically present in every part of the universe, giving form and qualities to material substances and guiding every event. Therefore, said Zeno, don't concern yourself with what you can't control, but make your own mind move in accordance with God's will. There is a plan, even if you don't understand it, and you are part of it. Athenian teachers brought this doctrine to Rome, where its emphasis on disciplined acceptance of fate appealed to the Roman instinct for social order and where it became popular.

For Stoics the same power that ordered human affairs also ordered the stars, and no one doubted that their influence reaches through all of life and nature. Though there was nothing you could do to change the future, there were advantages to knowing what was going to happen, and the key to that knowledge was in the sky. By that time, Rome was familiar with astrologers. They had arrived from the East with exotic costumes and curly beards, and Stoics listened eagerly to what they had to say. Some followed their teachings in a literal way while others thought it should be open to interpretation. The hard-liners were generally less educated; they awaited whatever fate had in store for them, ready to tough it out. For them, the stars' messages, properly interpreted, were the highest truth. Sophisticates spoke to one another in Greek and emphasized that though stars determine the conditions under which we live, we are free to choose how we will act. They smiled when simple people consulted the stars to see what was going to happen next, but in matters of imperial politics the distinction was less simple.

Some time in the 40s CE, when the emperor Tiberius knew he had little time left, he scanned the list of his possible heirs. All the promising candidates had died, and the only family members of suitable age were his grandson Gemellus and his grandnephew Gaius, nicknamed Caligula, an unstable young man given to cringing, fawning, and outbursts of rage. Instead of eliminating Caligula himself, Tiberius allowed the stars to do so, for the court astrologer told him Caligula's horoscope was so bad that he had as much chance of being emperor as of driving a chariot across the Gulf of Baiae. When Tiberius died and the empire's serpentine operations installed Caligula on the throne, the treasury was full. Four years later it was empty. Dio Cassius, senator, consul, and historian of Rome, writes a century and a half later that Caligula had spent it all to build a pontoon bridge 2½ miles long between Baiae and Puteoli, complete with stopping places for rest and refreshment. Across it he led a platoon of the Praetorian Guards, once in each direction; then it was broken up. Allowing that stories, like children, get taller as they age, something like this seems to have happened. Afterward, because the bills had to be paid, Caligula began to raid provincial treasuries, sell off imperial heirlooms, kill the rich to confiscate their estates, tax the middle classes into poverty, and dispose of anyone seen as lacking in respect. He had to be murdered, and two years later he was. This sad story shows again how stars could batter an empire.

• • •

Astrology was only one of the methods a Roman citizen had available when he wished to peek into the future, for all parts of his model of the world were linked together with a web of causality. If he wanted to know what kind of day he was going to have, there were many ways to find out. As he left his house in the morning and headed for the office, he carefully studied the scene. First the sky, for that is Jupiter's domain and where he caused his signs to appear. Different kinds of birds, depending on the direction of their flight, brought him messages from on high; odd appearance or behavior of an animal in the street warned or encouraged him. These were personal messages. To answer questions of state there were officials called augurs who, at midnight, solemnly sat facing south in a special area called a *templum* and waited for anything they could interpret as an omen. Shooting stars were noted. Lightning slanting downward from left to right was favorable; slanting the other way meant trouble, but the slant depended on where the augur was sitting and could be a tough call. An augur playing safe could close all public business for the day. Sometimes, after a riot or some other crisis that called for quiet negotiation, the augurs would announce that lightning had been seen and they would shut the city down. In the 70s BCE, after Cicero had served as an augur, he wrote that people often asked him how one augur could meet another in the street without smiling.

Cicero, in fact, devotes his book *On Divination* to refuting all arguments for it, but first he presents them clearly:

> According to the Stoic doctrine, the gods are not directly responsible for every fissure in the liver or every song of a bird; since, manifestly, that would not be seemly or proper in a god and furthermore is impossible. But in the beginning, the universe was so created that certain results would be preceded by certain signs, which are given sometimes by entrails and by birds, sometimes by lightnings, by portents, and by stars, sometimes by dreams and sometimes by utterances of persons in a frenzy.

The world embodied the future if one knew how to read it.

Cicero believed none of this and he tried to demolish every form of divination. I mention only two of his arguments against astrology. First, hereditary qualities, "the parental seed," are so obviously important to the makeup of a child that no forecast that ignores them can possibly be accurate. Second, note that twins, though born at the same time, often turn out to have very different characters and meet very different fates. A friend of Cicero's, an astrologer named Figulus, ventured the remark that the celestial sphere that supports the stars is so immense that, turning once a day, its circumference must be moving very fast. Thus, even if the second twin arrived only two minutes after the first, the stars would have moved a great distance in the interval. In about the year 400, Saint Augustine of Hippo (354–430) nailed that argument to the floor with the remark that if that is how things are, astrological forecasts become completely impossible since, though it is easy to measure the interval between two births, no one knows the exact moment of either. Besides, he says, there is a well-known case of twins born so close together that the second came out holding the foot of the first, and yet their lives took very different directions. Their names? Esau and Jacob, and their story is in the Book of Genesis.

So much for reading the cosmos to tell fortunes, but there is much more that can be done with it. If one knows how, one can take hold of the controls and make it obey. Often this wasn't even very difficult.

2.6 GUIDING HANDS

At dawn, the manager of a farm in the flatlands of Egypt stands on the shore of a lake. In an hour he will leave for Alexandria to sell a load of produce for as much as he can get in an uncertain market. He refreshes his memory from a scrap of papyrus, then his yell echoes over the still water: "Io, Tabao, Sokhom-moa, Okh-okh-khan-bouzanau, Aniesi, Ekomptho, Ketho, Sethouri, Thmila, Alouapokhri: whatever I set my hand to today, may it happen!" An early breeze ruffles the reeds where he stands, no answer comes, but something has changed: he has set a bit of the universe in motion.[3] There are hundreds of such names in a long magical papyrus of which part is in London and part in Leiden in the Netherlands. The Leiden papyrus, as it is called, turned up in Alexandria in the early 1800s. It dates from the third century CE and is one of the latest documents written in the old Egyptian language. Many of the grand names in the papyrus contain fragments from other languages, but because Alexandria was a Greek city and so few of the names are Greek, it is likely that the magic is older than the papyrus. It seems that most of the names the manager shouted out were only sounds that had brought someone luck when they were launched into the world.

It is a fact of history that nobody writes down the things everyone already knows. Therefore we have little information about any theory that might explain how shouted words can affect what happens, but we may be able to understand how Egyptians thought if we study some of the things they used to do.

The Leiden papyrus contains a great number of recipes classed as erotica whose purpose is to awaken a woman's interest in the man casting the spell. It may first be necessary to detach her from the man she is seeing at present, but that can be done. The other great desire that pervades the papyrus is to know the future, and there are many ways to do that. Finally there are medical recipes: against the sting of a scorpion, the bite of a dog, a bone in the throat. Some recipes call for items such as the blood of Hephaistus or the semen of Hermes that might be hard to come by, but commentaries attached to other magical books show that, for security reasons, priests encoded commonplace substances with fancy names. Blood of Hephaistus is really wormwood, semen of Hermes is dill, crocodile dung means Ethiopian soil, a snake's head is a leech, a man's bile is turnip juice, and so on. Some samples follow.

A recipe for a potion that will capture a woman's love begins: "Take a little shaving of the head of a man who has died a violent death, together with seven grains of barley that have been buried in the grave of a dead man; you pound them with ten *oipe* of apple-seeds; you add the blood of a worm of a black dog and a little of the blood of the second finger of your left hand (the one closest to your heart) and some of your semen, you pound them together and put them into a cup of wine . . . and you make the woman drink it." It doesn't say how you make her drink it.

---

[3] If the reader decides to try this I am not responsible for the consequences.

The papyrus also tells how to know the future. Here is part of a procedure that starts before dawn: "Fill a new dish with clean Oasis oil gradually so that it will not be cloudy; take a boy, pure, before he has gone with a woman. . . . Clothe him with a clean linen tunic, then make him lie on his belly with his head above the oil, eyes closed, while you utter an invocation [it seems to be missing] seven times. Then he opens his eyes and you ask him what you want to know." Several methods of divination involve the young boy and the oil. I suppose that in the dim light an excited child will look into the oil and see some picture in the faint reflection of the sky. If images do not come there is a prescription that will bring them by force: "Put the bile of a crocodile with pounded frankincense on the brazier. If you wish them to come quickly put stalks of anise on the brazier together with the shell of a crocodile's egg."

A few, but only a few, of the procedures are designed to damage someone. In one of them, a man desires the death of another. He first obtains the head of an ass. Then, in the early morning, he binds threads of palm fiber to his hand, his head, and his phallus. When the Sun is about to rise, he anoints his right foot with *set* stone from Syria and his left foot with clay. Facing the Sun, with the ass's head between his feet, he anoints his hands with its blood, then bends over and places his right hand in front and his left hand behind the head. In this awkward position he launches a long invocation toward the god Typhon Set of which the gist is as follows: "I invoke thee who are in the empty air, terrible, invisible, thou that hatest a household well established. I invoke thee by names thou canst not refuse to hear: Iopakerbeth, Iobolkhoseth, . . . [there are several more]: come to me and strike this man with frost and fire, for he has wronged me." This is done again in the evening, and the ritual is repeated every day for a week.

• • •

Egyptian writing faded away, but Greeks continued the tradition. We know less about Greek practices, but they were probably simpler than those just mentioned. For them, magic was a natural offshoot of religion. A person in trouble with a god would normally get a priest to offer prayer or sacrifice, but if that didn't work, or if the circumstances of the case weren't what a priest ought to hear, there were *pharmakides,* disreputable women who knew spells and procedures that got the same result. This was a dangerous business for the *pharmakis,* illegal because an angry god might punish a whole city. The crime was impiety and the penalty was death. One of Aesop's fables tells of such a woman. Her defense had failed, she was being led away, and a spectator at the trial called to her, "How was it that you were not able to persuade men, although you profess to be able to avert the anger of gods?"

Figure 2.3 *Defixio* intended to fix a chariot race. The left column reads "PHRIX PHÔXBEIABOU STÔKTA NEÔTER whether above the Earth or below DAMNÔ DAMNA LUKODAMNA MENIPPA PURIPIGANUX OREOBARZAGRA AKRAMMACHARI" and the rest is similar. (From Maricq 1952, 368.)

Greeks were especially fond of issuing curses and what are called binding spells, intended to prevent something from happening—to fix a sporting event or to make an accused person unable to find words to defend himself in court. Typically these were written on a thin sheet of lead, folded over, and secured with a nail to the ground near where they were to do their work. In this form they were called *katádesmoi* in Greek, *defixiones* in Latin. One *katádesmos* intended to fix a chariot race, is shown in figure 2.3. Circles show where the nails went through. A defixio from the Greek island of Amorgos rages against a man named Epaphroditus who has encouraged the plaintiff's slaves to run away:

> Lady Dimeter . . . see to it that the one who has put me in this condition finds no satisfaction, whether at rest or in motion, in body or spirit. Let him find no help from male or female slaves, from the small or the great. If he undertakes something let him not be able to accomplish it. May a binding spell seize his household. Let no child cry to him. Let him not set a happy table. Let no dog bark and no cock crow.

Another example shows how enchantments spread in the Roman empire. The emperor Tiberius appointed Germanicus, his nephew and Caligula's father, to supreme command over the empire's eastern provinces, but when Germanicus arrived at Antioch he was up against the local governor. At this point he fell sick, and suspecting that the governor was involved he had his house searched. Tacitus reports, "Explorations of the floor and walls brought to light the remains of human bodies, spells, curses, leaden tablets engraved with the name Germanicus, charred and blood-smeared ashes [of cremated bodies], and others of the implements of witchcraft by which it is believed that the living soul can be devoted to the powers of the grave." Germanicus died soon afterward.

Evidently *defixiones* were not the only controls for steering the world. But rather than weighing down this chapter with more conjuring, I will continue with a single cosmic tour-de-force that is first mentioned in a neo-Assyrian tablet from the seventh century BCE and had a long history afterward. The relevant passage reads: "As for the messengers whom the king my lord sent to Guzana, who would listen to the disparaging remarks of Tarasi and his wife? His wife Zazâ, and Tarasi himself, are not to be spared. . . . These women would bring down the Moon from Heaven!"

This is the earliest text that mentions this trick, but since there are several others let us digress for a moment. Plato mentions that women in Thessaly do the trick. A character in Aristophanes's *Clouds* who has to pay a debt at the end of the month thinks of a way to postpone his obligation: "Suppose I buy a Thessalian witch and have her draw down the Moon at night and shut it up in a round box, the kind you keep a mirror in." There are other classical allusions as well. One of the speakers in Plutarch's essay "On the Disappearance of Oracles" (c. 100 CE) tells how a woman named "Aglaonice, who was skilled in astronomy, always pretended at the time of an eclipse of the Moon that she was bewitching it and bringing it down." Five hundred years later, the Visigothic King Sisebut writes a poem to his friend Isidore, archbishop of Seville, in which he mentions a common superstition: the Moon is eclipsed because a witch has pulled it with a magic mirror into an underworld cave from which it can escape only when people frighten her by making a lot of noise.

What does all this mean? An eclipsed Moon stays right where it is; only some of its light is taken away. Nicias assumed that a god was responsible, but over many centuries the prevailing opinion seems to have been that it is something women do. In the Western world, fear of witchcraft climaxed in the seventeenth century and still lingers today. It rests on the belief that certain women have destructive powers. Much has been conjectured about the origin of this belief, but why women were supposed to be able to disrupt the cosmos in this particular way is still not understood.

• • •

It remains to glance briefly at a Roman assessment of magic as used in the pursuit of health and wealth. Pliny the Elder, Roman lawyer, admiral, statesman, and student, was born in about 23 CE and died on the beach

of the Bay of Naples during the great eruption of Vesuvius that buried
Pompeii and Herculaneum in 79. He could have sailed to safety but instead he went ashore to watch the eruption and try to rescue survivors.
Pliny is our most important source of information concerning the world
of his time. During his career he collected information concerning its
skies, weather, plants, animals, its peoples and their customs. It is said he
was never without a book in his hand, and he claimed that to write the
thirty-seven books of his great encyclopedia *Historia naturalis* he and his
secretaries combed 20,000 facts out of two thousand books by one hundred carefully chosen authorities. Why such an effort? Clearly it was in
his nature, but perhaps there was also a deeper reason. Greek imagination had encompassed the universe. Plato (see sec. 4.3) had told how it
came into being, the astronomer Hipparchus had mapped the heavens,
the geographer Pomponius Mela had written a treatise on the geography
of the known world—and what had the Romans been doing? They had
been thinking about their business, their politics, their empire. Pliny
showed them that there is a great world out there, full of wonders, and
that knowledge accumulated through centuries of study was in danger of
being lost. Knowledge in those days lasted only as long as the fragile roll
of paper on which it was written; after that, unless people were interested
and had it recopied, it went out with the trash. Pliny survived because he
was recopied again and again.

Here we are interested in magic. Pliny's chapters 20 to 37 feature medical procedures. He has no use for doctors. In his time almost all of them
were Greek, and their treatments were based on elaborate theories of
sympathy and antipathy. For Pliny the treatments are magic since they
have little to do with the Hippocratic principle that diseases have natural
causes and should be treated by natural means. Before the Greeks came,
he says, there were no doctors. Instead, Romans used remedies that often
go back to the dawn of civilization, but all are based on experience. They
used hundreds of herbs and other substances, applied in dozens of ways.
Romans treated a headache by drinking water in which willow tips had
been boiled. We use the same treatment. The salicylic acid in willow tips
is the effective ingredient in aspirin.

Here is a random selection of the kind of procedures in which Pliny
smelled magic even though they might actually work:

- For epilepsy, water drawn from the spring at night and drunk from the skull
  of a man who has been slain and whose body remains unburnt (Pliny 1938,
  28.2). There are some who recommend the patient to eat the heart of a black
  he-ass in the open air with bread, upon the first or second day of the moon
  (28.63).

- Applying to the forehead the mud obtained by pouring vinegar over a front door's hinges relieves headaches, as does also the rope used by a suicide if tied around the temples (28.12).

- The ashes of a burnt weasel, mixed with wax, are a cure for pains in the shoulders (30.13).

- Hollow teeth are stuffed with the ash of mouse dung or with dried lizards' liver. . . . [There are] some who recommend a mouse to be chewed up twice a month to prevent aches (30.8).

Some of the prescriptions may be very ancient (Pliny says they are), but of course most of their authority depends on what doctors call the *vix medicatrix naturae*—the body's wonderful ability, given time and rest, to cure itself.

Pliny's world was full of the unseen connections he referred to as magic. One unseen connection is conspicuously absent. In the Greek world daimones (and in the Roman, *genii*) were living, conscious, and generally benign presences. They might somehow have been involved with the procedures Pliny was describing, but I have not found them mentioned anywhere in the *Natural History*. As to malignant spirits, we shall see in section 6.3 that in the course of a few centuries, Western civilization was invaded by an army of them. They came from the East, the troops of Satan, working for the damnation of mankind.

So much for the great project of getting a grip on the fabric of nature, to hold it and read its signs in order to exercise some control over the destinies of people and kingdoms. But another way of thinking starts from the contemplation of nature. Of course, the Babylonians had contemplated stars and the livers of sheep for centuries, but always for practical reasons; there is no report about anyone doing it for pleasure. That form of entertainment seems to have started on the west coast of Asia Minor.

THREE

Guesswork

> They say that Anaxagoras was once asked . . . , "Why should anyone choose to be born rather than never to come into the world?" to which he replied, "For the sake of contemplating the universe and the order governing its parts."
>
> —Aristotle

THIS CHAPTER will start in fifth century Greece. It will touch events that occurred earlier and later and led to the discovery that one could take pleasure in speculating and writing about the mysteries of the natural world. Though Plato and Aristotle lived toward the end of this period, they will be little mentioned here, for they were not part of the discovery and will have most of the next chapter to themselves.

3.1 A MASS OF ROCK

In the previous chapter I mentioned the hard times Athens passed through during the fourth century. A hundred years earlier, the city had blossomed like a rose. It had led the forces that triumphed over an enormous Persian army, and talented people were gathering there to share in its glory and prosperity. A silver mine to the north of the city had struck it rich and money was plentiful. The Persians had sacked Athens in 480; now the ruins were being cleared, there was a theater, and each year in the Dionysiac festival the tragedies of Aeschylus, Sophocles, Euripides, and other authors now known only as names were presented before huge audiences. Originally, they were performed only by a chorus that sang and danced as it told the old tales of pride, doom, and the envy of the gods. Later, one actor was introduced; then Aeschylus used two, and in 468 Sophocles introduced a third. Now it became possible for the audience not only to hear about events but to see them happening.

In the following year or thereabouts, news reached Athens that a large meteorite had fallen near Aegospotami in the region of the Dardanelles. No contemporary records describe this event but later writers refer to it and Pliny, five hundred years later, says the stone was still on view. Almost anywhere in the regions around the Mediterranean, everyone would have regarded the meteorite as bad news, a sign of anger in the sky. But a well-known professor in Athens seems to have seen it in a different way. Anaxagoras (c. 500–c. 430) had come to Athens from Clazomenae, an island a few hundred yards off the Ionian coast. If the stories told about him are true, he reasoned that if a big piece of stone fell from the sky it had probably been up there for a while before it came down, and that there might be others. He had already startled Athenians by pointing out that we can explain the phases of the Moon by assuming that although it looks flat, it is actually a sphere with no light of its own, lit by the Sun. Further, to explain its changes of shape we need not think anything is happening to the Moon itself but only that each night, as we look at it from the Earth, it is being lit from a slightly different direction. He taught that eclipses of the Sun occur when the Moon casts its shadow on the Earth. He also taught that the Moon is eclipsed when something, perhaps the Earth, gets in the way of the Sun's light, but since everyone thought the Earth is flat, his discussion can't have been very convincing.

I suspect that few people at the time had thought of the Moon as a thing at all, much less a sphere. The heavenly bodies were regarded as phenomena, things that are seen, visible signs of divine beings. Those who thought of Sun and Moon in a spirit of religious awe considered it blasphemous to ask what these bodies are made of, or to say, as Anaxagoras did, "The Moon is made of Earth and has plains and ravines in it." And if the Moon is really like that, what is it doing up there? If a rock fell at Aegospotami, why doesn't the Moon come down? If the Moon is a sphere, why does it look flat? The Sun looks flat and everybody knew the Earth is flat. Besides, if the Moon were a sphere, why would it always show us the same face? Anaxagoras had only one argument in his favor: he could explain its phases.

Having cast off the lines that moored him to philosophy and religion, Anaxagoras headed into deeper water. He proposed that the Sun itself is nothing but a ball of rock, glowing hot and as big as the Peloponnese. This was altogether too much for respectable people. As long as one does not think about their physical nature, it is natural to worship Sun and Moon as gods and to tremble when one of them is eclipsed—but who prays to a piece of rock? Anaxagoras was thus prosecuted for impiety, the crime for which Socrates was later executed, and was saved only by the intervention of Pericles, his friend and former student. He went back to Ionia, where he died in Lampsacus. They say that the people there, awed by his brilliance and his reputation, gave him a public funeral. On his tomb was written:

Here lies Anaxagoras, who above all others
Passed through the limit of the world's truth.

In his honor, they erected two altars: one was inscribed TRUTH and the other, MIND.

This is the story that has come down to us. Nobody knows how much, if any of it, actually happened, but it implies that in Athens, and even in a small colony in Asia Minor, Anaxagoras had a respectful audience. Plato, about seventy-five years later, mentions rolls written by Anaxagoras.[1] He refers to Socrates' contempt for them, shared by many other Athenians, but no authentic fragment survives. Our most reliable information comes from Aristotle, a century later, who perhaps had the rolls on his table as he wrote but was not interested in biography. There are few further mentions of Anaxagoras before Cicero, 250 years later, and Plutarch, 150 years after that. Had either of them read anything Anaxagoras wrote? More than a millennium after Anaxagoras, Simplicius of Cilicia, one of the last pagan philosophers, wrote much about him, including some quotations that cannot possibly be accurate since they contradict one another.

From a modern perspective, Anaxagoras's innovation was to suppose that any object, including the Moon, has an existence of its own, separate from the sense impressions by which we know it and from the legends and associations that surround it. He encourages us, at certain moments, to ignore the thoughts that come to us when we see the Moon rise over a quiet hillside and to consider it as we consider a brick or any other familiar thing. A generation after him came the atomic theory of matter. We shall discuss this subject later, but to think of matter in terms of atoms—to imagine that atoms, and only atoms, are there—we must think like Anaxagoras. Nothing about a silver spoon—nothing about its appearance or its feel or its value or its associations with the family—suggests atoms at all.

How did it happen that a young Greek from the fringes of the Persian Empire arrived in Athens thinking thoughts that clashed so harshly with the Athenian view of the world that he barely escaped with his life? Yet when he returned to Ionia he was celebrated. Consider for a moment the earlier history of the region.

---

[1] In those days a book (*biblos* in Greek) was a roll of paper 20 to 40 feet long, made from the leaves of the papyrus plant (*byblos*), with writing on one side of it. These rolls were used to preserve legal agreements, religious rituals, and thoughts that someone wanted to keep. Literature was still in the minds and voices of orators and singers. To read a play by Aeschylus would have been a strange experience, words detached from their sound. Copyright was unknown; people copied what they chose and therefore "publication" did not exist. Herodotus wrote down his works, but made his living reciting them.

3.2 Ionians

Even before Homer's time, people who lived on Euboea and Attica, the
two long and arid peninsulas north of the Peloponnese, were discon-
tented. There were too many of them for the soil they cultivated, and they
were continually threatened and robbed by local warlords. To move
southward they would have had to displace other people already there,
but that meant war, and they took the easier course of planting colonies
in sparsely settled areas around the Mediterranean. Some sailed westward to Sicily and the coasts of what are now Italy, France, and Spain.
Others turned the other way and settled the deeply indented Aegean
shore of Asia Minor. By the eighth century BCE, Greeks had pushed back
the native populations of Lydians and Carians and occupied the whole
coastline. Nobody knows where Homer came from, but tradition—and a
whisper of evidence in his poems—suggests that it was from one of these
colonies. Clazomenae was another colony, but by far the richest and
strongest was Miletus, farther south. It traded with the Persians to the
east and all the Mediterranean to the west; it dominated coastwise bank-
ing and commerce, helped by a currency that, from the seventh century,
was backed by the state. It had colonies of its own, as many as ninety of
them, from Egypt to the Black Sea.

The Ionian colonists, whatever they may have intended when they set
out, ended with societies very different from one another and from those
they had left behind. When the Greeks landed in Asia Minor they were
not moving into a vacuum but into the Persian Empire, and from the beginning they mixed with local populations. As evidence of this, Herodotus mentions that in his time, a few generations later, the Milesians,
who had mixed with Carians, spoke a language completely different
from the Lydian-Greek of Ephesus and Clazomenae to the north, and
that neither resembled the Greek of Athens. Most important, the colonists'
political institutions reflected both their alienation from the mother cities
and their contempt for the despotism of the empire. In their coastal towns
and cities they felt like a new and specially favored part of humanity. We
can best understand Anaxagoras if we look at some of the men and ideas
that by his time had already sprung from that rocky soil.

• • •

In Miletus, more than a century before Anaxagoras, lived a man named
Thales (ca. 624–546 BCE), known as a student of nature. Aristotle tells a
story about him. Once Thales somehow deduced that "there would be a

great harvest of olives in the coming year; so, having a little money, he gave deposits for the use of all the olive presses in Chios and Miletus, which he hired at a low price because no one bid against him. When the harvest-time came, and many were wanted all at once and of a sudden, he let them out at any rate he pleased, and made a quantity of money." I mention this anecdote because it shows one way a Greek colonial could get himself respected and listened to. Rich men, successful generals, and political leaders sometimes had thoughts about the world and how life in it should be lived, which the public respected. Some were written down. No books survive, but later writers quote passages from them to illustrate a point. By now, scholars have dug out hundreds and arranged them, cataloged them, and tried to figure out what they mean.

Tradition names Thales of Miletus as the first Greek philosopher and surrounds him with legends. He is said to have proved some simple and obvious theorems of geometry, as well as one that is not so obvious, as follows. Draw a semicircle and its diameter, D; now choose any point on the semicircle and draw lines from it to the two ends of the diameter. Theorem: The triangle formed in this way is always a right triangle. The proof (starting from simple notions that no reasonable person would doubt) is left as an exercise for the reader. The story raises an important point whether or not Thales invented the proof. Babylonians and Egyptians had a number of mathematical tricks. For example, Babylonians knew this proposition, as well as the Pythagorean theorem, a thousand years before Thales and Pythagoras found them. If they were known they must have been proved, but there is no sign that anyone thought the proofs were important enough to preserve. Whoever set the process of proof at the center of the stage is the founder of all the mathematics since then, and if it was not Thales it was someone who lived not long afterward.

Thales' model of the world, the way it is situated in the universe, and the materials of which it is made have fared less well, but (if correctly reported) they are worth noting as examples of the way someone could think at that time. For Thales, as for the Babylonians, the Earth was a disk. Why didn't it fall straight down? Because it floated on water. This idea too can be found in Egyptian and Babylonian texts. In winter we see less of the Sun; it slips farther south and the days are shorter. Remember (fig. 1.6) that if we place the Earth's poles straight up and down, the Sun goes around the Earth in a path tilted with respect to the equator. It is at the lowest part in winter. Thales is said to have taught that the lower part of the Sun's path dips under water. Even then, this was not an original notion. Whoever laid out the zodiac long before had given watery names to three of the winter zodiacal signs: Capricorn the goat-fish, Aquarius the water carrier, and Pisces the fish, and near them are a Dolphin and a Whale.[2] Since no constellation except perhaps Orion resembles whatever it was named after, the names probably originated in some vast cosmological myth.

Thales' second contribution refers to the nature of matter. We are used to seeing matter change its form. Water becomes ice or steam. One can hard-boil an egg. An ox eats grass and turns it into flesh and bone and hair. Are we to assume that the grass ceases to exist and the other substances are created, or is there some continuity? If so, what stays the same when transformations like these occur? Aristotle mentions early thinkers who taught that matter is not created or destroyed, but only changed. "So they say nothing comes to be or ceases to be; for there must be some entity—either one or more than one—from which, since it is conserved, all other things come to be. . . . Thales, the founder of this school of philosophy, says the principle is water (for which reason he declared that the Earth rests on water), getting the notion perhaps from seeing that what nourishes all things is moist." He means all living things, of course, but we shall see that even for Aristotle, who did not agree with what he had just reported, the whole world was, in some sense, alive. Water was the only substance known in Thales' time that exists in all three forms of matter: solid, liquid, and gas. He would probably not have been very surprised if someone had shown him that his guess was wrong, for as Aristotle realized, his main point was that *something* about matter is constant, that when forms change this something remains. Whatever it is, the search for it still guides the research of physicists who study the most fundamental forms of matter.

So much for what things are made of, but the world is more than a collection of objects, for it *happens*. What makes something happen? Soul, will, intelligence make it happen. According to Aristotle, Thales said that because a magnet moves iron, it has a soul. He also said "All things are full of gods," which Aristotle takes to mean that life and soul pervade the universe. If stars and trees and rivers have life and soul, why not sticks and stones? Of course, the mere existence of soul has little power to explain what we see in the natural world. There is something arbitrary about will and intelligence; no one can predict what they are going to do. But in inanimate nature, at least, many things happen in a predictable

---

[2] In the first century CE, the Chinese astronomer Chang Heng describes the heavens as a sphere half full of water with the Earth floating in the middle (Needham 1954, vol. 3, 217). In the fourteenth century the Arab historian Ibn Khaldun compares the Earth to a floating grape (1958, I.1). The wide distribution of this idea suggests that it, like some names of constellations, may be very old.

way. It would be impossible to make a pot or cook a good meal if the materials behaved differently every time one used them. Some general principles are waiting to be found.

3.3 Earth, Sun, Moon, and Law

The ingenious Anaximander (c. 610–c. 546), a few years younger than Thales, also lived in Miletus. I will start with his picture of the solar system. It is a picturesque contraption, but while reading the following pages please note that we have nothing firsthand from him, not even a fragment, and almost all of what follows comes from authors who wrote from five to eleven centuries later. Anaximander seems to have taught that the Earth on which we live is shaped like a drum, with its depth one-third its diameter. It is situated in the exact center of the spherical cosmos, and it stays there because there is no reason why it should move in one direction rather than another.[3] It doesn't need to float on water. We are told that he was the first Greek who ever drew a map of the inhabited lands that sit on this drum. In the middle is Greece, in the middle of Greece is Delphi, and around the rim flows Oceanus. It is the Babylonian map of figure 1.4, transplanted and reduced in scale.

Over and under the drum rotate two vertical hoops of fire. Each is sheathed in smoke or mist except at one place, where there is a break and its light shines out. These breaks are the Sun and Moon. The Moon-hole's shape changes during a month, and occasionally a hole gets blocked up for a while: an eclipse. The diameters of the Sun's and Moon's circles are twenty-eight and eighteen times, respectively, that of the Earth, the planets have their own hoops, and the stars are closest of all. There is more, but this will do. Implausible as it is, it suggests a possibility that has been very fruitful for science: fire up there is assumed to be fire, smoke is smoke. The lights in the sky are not just lights, they may be things that are open to understanding.

Thales and his pupil Anaximander have raised a new kind of question: What is the world made of? Their attempted answers are often dismissed as naive and amusing, but the two Ionians, out there at the edge of what considered itself the civilized world, deserve a certain honor for being the first in recorded history to raise such questions at all. Inevitably another question had to arise: The world does not just sit there; what does it do? Here again, Anaximander offers a wild guess.

---

[3] The principle invoked by this argument is today called the principle of sufficient reason. It is illustrated by the story, attributed to the fourteenth–century theologian Jean Buridan, of the donkey that starved to death because it couldn't choose between two identical bundles of hay.

Anaximander, like Thales, tries to imagine one existent thing that is responsible for the qualities of the world we know: material qualities, but also others like purpose, energy, and tendency to change in predictable ways. This thing has no beginning in time but is the beginning and cause of everything else. It cannot be any familiar substance because it contains all the opposites: hot and cold, wet and dry, living and dead; Anaximander calls it *to apeiron,* translated as the Boundless, meaning that it is not limited in space or time or in any other way. According to a later account, "He says that it is neither water nor any other of the so–called elements, but some other apeiron nature, from which come into being all the heavens and the worlds in them." What does this actually mean? The fragments don't say, but in section 10.2, still a long way from here, we shall see that modern physics offers a glimpse of something similar.

Anaximander's next contribution needs a little introduction. In the Greek homeland, and probably throughout the colonies, there was frequent political turmoil as competing groups—usually the aristocrats and the rich—struggled for power. The sufferers were the people, and among them the institutions of government with the assent of the majority slowly began to emerge. A crucial step in this process was the idea of a code of laws enforced equally on all. In Babylonia, King Hammurabi had enacted such a code in about 1900 BCE; in Greece it began with statutes drafted by Draco in 621. These codes were extremely severe, punishing those committing relatively slight infractions with death (hence our word *draconian*). They were replaced in 594 by Solon's code in which punishment reflected the severity of the offense. It was in the generation after this that Anaximander groped toward a way of saying that nature, too, is not allowed to behave capriciously. We have no text from his own time, but later writers report words something like this:

> The source from which existing things derive their existence is also that to which they return at their destruction, for they pay penalty and retribution to each other for their injustice according to the assessment of Time.

The words sound poetic but the important ones are legal terms. Anaximander is telling us that the world is just. The seasons alternate, and generations of plants, animals, and people rise and die away. Whatever disturbs the balance of nature—flood or pestilence, locusts, or even something smaller like the ripple in a pond after a stone drops in—endures for a time and is gone; nature, like us, lives under the commandment, "Nothing too much." The Boundless accommodates its conflicting tendencies in the manner of a court of law, not a battlefield, and Time is the judge.

If the Boundless is eternal, how did living creatures come into existence? Anaximander proposes that the Earth was originally covered with water and that land emerged as the Sun dried it. An early Christian commentator named Aëtius reports: "Anaximander said that the first living creatures were born in moisture, enclosed in thorny barks, and that as their age increased they came forth onto drier places and, when the bark had broken off, they lived a different kind of life for a while." Another commentator mentions a very acute deduction: "He says that in the beginning man was born from creatures of a different kind, because other creatures are soon self-supporting but man alone needs prolonged nursing. For this reason he would not have survived if this had been his original form." In fact, he says, "Man was originally similar to another creature—that is, to a fish." Like Darwin and Wallace, he groped his way toward a theory of evolution. He makes some other fanciful speculations about these changes, but then, even today, though most people understand that evolution happens, no one can say in plain language how it happens.

Anaximander was born, thought, and died. He lived in a precarious place at a precarious time. In front of the Milesians as they looked out over the sea were Greeks with whom they felt little affinity and from whom, in case of trouble, they could expect little support. Behind them was the Persian Empire. In 546 the Persians conquered Caria, and in that year Anaximander died. We do not know whether he was tall or short, fat or thin, or preferred red wine to white, but he succeeded in untying the ropes that held his mind to experiences of its own time and place, and his thought took off like a balloon.

3.4 A World Made of Numbers

The Babylonians were great calculators. They covered clay tablets with multiplication and division tables, and they set and solved problems that involved quadratic and cubic equations. They knew the formula $a^2 + b^2 = c^2$ that relates the sides of a right triangle; $3^2 + 4^2 = 5^2$ is a familiar example, but what about $3367^2 + 3456^2 = 4825^2$? Somebody invented a system for finding numbers like these, and clay tablets show Babylonians training themselves with examples that use the system, but there is little sign, over the centuries, of an expanding and a deepening knowledge. It was different in the Greek world.

Consider the diagram of figure 3.1. It shows pebbles laid out on a table; I have drawn lines to guide the eye. Look at the large squares with $1, 4, 9, \ldots$ pebbles, built up by adding more along the borders. Counting shows that

$$1 + 3 = 4, 1 + 3 + 5 = 9, 1 + 3 + 5 + 7 = 16,$$

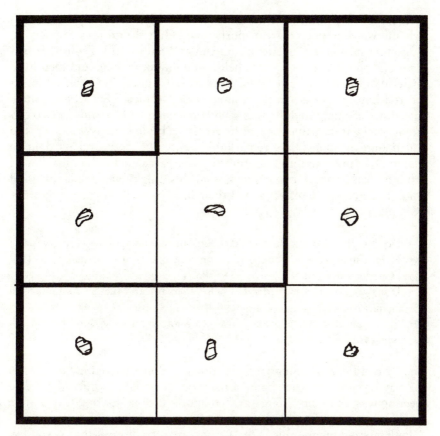

Figure 3.1   Pebbles arranged on a table illustrate the geometry of numbers.

and so on, a fact about numbers that is not quite obvious until you look at them the right way. You can see why 1, 4, 9, . . . are called squares. There are rectangular numbers. Draw some; you may discover that

$$2 + 4 = 2 \times 3, 2 + 4 + 6 = 3 \times 4, 2 + 4 + 6 + 8 = 4 \times 5,$$

and so on, and one can also arrange pebbles in triangles: 1, 3, 6, 10, ...

Pythagoras, c. 580–c. 500 BCE, in his lifetime but more so later, is enveloped in myth. He was probably born on the Aegean island of Samos, and legend says that as a youth "the long-haired Samian" radiated wisdom and serenity. It says he studied with Thales in Miletus, that he spent years in Egypt and among the Chaldeans. He returned to Samos, but in the middle of his life, because of turmoil on the island and because he wanted to reinvent human society, he and a party of followers went off to a remote Greek colony called Croton that had been established about a century earlier in southern Italy. By now it was rich and was known as a great sports town. Here the group set up a commune that recalls settlements in the United States in the 1960s: life was plain and simple, property was held in common, women and men lived on terms of equality, and Pythagoras was the guru. He believed in reincarnation up and down the animal scale, so meat was seldom eaten. Anecdotes accumulated: he was seen conversing with friends in two different places at the same time; as he crossed a bridge the river rose up and hailed him by name (less remarkable, since the river and its daimon shared a common nature); another time he persuaded a doubter that he actually was Apollo by showing that his thigh was made of solid gold. The Pythagoreans regulated their lives by oracular principles. For example,

- Don't poke the fire with a sword. [Don't further antagonize an angry man.]
- Help a man who is loading a wagon, not one who is unloading it. [Encourage effort, not rest.]
- Don't lay down your burden but help others to carry theirs. [An injunction against suicide.]
- Once you have started on a journey, do not turn back. [Don't fight death when your time comes.]

The Pythagoreans dominated the political and intellectual life of Croton for a few years, and then came a reaction. A house in which they were meeting was set on fire. Some say Pythagoras died in it; others say that he lived and taught a few years more in Metapontum, not far away.

Pythagoras taught—this is the Pythagorean tradition—that if you examine the natural world closely enough, if you throw out everything that is temporary and contingent and pertains only to appearance, if you ask what this world really is, the answer is number. Number is more than just a convenient way of expressing a quantity. I have illustrated this at a simple level by showing that one can learn something by representing numbers as patterns of little stones, and of course are many more such tricks. For a Pythagorean, the series of numbers from 1 to 100 is like a line with a hundred people in it. No one would think that the only difference between the people numbered 31 and 32 is their order in the line, for every one has an individual nature. The ancients were sensitive to the individualities of numbers. Numbers are even or odd. Some are prime, others composite. There are squares and triangles and cubes. If a number can be both square and triangular, there must be something special about it. Try 36, and take my word that the next one is 1225. There are so-called perfect numbers; 6 is the smallest one. It is divisible by 1, 2, and 3, and $1 + 2 + 3 = 6$. The next one is 28, divisible by 1, 2, 4, 7, and 14, whose sum is 28. Like perfection itself, perfect numbers are rare. The next is 496, then comes 8128, and the one after that is in the millions. In the Middle Ages it was not hard to explain why God took six days to create the world and not five or seven.

For another view of the Pythagorean idea, consider the number 1, for example, one egg. But now things get complicated. The egg has a top and a bottom, a right side and a left, and furthermore there is something inside. But if one egg isn't an example of unity, or oneness, what is? Euclid (c. 300 BCE) realized that if we want to say anything about oneness we must first say what we mean. He writes his definition: "One is that according to which each existing thing is said to be one." One egg, one city, one army. If we say 12,000 soldiers they are being counted as individuals; if we call them an army we are speaking of the same men in a different way.

One is more than just the first integer; its nature sets it apart from all other numbers. Then comes two, and with it the possibility of talking about relations involving more than one thing, action, or idea. If there were no two, nothing could ever happen. Look at the world this way, keep on attributing special qualities to the integers, and you can see why someone could say that the world is composed of numbers.

The Mesopotamians and Jews also gave numbers special connotations, many of which are found in the Bible. As a single example, consider the number 40. It was bad news in Mesopotamia, and in the Scriptures it is associated with danger or deprivation: the days of rain that made the Flood, the Jews' years in the wilderness, the temptation of Christ, and later, the days of Lent. Scheherazade tells of forty thieves. Even today the word quarantine, from the French *quarantaine,* is associated with illness and isolation. Consider what these and other examples reveal if the Bible is taken as literal truth: that God in the beginning created not only the world but an unalterable plan for its history, so that with the certainty of mathematics specific themes will always be identified with specific numbers. History, if you look carefully enough, will seem to be built out of numbers. Until the Renaissance, the word "arithmetic" referred to these special meanings. What we call arithmetic was known as logistics and was regarded by most educated people as a study fit for grocers and carpenters.[4]

---

[4] In 1664 the diarist Samuel Pepys, settling into a top post in the Royal Navy, hired the mate of a naval vessel to teach him the multiplication table, which had not been part of his studies at St. Paul's School or Cambridge University.

• • •

The Pythagoreans set out to relate their understanding of number to actual experience. The world they imagined was mathematical in nature, a thought rich in implications even if one has a hard time saying what it means. They had to start somewhere, and very wisely they began with a thing everyone notices: the contrast between discordant and harmonious sounds. To study it, they made themselves an instrument called a monochord, a string stretched over a flat box that serves to amplify the sound when the string is plucked. There is also a bridge that can be moved under the string so that the length of the vibrating part can be changed without changing the tension. Twang the string of a monochord, then move the bridge so that the vibrating part is just half as long as it was and twang it again. The sound is just an octave higher. The Pythagoreans didn't know anything more than that about octaves, but they realized that the two notes were related. Furthermore, if you make the length of the vibrating part just a little different from half, the two notes fight. Clearly, there is something special about the half. Move the bridge so that the vibrating part is ⅔ as long. Again, the notes sound best if the ⅔ is exact. Today we would say that the harmony of these simple numerical relationships says something about the human auditory system; but from the time of Pythagoras until the seventeenth century, many educated people thought it says something about the objective nature of the universe.

The Pythagoreans saw numbers everywhere. Aristotle gives a page of examples: they found that justice, soul, reason, opportunity, and many more were, fundamentally, integers. Encouraged by this, since "all other things seemed in their nature to be modelled after numbers, and numbers seemed to be the first things in the whole of nature, they supposed the elements of numbers to the elements of all things, and the whole Heaven to be a musical scale and a number." Later, Aristotle tells us that the Pythagoreans even considered that numbers occupy space and form matter, but of course, he says, it is not possible that solid bodies are composed of numbers.

Of course it isn't possible, if we take the words literally, but I have already tried to explain that people in High and Far-off Times spoke a language that leaks meaning if taken literally. Pythagoras is perhaps two-thirds myth, and we shall see in the next chapter that Plato, Aristotle's teacher, chose the language of myth to say what he had to say about the Grand Contraption. Numbers denote permanence and stability, a world governed physically and ethically by mathematical principles that do not reflect human choice or the whims of a god,

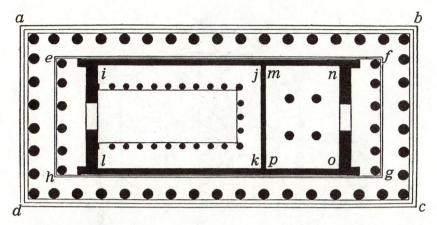

Figure 3.2   Pythagorean ratios in the plan of the Parthenon. The length-to-width ratios of the marked rectangles are *abcd*: 9/5, *efgh*: 14/5, *ijkl*: 8/5, *mnop*: 7/5; *inol*: 7/3. (Adapted from Dinsmoor 1950, 161.)

that are immune to change and that cannot be different from what they are.

Even if most people didn't go as far as the Pythagoreans, the relation of musical intervals to ratios of integers was probably well known, and it is not surprising that similar ratios show up in the architecture of the period. Figure 3.2 shows the plan of the Parthenon in Athens with a few length-to-width ratios of the rectangles it contains.[5] Except near the corners, 9/4 is the ratio of the spacing between columns, measured center to center, to the diameter of a column at its base, and it is also the ratio of the width of the stylobate, the platform on which the columns stand, to the temple's height measured from the stylobate to the top of the cornice. I doubt if anyone thought at the time that the exactness of these ratios improved the look of the building, but they improved its fit into a cosmic scheme.

Figure 3.3 shows the ground plan and vaulting of the late-Gothic church of St. Wolfgang in Schneeberg, Germany, built two thousand years later. The side walls are just twice as long as the façade, and the vaulting is a paradise of geometry governed by small integers that represent in stone the perfection of God's work.

[5] Of course, any such ratio in a classical building may have occurred by chance; the Parthenon's 9/4 probably did not.

Figure 3.3   Plan and vaulting of St. Wolfgang's church in Schneeberg, Germany, fifteenth century.

3.5 CHANGE AND ETERNITY

A while back I mentioned the obvious point that the world doesn't just sit there; it functions. What does it do? Pythagoras used the language of numbers to suggest how it is structured, but he missed half the story, for numbers don't do anything. Anaximander proposed a sort of cosmic equity resembling that of a law court: as justice requires that an injured party be recompensed by the person who produced the injury, so in nature darkness is followed by light and drought by rain; nature is full of these compensations. He spoke his sentence and fell silent. Sixty years later an enigmatic thinker took up the story.

Heraclitus (c. 550–c. 475 BCE) was born in Ephesus, and unlike other Ionians mentioned here he seems to have spent his life where he was born. Legend says that he presented himself as an aristocrat stuck in a frontier town; this pretension was not appreciated by the townfolk and he lived alone. We know nothing of the book he is said to have written, but we can guess that it consisted of cryptic utterances, directed at the wise and puzzling to ordinary readers. More than a hundred of these aphorisms have been preserved; probably they don't give a balanced version of what he wrote, but I will try to convey at least some of his thought by quoting a few (for more, consult Freeman 1978).

Let us start with the word *logos.* This is a very rich term: often it means word, but the small Greek dictionary on my desk gives fifty-four meanings. In Homer, *logoi* are stories. For Heraclitus it is the Reason, perhaps the Design, that pervades and governs the universe. He doesn't define it—perhaps it can't be defined—but the following fragments will show some sides of it. Heraclitus wanted his readers to puzzle out his sayings for themselves, so I will economize on interpretations.

- When you have listened, not to me but to the *logos,* it is wise to agree that all things are One.
- God is day-night, winter-summer, war-peace, satiety-famine.
- We would not know right if we did not also know wrong.
- Into the same river we both step and do not step, we are and we are not.

(Socrates, in Plato's *Cratylus* 402a, interprets the last one as saying that everything is in process and nothing is at rest, and other commentators said it in two words, *panta rhei,* everything flows. This formula will be recalled in sec. 11.4.)

- One should know that war is common and right is strife, and everything happens by strife and necessity.
- Fire lives the death of earth, and air lives the death of fire; water lives the death of air, earth that of water.

(Heraclitus associates the *logos* with fire, which seems to be his fundamental element. This may tell us what *logos* means to him, for a flame is a process and not a thing.)

- The ordered cosmos, which is the same for all, was not created by any one of the gods or by mankind, but it was ever and is and shall be ever-living fire, kindled in measure and quenched in measure.

- The sun will not transgress his measures; otherwise the Furies, ministers of Justice, will find him out.

Finally,

- Immortals are mortal, mortals are immortal: each lives the death of the other, and dies their life.

It is said that Socrates once remarked to Euripides, "What I understood [of Heraclitus] is noble, and also, I think, what I did not understand." For myself, I read Heraclitus like a poem and do not try to squeeze his words into ordinary discourse.

The word *logos* has another meaning that may be unexpected. It means the ratio between two numbers. Similarly in Latin, *ratio* means both reason and ratio, and even in modern French a word for ratio is *raison*. The connection is Pythagorean. For Heraclitus it was logos that pervades and governs the universe; for Pythagoras it was harmony, but both were talking about the same thing.

• • •

If Heraclitus's book traveled as far as Athens, it may also have reached Italy, where Parmenides (c. 515–c. 445) lived and taught in the city of Elea, a few miles south of Naples. He was born into one of the ruling families and must have governed the city for a while, for Plutarch reports that he "set his own state in order with such admirable laws that every year the government makes its citizens swear to abide by them." Nevertheless, he seems to have left public life to become a recluse and ponder the most fundamental of all questions: Does anything actually exist, and if so, what? His surprising answers are contained in a poem that gains authority from being composed in the style of Homeric epic. Time has been unusually generous to this poem, and much of it has survived. The Prologue tells how Parmenides enters the domain of the goddess Justice and how she teaches him the way of Truth. Her lesson can be compressed into a series of propositions, each with its own proof:

1. What one can think about exists. PROOF: Assume that something that does not exist can be thought about. "But this is a path that cannot be explored, for you could neither recognize what does not exist nor think anything about it." Therefore what does not exist cannot be thought about, and it follows logically that what can be thought about exists.
2. What exists is eternal. PROOF: Assume the contrary. This means that at some time in our past or our future it did not or will not exist, but since we are thinking about it, this is impossible.

3. Further, what exists does not change, since if it did, something about it would have come into existence or passed away.

4. What exists has no gaps, for if it had there would be regions of nonexistence. "But it is motionless in the limits of mighty bonds, without beginning, without end, since becoming and destruction have been driven very far away, and true conviction has rejected them."

This reasoning has taken us far from the original question and presented us with an unexpected conclusion: Yes, the world exists, but it is not like the one we think we know and live in, for that one changes all the time. Therefore the world our senses report does not exist. We shall have to struggle with this claim, but let us continue with Parmenides: "And remaining the same in the same place it rests by itself and remains fixed, for strong Necessity holds it within the bonds of a limit that constrains it round about, because it is decreed by divine law that existence shall not be unbounded."

The logic is not very compelling. Proposition 1 says you can't think about a four-legged bird, but you will find it easy if you try. And the conclusion is not surprising: existence exists, everywhere, and nonexistence does not. More than a century would pass before Aristotle began to show systematically what we can and cannot learn from logical arguments. Existence cannot be explained in terms of anything else. But having said that, let us look past the panoply of proof and consider the vision of the world expressed by Parmenides' words.

Something, "what exists," pervades the universe. Anything else? Parmenides mentions Necessity and Justice; they enter his poem as goddesses. Both can be thought about, therefore both exist. Both are everywhere, and Necessity is at the center of the revolving world. She is the source of its energy, and Justice guides her. Justice, he says, does not allow anything to come into being or to pass away, but holds it fast. In Parmenides' mythic language, something important is being said. As I interpret it, Necessity insures that the world is lawful: the Sun sets in the west every night; things fall down and not up; if the right procedures are followed, smelted copper will be pure. And Justice guarantees that the world is also fair. The operations of nature do not favor one person over another. Anaximander made the same claim, and Heraclitus taught that the *logos* tends toward a balance of opposites. We may if we wish see in this an early intuition of a principle of conservation: if something disappears in one place it will turn up somewhere else.

The Way of Truth opened by Parmenides has inspired a few religious minds in every age to believe that, when properly understood, the world is One; and Love, Justice, and Necessity and many more, since they do not differ from one part of it to another, are also One. Parmenides has also taught us that the worlds of thought and of sense are not the same, for none of the connections and interpretations we make, none of the meaning we find when we look at the world, is directly present in what our senses report. What then is the relation between the two worlds? Many Western philosophers have tried to answer this question. Maybe it is just too hard for us.

• • •

Early writers on the history of science did not take the views of pre-Socratic philosophers very seriously. After all, justice, or a bowl of soup, are not numbers; not everything changes all the time; and yet some things really do change. How can those great sages have failed to notice such obvious facts? In the first place, they were beating on the doors of the cosmos with the only weapon they had—pure thought—and also they were writing in the only language they knew that was large, general, and dignified enough to express their thoughts: the language of myth. Parmenides was explicit, but Pythagoras's numbers and Heraclitus's fire are no less mythical. One may say that they would have spent their time better in experiments than in making exaggerated claims, but there aren't really any experiments that would either support or refute what they say. The bedrock of the natural world is experience: of fire and water, of mountains, clouds, and seasons, of events that are predictable and also of events that are not, and in truth not much can be deduced from this experience about how they all fit together.

From this perspective we see that Pythagoras and Parmenides were looking for fixed principles that control and limit the disorder of the world around us. Numbers do not change, nor does Parmenides' motionless sphere of reality. Winds and wars do not blow them away. On the other hand, nothing we know is exempt from change, and Heraclitus claims it is the *logos* that decrees what will happen. The next step was to focus more narrowly on the qualities of things in the effort to see, at least in simple cases, how change comes about.

3.6 THEORIES OF MATTER

In southern Sicily the city of Agrigento stands on a rise of ground that was once the acropolis of the Greek Akragas. The poet Pindar, who lived there, called it the fairest city of men. From the acropolis, walk down toward the sea through groves of almond trees to the remains of seven Doric temples of tawny limestone. The dust of the easternmost of these, the one known as the Temple of Hera, bears the footprints of Empedocles (c. 500–430), for it was built when he was a young man. He was a great orator who in 470 led a revolt that threw down a tyrant; he was also a successful doctor. He saves us the trouble of guessing his status in the community by describing himself: "An immortal god, mortal no more, I go about honored by all, as is fitting, crowned with ribbons and fresh garlands. . . . [People] follow me in their thousands, asking where lies the road to profit, some desiring prophesies, while others ask to hear the words of healing for every kind of illness." Empedocles was friends with Parmenides and members of the Pythagorean group, and he surely knew about Heraclitus. The first claimed that nothing ever changes and the second claimed that everything changes all the time. These are extreme positions, but there is truth in each. Can they be combined into a single theory of matter and process that makes sense? Empedocles wrote two books in Homeric style called *On Physics* and *Purifications,* both lost. They must have been just as abstract as Heraclitus and the rest, but the abstractions are of a new kind. Thales had suggested that water is somehow the basic constituent of every form of matter; Empedocles saw that though words like these might explain something about matter, they had nothing to say about the oppositions we sense in our world. Oppositions suggest more than one basic substance, and he suggested four, calling them roots: "Shining Zeus, life-bringing Hera, Aidoneus, and Nestis who with her tears waters mortal springs." Later on they were known more prosaically as Earth, Air, Fire, and Water. Obviously the weeping Nestis was water, but even in antiquity no one was sure which of the others was which. It's not important, since those names are the ones that have stuck, and we say elements, not roots. Things around us contain all of them, but the mythological names and capital letters serve to remind us that they are abstractions. The element called Water is not what comes out of the tap, and I don't think Fire is always hot. "From these elements are all things fitted and fixed together, and by means of these do men think, and feel pleasure and sorrow." It is not surprising that this first announcement of a material theory of mind came from a doctor.

The mix of elements determines properties, but the elements themselves participate in the dance of the changing world. The dance is propelled by two opposing forces: Love, which draws them together; and Hate, which drives them apart. The elements are what is permanent, for, echoing Parmenides, "It is impossible for anything to come to be from what is not, and it is impossible that what is should be utterly destroyed."

The existence of the Earth and the nearby cosmos shows that in our time Love dominates. Empedocles tells how Love drew the elements together, but the ancient principle of balance requires that Hate will one day break this unity apart until Love once more brings it together. Myth is a language. Does that mean it is not a belief? Well, children believe in Santa Claus; their parents don't.

Empedocles believed in a kind of evolution. In the beginning of our epoch, as Love began to draw the elements together, "Here sprang up many faces without necks, arms wandered without shoulders, unattached, and eyes strayed alone, in need of foreheads." Later, in the second generation, the pieces began to unite, and "many creatures were born with faces and breasts on both sides, man-faced ox-progeny." And what finally survived? The fittest, of course—or so Empedocles thought.

The explanation of matter in terms of four elements had its problems. Silver, for example, must be mostly Earth, the only solid element, but heat it over a fire and it becomes a liquid, then it is mostly Water. Where did Earth go, and where did Water come from? Surely not from Fire. Nevertheless, the four-element theory had an extraordinarily long life. It was still being taught in some European universities in the early sixteenth century, and it was only after that, when people decided to get their hands dirty with experiments, that they began to get some idea of how many elements there really are.

•••

Anaxagoras and Empedocles were both born about 500 BCE, but their styles were different. Empedocles used mythical names and abstract tendencies to explain the structure and behavior of matter. Anaxagoras identified the meteorite as a stone that fell from the sky. He did not mention Zeus or speculate about the purpose of the meteorite. For him it was a stone and, we gather, that was all.

Anaxagoras's ideas concerning the nature of matter have the same specificity. In order to be fairly sure of reporting him correctly, I am taking as much as possible from almost seventy references scattered throughout Aristotle's works.

Suppose you drop a grain of wheat into good soil, and a little stalk emerges and grows larger. Assume that matter is not being created out of nothing. Evidently the plant is sucking matter out of the soil, but something in that first little grain is making sure that what grows will be wheat. When the plant stops growing it contains much more wheat-substance than was in the seed, and if we assume the wheat-substance is neither created nor destroyed, there must already have been some of it in the soil; also, since a seed will grow for a while in nothing but water, there must be some in water too. There is a remark in a book formerly attributed to Aristotle: "A plant does not breathe. Anaxagoras, however, held that it does," so he may have thought there is some plant-substance in air. Anaxagoras next assumed that any sample of matter consists of invisibly small particles he calls seeds, and that every seed contains substances of every kind. A seed also contains pairs of certain basic qualities: hot and cold, wet and dry. When seeds enter an organism the processes of life sort out what it needs and reject the rest.

This theory of matter accounts for the changes that are seen in every form of life, but it does not say, for example, what makes the wheat stalk extract wheat-substance from the environment. Anaxagoras says it is Mind (*nous*): "All things contain something of everything except Mind, and there are certain things that also contain Mind." This is "the most insubstantial of all things, and the most pure," and it exists separately in matter, not compounded into any seeds. Mind governs the worlds and their formation. Our world began as air, ether, and a great mass consisting of an exactly equal mixture of all the material substances. Mind packaged these into seeds. Every seed contained every substance, but in each one some substances and qualities predominated, so that different kinds of matter could emerge. Then the heavens began to turn about the North Pole in the sky, and the flat disk of the Earth was formed. Anaxagoras suggests that our cosmos is not unique and that the whole process has repeated itself here and there in infinite space and time. More than one cosmos at a time? Metrodorus of Chios said, "It is strange for one stalk of wheat to stand in a great plain, and for one world to exist in boundless space."

All this is guesswork, and so are the seeds. Anaxagoras's myth is fanciful and does not tell how Mind makes anything happen, but it fills some gaps in Empedocles' equally mythical scheme. It accounts for the miracles of reproduction and growth, and it incorporates Anaximander's intuition that what happens is not at random but is governed by an intelligible ordering principle. And like Anaximander, Anaxagoras talks about heavenly bodies, as well as the meteorite that got him into trouble, in earthy language. Perhaps in the long run, that is his main contribution.

One thing lacking in Anaxagoras's account is an adequate description of the seeds. "Our senses are weak and we cannot judge the truth," he writes, but "what we can see gives a glimpse of the unseen." What distinguishes one seed from another, and what do seeds actually do? For the next version of Anaxagoras's theory we return to Miletus.

• • •

During Anaxagoras's last days in Lampsacus, tradition says he was visited by a young Milesian named Leucippus. Miletus had seen good times and bad since the days of Thales and Anaximander. It had fallen under Persian rule and organized a revolt that developed into a great Greco-Persian War. The war finally liberated all the Ionian colonies, but not before, in 494 BCE the Persians had captured Miletus again and burned it. At once, the inhabitants started to rebuild but the fighting went on; in 480 the Persian fleet was destroyed at Salamis, and the next year came a land victory at Plataea. The war was not yet over, but Ionians felt a new freedom and Miletus prospered. This was the time when Leucippus was born. As to how he lived and what he did, we know only that he seems to be the one who invented the atomic theory of matter. Only one fragment of his works remains, which I will quote in a minute; most of what we know of his idea is thanks to Aristotle, who explained it with his usual precision a hundred years later but didn't believe a word of it.

Leucippus had a student, about twenty years younger, named Democritus, who came from Abdera, a colony at the top of the Aegean Sea that seems to have been an intellectual center. Democritus traveled widely, knew everybody, and wrote dozens of books. Three hundred fragments remain, almost none relating to atoms, but he so overshadowed his teacher that he is often given the credit for being the first atomist.

According to Aristotle's account, Leucippus's atoms resemble Anaxagoras's seeds except that they are all made of the same substance; they differ only in size and shape. All atoms of iron, for example, are exactly the same. Atoms are what is permanent in the changing world. Their number and variety are enormous. The universe consists of atoms and space for them to move around in, and that's all there is. Simple natural forces determine whether atoms move or stay where they are; there is no need for overarching principles such as Love and Hate; there is no Mind. This picture, simple and mundane, explains common experiences. If a wet shirt dries, the water in it does not cease to exist or turn into air; it is just that the particles of it blow away, one at a time. In the rotten stump of a dead tree a sapling springs up; the new tree is made of the atoms of the old tree, plus water, but some vital process is rearranging them. In this way, once again, a balance between Parmenides and Heraclitus is achieved: the world's constituents never change but its aspect is always new.

What properties does an atom have? Only size, shape, position, and motion. What is the taste of an atom of salt? What is the color of an atom of red light? An atom, Democritus says, has no taste or color, only the four properties, and: "Sweet exists by convention, bitter by convention, color by convention; only atoms and emptiness exist in reality. We know nothing for certain except what changes according to the body's condition and the constitution of the things that enter it and press upon it." We recognize salt by its taste, but taste does not reside in atoms. Today we say that sensation originates in the brain. If our eye sees a red flash or our tongue detects salt, each sends its message along a path of nerves. The electric impulses are essentially identical; they say nothing about color or taste, but they go to different places in the brain: one registers the color red, the other the taste of salt. I mention this to show how deeply Democritus understood the implications of an atomic theory. The subjective nature of physical sensations is illustrated by an old paradox often used to refute the objective nature of physical sensation. Prepare three bowls of water, one hot, one tepid, one cold. For a couple of minutes, soak your right hand in hot water and your left in cold, then test the tepid water with each hand. The right one reports it is cool and the left hand reports it is warm. When Democritus mentions the body's condition in the fragment just quoted, he is referring to paradoxes of this kind.

How many different kinds of atom are there? Countless. Consider only the atoms of sound. No two voices are quite the same, and for each voice every pitch has a different atom. Consider all the nuances of colored light, of odor, of taste—and these are only some atoms that affect our senses.

If atoms move, there must be empty space for them to move into. Parmenides objects that nonexistence cannot possibly exist, but Democritus contradicts him: "The existence of nothing is just as real as the existence of thing." Aristotle is unconvinced: "To maintain that [the material world] is divisible at some points but not at others looks like an arbitrary fiction," but as usual what he had in mind was subtle. Choose any substance (he chooses flesh) and start cutting it smaller and smaller. This can be done ad infinitum, he says, but after a certain definite point the pieces are no longer flesh. In this sense, each substance has a smallest particle, but even it can still be cut.

If the whole world is made of atoms, and if, as it surely seemed to Democritus, these are so small that we can never know anything about them, then the chance of understanding what the world is and how it functions must have seemed very remote. Thus his most famous saying: "We know nothing in reality, for truth lies in the abyss." I think, though, that he was thinking of more than the mysteries of matter, for there are mysteries of the cosmos, of life, of human thought, of the divine that must have seemed to him as impenetrable as they do to most of us today.

The atomists' cosmos is infinite and comprises an infinity of worlds. Their version of how a world is formed, as related by an amateurish but valuable historian named Diogenes Laertius, echoes Anaxagoras: "Many bodies of all sorts of shapes move . . . into a great void; they come together there and produce a single whirl, in which, colliding with one another and revolving in all manner of ways, they begin to separate apart, like to like." The lighter particles migrate outward, while heavier ones stick together to form the disk of Earth at the center of the whirl and, circling about it, the heavenly bodies. "In some worlds there are no Sun and Moon; in others they are larger than in our world, and in others more numerous. . . . They are destroyed by collision with one another," and the cycles begin again. The idea that worlds pass through cycles of creation and destruction is an ancient one; we have seen it in the teachings of Anaximander, Anaxagoras, and Empedocles, and Aristotle says that Heraclitus thought the same.

If Diogenes Laertius has it right, the atomists have unloosed a new idea: worlds come into existence at random. Anaxagoras taught that Mind guided the beginning of a world, but there is no mind or god in the atomic picture. Instead, there is chance. In the Hebrew tradition, exactly the opposite was the case, for God had planned every detail. It is hard to imagine two ideas more sharply in conflict, either in their pictures of what happened in the beginning or in their interpretations of the world in terms of purpose and meaning. Two very different models of the world emerge, and we shall see them struggling for dominion. We shall follow the chance-governed model quite closely when it begins to develop, but after the Classical age the idea stood still for about fifteen centuries. Is chance the same as anarchy? Here are the only words we have from that quiet genius, Leucippus: "Nothing happens at random; everything happens out of reason [*logos*] and by necessity." It is the exceptional simplicity of his atomic picture of the world that allows him to make such a simple claim. There are very few things an atom can do. It can go where it is pushed, or it can fall freely; when two atoms collide, they can bounce or, if they are made to stick together, they do that, but if a mass of them is hit hard enough it comes apart. Leucippus is saying that these simple processes are rule-governed; they happen the same way every time. And where in this picture does the human animal fit? The first atomists seem to have been as strict as Empedocles, who said that it is by means of the four elements that "men think, and feel pleasure and sorrow." Leucippus lived in Miletus at the same time as Sophocles, in Athens, was writing "There are many wonders, but mankind is most wonderful of all." Their words express different views of humanity, but while we may not like to think of ourselves as automata, there may be some advantages to living in a mechanical universe. Let's see.

3.7 ATOMS AND THE PURSUIT OF HAPPINESS

Epicurus (c. 341–c. 270) was born to an Athenian family on the island of Samos. He showed an early talent for philosophy and studied for a while with a slave who had been a pupil of Democritus. When he could leave home, he gravitated to Athens. He was there when Alexander died, and probably because he foresaw the troubles to follow he went home for a few years. Then in 306 he returned, bought a house and garden, and established a school, a group of people who met in the garden to listen to Epicurus and discuss great questions. Women, members of a society in which they received little education and no respect and rarely ventured out of the house, were welcome in the garden on terms of equality. The circle even included a slave.[6] What was most different about this school was what they talked about: metaphysics and politics and ethics were replaced by topics that Epicurus considered loftier still—life and the search for personal happiness. Most in the circle seem to have been people of property and leisure, with slaves to attend to their wants and ease their lives. What did they know of pain or limitations? In fact, quite a lot, for life in the Classical world was harder than ours, and at this time Athens was caught in the breakup of the Macedonian Empire. During the siege of 295, when hundreds starved, Epicurus counted out the rations of his followers from their small common supply, bean by bean.

The lifestyle that Epicurus urged on his friends can be summarized as frugality without worry. But aside from the dangers of war, what was there to worry about? For one thing, the common explanation of epidemic diseases was that they were sent from above. A pestilence during the Peloponnesian War had carried off Pericles and thousands of others, and no one knew when the gods might strike again. They also sent omens for people to worry about; remember the one that defeated the Athenians at Syracuse. Messages from an incomprehensible elsewhere poured in on everyone, all the time. When Epicurus settled in Athens the immortals guided the daily activities of most citizens, as well as the policies of government. Did health and sickness, the success and failure of enterprises, really depend on the whim of invisible beings? So people believed, but Epicurus thought not. First, as to the gods, "A blessed and eternal being has no trouble himself and brings no trouble upon any other being; hence he is exempt from movements of anger and partiality, for every such movement implies weakness." In reading this, we need not think of Zeus, Athena, and the other Olympians. Taking one's troubles to them was like going to the governor to complain about a traffic ticket. We should think of the lesser divinities, hostile *daimones* who bring bad luck, or friendly ones to whom a family has prayed and sacrificed for generations. But did anything really result from this attention? Perhaps it was time to reexamine the relations between gods and men.

If the gods are not responsible for the good and evil that befall us, what is? To answer, Epicurus turns to the physics he had learned, which taught that the thunder and lightning which threaten us, like the rain for which

---

[6] Writers who praise Athenian culture sometimes forget to mention that the leisure which allowed so much fine thinking was afforded by the work of slaves, who at this time composed about a third of the population.

we offer prayers and sacrifices, occur "out of reason and by necessity"; they happen or they don't, and we should not take them personally. Necessity was a goddess in Parmenides' poem, but Epicurus, like Leucippus, sees it as whatever has to happen in the ordinary course of things: if you throw a piece of paper into the fire it burns, and nobody can do anything about it.

Epicurus's students learned that there is no divine plan, that what happens happens, and there is no reason why, in the circumstances in which they found themselves, they should not try to live lives of modest enjoyment.

• • •

Having thrown off some of life's heaviest burdens, Epicurus moves on to an unavoidable question: This feeling of liberation is very nice in practice, but how does it work out in theory? Epicurus widened his philosophy of living into a philosophy of nature. I will sketch it here, though it flowered later.

In his cosmology Epicurus agrees with the atomists: the universe we inhabit consists of matter that behaves like matter. Objects and space exist, nothing else. Planetary motions are purely mechanical. Space must be infinite; the only way it could end would be with a wall of some kind, but this is absurd since the wall itself would occupy space. Infinite space must contain infinite matter and an infinite number of worlds; otherwise all would disperse and be lost. In this boundless space worlds form and disperse at random, as Anaxagoras and others had taught.

The human soul is made of atoms, very small and spread through our bodies. They are not soul as we might use the word, for they are also the path by which sensory impressions reach consciousness. When we die these atoms disperse and that is the end of the soul, so there can be no afterlife, no punishment to fear. The gods are real; we know this because people of every race feel their presence, but they are made of the finest matter and live far off in the space between worlds. They do not concern themselves with our affairs.

If every atom goes where it is pushed, one could say that the universe is preprogrammed. Chance would never enter such a universe, and neither human nor animal would have a will of its own. But Epicurus seems to have taught that that is not how it is: any atom at any time may swerve a little out of its path, not as if it had a mind but more as if its motion were not strictly rule-bound. In such a world, what a mass of atoms does cannot always be predicted. Our minds and bodies are not rule-bound; if they were, Epicurus would waste his time trying to teach us how to live. Diogenes Laertius says that among the three hundred rolls Epicurus wrote was one called *On Choice and Avoidance* and another called *On the Angle in the Atom* in which, I suppose, all this is explained for us, but no word of them remains.

Epicurus says clearly that the axioms of physics should come not from imagination but from observation. Nevertheless, in the existing writings, the evidence he gives to back up his atomic theory is scanty. He claims that when we smell a rose we are detecting atoms it gives off. This is reasonable, but to explain how water freezes he needs two kinds of particles, round and angular. Cold adds more angular particles and drives the round ones out. Where do they go? It was three hundred years before a Latin poet put together a serious collection of evidence that might have changed a skeptic's mind.

• • •

We know little of Titus Lucretius Carus (c. 95–50) except for a rumor that he was subject to fits of insanity and finally took his own life. One or two copies of his only surviving poem, *De rerum natura,* "On the Nature of Things," drifted into monastery libraries in northern Europe and were themselves copied from time to time; the best remaining manuscripts date from the ninth century. The book-length poem is a manifesto of Epicurean philosophy and has placed its author with Vergil and Catullus at the summit of classical Latin poetry. I will not discuss him as a poet; luckily, there is a fine verse translation by Rolfe Humphries (1968), from which you can get some idea of his work.

Like many educated Romans, Lucretius disdains the state religion. Evidently he thinks the Stoics are too stoical, and he welcomes Epicurus's prescription for the good life. It was time for that prescription to be stated clearly once again. First, an increasing number of people were interpreting it as permission to abandon the austere Roman style in favor of coarse sensuality. Also, some seemed to be accepting and believing Epicurus' message without leaving their superstitions behind.

Lucretius explains the atomic theory with arguments based on experience that try to show how the world can be understood without resorting to the supernatural, adding a poet's specificity to the philosophical arguments of Epicurus. He shows that the physics of Heraclitus, Empedocles, and Anaxagoras is mostly empty talk that really explains nothing, while the atomic theory helps us understand not only what goes on in our world but the rest of the cosmos as well. A few examples. I have already mentioned how laundry dries, atom by atom. Light passes through the sides of a lantern made of horn while rain does not; therefore an atom of light is smaller than an atom of rain. Wine flows more quickly than olive oil because the atoms of oil tend to stick together. Honey and wormwood taste different because the atoms of honey are smooth and round while those of wormwood have barbs and hooks that tear the tongue. Though atoms are eternal, Empedocles' four elements are not: Earth dries and blows away as dust, Water evaporates, Air continually exchanges with other forms of matter, Fire goes out unless it is fed.

Like Epicurus, Lucretius writes of infinite worlds in an infinite universe, and the gods have nothing to do with any of it. After all, if there had been a divine plan, the Earth would not contain regions made uninhabitable by heat, cold, or lack of rain; there would be no mosquitoes and no disease. And judging from the evidence, our world is not very old. There is no history of events earlier than the Trojan War. Since weather wears down old carved stones, stones much older than that would be very worn by now, and we do not see any. Lucretius's poem was admired as an inspiration to throw off superstitions and fear of the hereafter.

At about the same time, the architect Vitruvius explained the properties of building materials in terms of their atoms in his book *De architectura,* which was the standard treatise on the subject until the Renaissance, but readers took little note of atoms there or in Lucretius. In Roman times the flame of scientific curiosity burned very low, and in Europe for long after that it burned lower still. We shall see that in the sixteenth century, when thought of atoms began to interest the educated public, it was Epicurus's version, with its godlessness, of course, neatly cut away, that they preferred.

The original Epicureans were mostly ladies and gentlemen of leisure who could afford long hours in the Athenian garden. What interested them could be called Philosophy Lite. Sections 4.3 and 4.4 of the next chapter will describe University Phil., the heavier stuff, studied at a high level in Athenian schools called the Academy and the Lyceum by rising men who would become administrators, politicians, generals, and professors. The names of Plato and Aristotle have reverberated through the centuries, and thousands of students in European and Near Eastern lands have struggled to understand what they said, but when the Renaissance came and people got interested in finding ways to understand the world around them, the mechanized version taught in Philosophy Lite turned out to have more of the future in it. I think there are more Epicureans than Platonists or Aristotelians in the general population today.

✳ FOUR ✳

Earth and Heaven

> The primary science (*epistēmē*) deals with things that are both
> separable and unchangeable.[1] Now all causes must be eternal,
> but especially these; for they are the causes of so much of the
> divine as appears to us. There must, then, be three theoretical
> philosophies: mathematics, the science of nature (*physis*), and
> theology, since it is obvious that if the divine is present
> anywhere, it is present in things of this sort.
>
> —Aristotle

WE HAVE LOOKED at several cosmological models with which people represented the world and their places in it, but a cosmos is more than a layout; it must also explain at least some of what happens. Even storms and floods don't happen at random; they are anchored to the seasons, and seasons are anchored to the annual motion of the Sun so that they tend to recur in a certain rhythm, year after year. The Chumash noticed this (though they were not quite sure and played safe); so did everyone else. In many cultures, all over the world, there seems to have been a belief that the world looks kindly on us black-headed ones, that though it can occasionally be provoked into flood or hailstorm, when things are running normally our lives are safe from acts of nature and our needs are met. How does this happen, and why?

4.1 LAW AND NATURE

Heraclitus wrote, "The Sun will not transgress his measures or the Furies, ministers of justice, will find him out." The measures? Not too bright or too dim, of course, but also the Sun follows exactly the same path among the fixed stars every year, and at the same rate. This isn't obvious, for when you look toward the Sun you don't see any stars. First you have to observe the stars at night and make a chart of the twelve zodiacal constellations. Watch these as they set, one after another, about every two

---

[1] It is clear from what Aristotle says about science that he is not using the word as it is used today. His conception of science is discussed in section 4.4.

hours. An hour after sunset, a certain constellation is visible on the horizon. You know then that the Sun is about half a constellation ahead of the one you see. This is very rough, but it enables you to plot the Sun's path as it moves, at a rate of about two of its diameters every night, along the zodiacal band of your chart. Also it is easy, observing where the Sun breaks the horizon at sunrise and sunset, to trace its journey north and south. Keep track of the dates and you will find that the Sun's path is always the same—the Chumash tightrope—and that a year ago it was where it is today. It had better be back next year.

I sense in Heraclitus's words the uneasiness expressed in the Chumash ceremonies: before December ends, the Sun *must* turn northward. For Heraclitus, the Sun's measures originate in the *logos,* which I have called the Reason, perhaps the Design, that pervades the universe. There is of course much more to *logos* than the establishment of astronomical measures. Sometimes its control is exact; I single out astronomical regularities only because they illustrate the exactness on a huge scale. Heraclitus's warning suggests a universe that would do more harm than it does if it were not limited by a greater power. He does not say what that is, but Anaximander has suggested that it is Time, which compensates nature's excesses just as human law seeks to compensate excesses of behavior. Let us look at some other principles of moderation from ancient cultures.

Rta.   This is the Hindu version of *logos,* a divine, impersonal principle that orders the cosmos as well as human affairs. It first appears in the collection of Sanskrit texts known as the Vedas. The oldest are the hymns of the *Rig Veda,* some of which probably date from the thirteenth century BCE, though they were written down much later. A few words tell how the seasons follow their regular course: "The twelve-spoked wheel of *rta* rolls around and around the sky and never wears out."

Ma'at.   The Egyptian world was permeated by an abstract principle called *Ma'at* which, though sometimes personified as a goddess, exerted an influence wider and more pervasive than any divinity. It is a principle of natural and social order, of truth and justice that inhere in the nature of the world without being imposed from outside or from above. Illustrations that accompany the Book of the Dead show Ma'at identified by a feather in her headdress. She places it in one pan of the scales used to weigh a dead man's heart, for that is where his conscience is kept. If the sins in his heart outweigh the feather, he goes down to join the legion of the damned. In nature, Ma'at sees to it that no drought or famine lasts forever, and that an eclipsed Sun brightens again. The Book of the Dead says that "Ra [the Sun] lives by Ma'at the beautiful" and that Ma'at and

Thoth, a god of wisdom, plot out every day the course of the boat in which Ra rides across the sky.

Anankē.   Anaximander's principle of cosmic justice was enforced by Time. A century later, Leucippus imagined the world as if through a microscope and he saw atoms. When he wrote "Everything happens out of reason (*logos*) and by necessity," the word translated as necessity is *anankē,* but its sense in Greek is stronger, more like compulsion backed up by force. Parmenides makes Necessity a goddess representing an abstract principle, but for atomists the word's meaning was simple. We may not know the details, but every physical change is the result of atoms changing their positions and nothing else. And each atom, struck from one direction or another, goes where it has to. It has no choice. For Leucippus, logos is the legislator and anankē the enforcer.

Number.   We have seen in section 3.4 that the Pythagoreans, encouraged by the success of their efforts to reduce musical harmony to simple fractions, found that numerical harmony pervades the universe. Some Pythagoreans claimed that numbers have size and weight and that matter is made of them. Plato and Aristotle both rejected the idea as absurd, but a thousand years after Pythagoras the Greek philosopher Proclus understood what was meant: "What Pythagoras in his enigmatic style conceived as number was the intelligible order of the universe." Number does not need to legislate or enforce; it corresponds to the way things actually are.

The Tao.   In China the guiding principle was called the *Tao,* literally the Road, or the Way; it adds purpose and meaning to the idea of natural and social regularity. As to what the Tao actually is, the meditations of the *Tao-te Ching,* the book ascribed to Lao-Tse (Anaximander's contemporary), begins,

> The Tao that can be told is not the eternal Tao.
> The names that can be named are not eternal names.
> The nameless is the origin of Heaven and Earth,
> The named is the mother that rears ten thousand creatures.
> Those who rid themselves of desire can see the secret essences;
> Those who desire see only what is visible.

Tao is a formless and wordless principle that existed before Heaven and Earth; it distinguishes right from wrong and it guides, without controlling, the world's physical, moral, and social development. A principle called *te* represents efficacy, with suggestions of force. Tao-te rules the universe. Embodied in a king it rules a country. Joseph Needham, the great historian of Chinese science, offers the doctrine of Tao as the reason why the Chinese, who developed medicine, printing, and many other arts and sciences centuries ahead of the West, never produced anything like a physical theory. Tao governs how things happen and should happen; nothing more need be said.

Wisdom.   In the book of Proverbs, Wisdom demands to be listened to: "She takes her stand at the crossroads, by the wayside, at the top of the hill, ... she cries aloud. . . . Listen! For I shall speak clearly, you will have plain speech from me, for I speak nothing but truth." She aims to teach people to live sensibly and love God, but first she shows them that she is established in the fundamental nature of things: "The Lord created me the first of his works long ago, before all else that he made. I was formed in earliest times, at the beginning, before Earth itself. I was born when there was yet no ocean, when there were no springs brimming with water. . . . When he set the heavens in place I was there, when he girdled the ocean with the horizon, when he fixed the canopy of clouds overhead and confined the springs of the deep." She claims she was the Lord's "darling and delight, . . . while [her] delight was in mankind." She says: "Through me kings hold sway, and governors enact just laws," but her invitation is to all people, especially the simple, to approach her and learn to be wise. The book of the Wisdom of Solomon, from the first century BCE, says that she is "the breath of the power of God, and a pure influence flowing from the power of the Almighty." A century later, Philo of Alexandria, a Jewish philosopher in the Platonic tradition, says, allegorically of course, that she is God's wife, who received the seed of God's plan for Creation and made from it the world and all its inhabitants.

Having made humanity, Wisdom cherished it and tried to protect it from the severity of God's justice. As Christianity evolved, the creative power of Wisdom changed its gender and became Christ. Catholics transferred the nurturing part to Mary, though she plays a very minor role in the gospels, while the Eastern Orthodox Church remembered that Wisdom's part in the drama of salvation preceded Mary's and in some churches she is worshipped as before.

Where did Wisdom come from? There is a much damaged Aramaic papyrus that dates from the fifth century BCE, but it refers to Akkadian times and suggests a similar feminine power: "To the gods she is dear. F[or all time] the kingdom is [hers]. In he[av]en she is established, for the lord of holy ones has exalted her." That is all that remains.

•••

I have gathered these principles of order and rightness into a few paragraphs that make them sound more similar than they really are. The *Tao* seems to be the condition that makes it possible for natural and moral law to exist. Ma'at, rta, logos, Wisdom, and Number refer to principles of justice and proportion and beauty. With them there is little or no sense of enforcement, but te and ananke are stronger: they prescribe and enforce. Because we are talking of matters that cannot be seen or directly experienced, we float among metaphors, but always, in one form or another, the idea of law recurs. Allowing for differences in expression, it seems the world is made so that behind what may seem to be random acts of nature or caprices of gods and men, there are guiding principles that should, or must, be obeyed. And for many people (though not for someone who thinks only atomistically) this law has something to do not only with necessity but also with justice and moderation, or even with beauty and fitness, as in musical harmony. If it is strange that a law that governs equity in human affairs can at the same time tell the planets where to go, remember that in ancient times all matter was perceived as being in some sense alive. Cicero quotes Aristotle as writing in a work that has otherwise vanished, "What remains, then, is that the movement of the stars is voluntary." In section 4.4 we shall examine some of the reasons that led to this claim, which most Arabs, Jews, and Europeans who thought about such things still believed two thousand years later.

Rereading the last few pages I am reminded of a line I copied from the Rig Veda long ago: "Truth is one; the wise call it by different names."

4.2 MEASURING MONTHS AND YEARS

The movement of the stars may be voluntary, but if so they choose to follow an exact schedule. The degree of that exactness interested astronomers very much and led to ingenious ways of measuring it.

Many of the world's laws and ceremonies, in old times and today, are timed by the Moon's phases, which repeat about every 29½ days. This number ties the lunar calendar to the calendar of years and seasons. By the fourth century BCE the Babylonians had a number for the length of a month: 29 days, 12 hours, 44 minutes, and 4 seconds, or 29.5306 days, agreeing with modern measurements. How, using naked-eye observations, could they have arrived at such a figure? It took time, and because the astronomers' notebooks did not survive, I can only speculate how they could have done it.

It is hard to tell just when the Moon is full, much easier when it is half-full. Suppose that under ideal conditions, by aligning the Moon's shadow line with a straight edge and observing it several times during a night of the half-Moon, one can estimate the moment when the Moon is half-full with an uncertainty of 4 hours. Suppose also that lunar records have been kept for 500 years and you know that in a period of 182,617 days, plus an estimated 6 hours, there have been exactly 6,184 lunar months. Dividing these numbers yields the Babylonian figure quoted above. With the 4-hour uncertainty spread over 6,184 months, the uncertainty in the length of one month is 1/1,546 hour, which works out as a little over 2 seconds.

By the thirteenth century BCE, Chinese astronomers of the Shang dynasty who were trying to adjust the calendar had measured the year and the lunar month as 365.25 and 29.53 days, respectively. (Modern values are 365.2564 and 29.5306.) Babylonian astronomers certainly measured the year, but I can find no number. If they knew the number of days between two summer or winter solstices that are about 500 years apart, even if the uncertainty in the number was as great as a whole day, the resulting figure would have had an error of only 3 minutes.

As centuries went on, the measurements improved but the numbers changed only slightly, and in the second century BCE the Greek astronomer Hipparchus declared that the number of days in a month and in a year, are, to a precision measured in seconds, constant, year after year.

• • •

We live in a world of unimaginable complexity. Until very recently, few people enjoyed what we would call comfort, yet it was possible, with luck and foresight, to live a long and happy life. This suggests that the world has some kind of rational order—one couldn't live well in a world of chaos. Any suggestion that the cosmos came together by chance and continues to function that way has to explain this order, one of whose noticeable qualities is its extraordinarily exact performance. Questions demand an answer: What is the purpose of this vast machine? And what are *we* doing inside it? Let us now look at the last of the world's great myths, a single story created by a single man, which aims to answer those questions.

4.3 PLATO'S FANTASY

Plato (c. 427–347) grew up in Athens in a rich and aristocratic family. His father claimed descent from Poseidon, which must have set him near the head of anyone's table. While in his twenties Plato traveled in the Greek world; later he cultivated friendships among Athenian intellectuals. When he was about forty, he bought an old sports ground known as Academē on the outskirts of Athens, a pleasant place with a grove of olive trees. Here he established a school, or it might be better to say, the world's first version of a university, known as the Academy. When Justinian closed it in 529 CE it had been running with occasional breaks for 915 years. The main studies were philosophy, rhetoric, mathematics, law, and politics.

One can imagine Plato staffing and running this place, lecturing, writing, and, to use modern words in an old context, conducting seminars. He wrote a few philosophical treatises. They have not survived, but perhaps this would not have bothered him, as everyone agreed that philosophy is better talked than written. What remains are popular works in the form of conversations on a great range of subjects, and in them his teacher Socrates plays a leading part. When Plato was sixty, he was invited to Syracuse to teach wise government to the son of the local ruler. The effort was not a success, and a second visit several years later almost ended in disaster. He returned to Athens, where he died in 347.

Shortly I will say something about Plato's mythical explanation of the universe, but it will make more sense if I put it into a wider context. It was Socrates' habit, when he joined a discussion in Athens's public square called the Agora, to insist that people explain exactly what they meant, and especially that they can define the words they used. If, for instance, someone was praising so-and-so as being very wise, Socrates would interrupt to ask, "What is wisdom, anyhow?" Someone would answer with a couple of examples, and Socrates asked what they had in common. Everyone would have a try at answering, but as the examples and opinions accumulated they became more diverse, until the discussion broke up and people walked away. His aim was more moral than intellectual: he did not see how it was possible for a person or a state to behave morally unless the standards to be followed were clearly stated and understood.

Plato continued this quest for clarity by turning it upside down. He asked, "Supposing that Socrates has managed to elicit the definition of a quality called Wisdom, what is it, exactly, that has been defined? Surely, not just a word." Perhaps it pertains in some way to events as they actually happen, but a definition, once it has been formulated, may not correspond to any particular experience. The definition is fixed and timeless, while experiences happen and come to an end, and no two are alike. If experience belongs to the world of anecdote, what does an exact definition really contribute? Old Parmenides said that since what really exists is solid, changeless, and above all, One, then the disorderly spectacle that we see as we look around us is not the real world at all but some rough imitation of it. Thus Plato, in his search for clarity, was driven to anchor his definitions, not in our shifting world, but in the changeless one which, according to Parmenides, is the one that really exists.

The myth starts there. There are two worlds, and the one we live in is in every sense the lower one. The "Heaven above heaven" is where reality is. It is the domain of Ideas, or Forms (using capitals when the words are used in this special sense), which have an existence independent of any earthly examples. Wisdom exists in that world; in our world, events occur that illustrate wisdom, but mostly with some condition attached that makes them not quite the real thing. A carpenter making a window knows what the right window for this spot would be. He works his window as close to that as possible, within limits set by skill, materials, and time, but it is not the window he imagined. Ideas are relevant for us, Plato says, because it is possible for a properly prepared mind, after long study, to catch some notion of them and become wiser in thought and action.

And what, if one were to ascend to the world of Ideas, would we see if we again looked upward? In the sixth book of *The Republic* he tells us it is the Idea of the Good. "The Good is not in itself a state of existence but something far beyond it in dignity and power." It is the reason Ideas exist at all, and it is the principle that unifies them. He uses this mythical setting—which I think we are meant to take seriously but not literally—for his dialogues on love and politics and the human condition. In one of the last of them it serves to answer two questions asked at the end of the last chapter: Why was the world made, and what are we doing in it?

• • •

Plato's dialogue *Timaeus* is named after the fictional character who does most of the talking.[2] It was planned as the first of a trilogy. The second breaks off in midsentence and the third was never written. Cosmology in Century 21 contains no trace of the cosmology in *Timaeus,* but an accident of history made it a standard text in the European Middle Ages. In the fourth century CE, the first two-thirds of it was translated into Latin by Calcidius of Tyre, and for the next eight centuries it was, except for some logical works by Aristotle, the one Greek philosophical text available in Europe and the only window that gave a glimpse of Plato.

Timaeus is a long and complicated work. At the risk of giving a distorted impression of it, I will write mostly about the first half, which, in terms that are no less mythical for having been carefully thought out, describes Plato's Grand Contraption. Its history stretches to infinity in each direction, but to make his account easier to understand he describes it in terms of a myth within the myth in which it is put together, part by part.

---

[2] There is a fine book on this dialogue, Cornford 1937, which translates Plato's text and comments on it. To make life easier for anybody who may want to go beyond my short summary of a long work, I will quote from Cornford's translation.

The opening scene recalls the one in Genesis: a god looks out over a formless chaos and considers how it can be made into something good. It cannot be made exactly like the heaven above heaven, but he resolves to make it as much like it as he can (30a),[3] and he sets to work like a divine craftsman who works from an ideal design to create the best world he can out of available materials. Timaeus does not claim to know the details or to get everything right, but here is a likely story, he says, that covers the main points. First, what is the god's purpose?

> Desiring that all things should be good and, so far as might be, nothing imperfect, the god took over all that is visible—not at rest, but in discordant and unordered motion—and brought it from disorder into order, since he judged that order was in every way the better. . . . Taking thought, therefore, he found that, among things that are by nature visible, no work that is without intelligence can ever be better than one that has intelligence. . . . In virtue of this reasoning, when he framed the universe, he fashioned reason within soul and soul within body. . . . This, then, is how we must say, according to the likely account, that this world came to be, by the god's providence, in very truth a living creature with soul and reason.

Plato, who often leaves his main thoughts unspoken, clearly intends that the reason, harmony, and beauty of the cosmos he is about to describe will be reflected in the souls of the people who will later inhabit it.

Material substances, "things that are by nature visible," are without intelligence or soul, and so the god must deal with them as best he can. Reason must guide necessity, but his meaning of necessity is not that of the atomists. The atomists imagine a world they cannot see in which atoms move as they must, guided by principles that are never violated. Plato's ananké is different. The god's material stockpile consists of huge amounts of each of the four elements. He puts them together to form all the world's substances, complete with the roughness and unpredictability that we know well: the split in the wood, the defect in the stone. These are necessities of the material world. A craftsman can only do the best that is possible, and what this one made is the world we live in.

Our cosmos, like Parmenides' stately world, is a perfect sphere, and because if it just sat there nothing would ever happen, he gives it the motion most appropriate to its shape: he sets it turning and then he gives it a mind. As the world turns it thinks, but slow rotation is a changeless motion, and there are many other kinds of change. To make these possible, the god creates time, "eternal but moving according to number" as an image of the higher world.

---

[3] The 30 refers to the page number in the first printed edition of Plato. Each page is divided *a* to *e*, so that 30a starts at the top of p. 30. Every scholarly edition carries these numbers.

Still, nothing that has been created actually changes, so "in order that time might be brought into being, Sun and Moon and the other five stars—'wanderers,' as they are called—are made to define and preserve the numbers of time."[4] Having created them, he installs them moving in appropriate orbits in the world's soul. There, "as bodies bound together with living bonds, they [became] living creatures and learned their appointed task."

There is an inconsistency here: time has just been created, but for a while before that the god has been putting his universe together, piece by piece. This is how we know that the craftsman god, whose nature and origin has never been made clear, is only a literary device that enables Plato to explain the purpose of each part of the created world.

Souls are not enough. The universe so created wants life that shows itself in action, and so the craftsman creates gods, one for every planet and fixed star. They are spherical, made mostly of Fire, the gods "that revolve before our eyes." As for those other gods celebrated in poem and story, "to know and declare their generation is too high a task for us; we must trust those who have declared it in former times."

Plato's god now makes as many souls as there are stars, and assigns each soul to a star. He shows them the nature of the universe and the law that governs their destiny. Any soul that lives its life well will be allowed to return to its star. A male soul that lives badly will come back to Earth as a woman, and if it persists in its bad ways it will descend the scale of existence until it is somehow rescued and returned to its original state. This doctrine, called the transmigration of souls, implies that every creature, every insect has a soul, so that in all there are a huge number of them. Look closely at a beam of light in a dusty room: thousands of particles drifting and slowly settling, and thousands more where the light does not shine. One Pythagorean philosopher suggested that every particle is a soul drifting to earth from its star to be born again, to live and die and return home.

Finally, Plato's divine craftsman plants each soul where it will live for a while, some on the Earth, some in the Moon, and some in "all the other instruments of time." It is then up to the freshly minted gods to add whatever was necessary to these individual souls and use the four material elements to make bodies for them. And so "they confined the circuits of the immortal soul within the flowing and ebbing tide of the body."

It wasn't easy, for just as the sculptor's chisel reveals unexpected flaws in the stone, nothing in the created world ever conforms to the divine plan or functions exactly as it should. Nevertheless, the gods create human bodies which, whatever their faults, are able to move from one

---

[4] See section 2.4 for the way in which planets "define and preserve the numbers of time."

place to another, to see and hear and speak. But Plato has not told us what anything is made of, and now his story turns back to describe the structure of matter.

• • •

A theory of matter has to account for its many forms and ways of changing. Fire moves out of the tip of a flame and becomes air; air condenses into cloud and moisture. Evidently the four elements are not substances of a fixed and definite kind, but if they are not, what is? Geometers of Plato's time knew that there are just five ideal solid mathematical forms, now often called the Platonic solids, defined as shapes bounded by regular equal-sided polygons that are exactly alike. They are shown in figure 4.1. The tetrahedron, octahedron, and icosahedron are bounded by 4, 8, and 20 equilateral triangles, respectively, the cube by 6 squares, and the dodecahedron by 12 pentagons. The god makes Fire, Air, and Water out of the tetrahedron, octahedron, and icosahedron, the first because it is the smallest and has the sharpest edges, and the other two follow naturally. Earth is the cube, assigned because it is less subject to change than the others,[5] and each kind (including Earth, if one cuts the squares along the diagonal) can be taken apart into component triangles and reassembled into other shapes.

It is hard, reading this account, to take it literally; it is so facile and explains so little. Do these geometrical figures perhaps belong to the realm of ideas, abstract archetypes of more earthy materials here below? I think not; they are more like real particles. They disassemble and reassemble, and Plato says "We must imagine all these to be so small that no single particle of any of the four kinds is seen by us on account of their smallness, but when many of them are collected together their aggregates are seen." The implication is clear: if one of them were by magic made large enough, we could see it and play catch with it.

The story goes on. Timaeus says, "There still remained one construction, the fifth; and the god used it for the whole, making a pattern of animal figures thereon." Clearly, this one with its twelve sides stands for the cosmos, scattered with animal constellations, but other than that Plato says nothing about stars.

Timaeus now goes on at great length to explain how his "likely story" accounts for the properties and transformations of matter, for living tissue and the architecture of the human body, for health and disease. The gods responsible for this work have their eye on the necessities of living. The

---

[5] This corresponce of solids to elements may be older than Plato: Aëtius (*Opinions* III.11.3) attributes it to Philolaos, a member of the Pythagorean circle in South Italy.

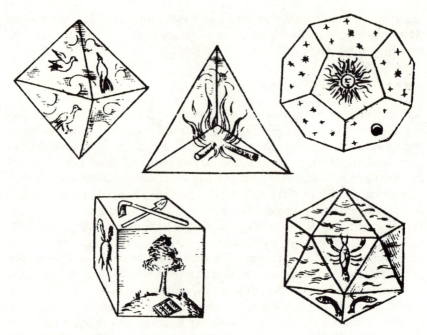

Figure 4.1   The five Platonic solids, decorated with symbols of the four elements and of the cosmos. (From Johannes Kepler's *Harmonices mundi,* 1629.)

divine soul lives inside the head, but now the gods create two more souls: one, in the heart, for courage and energy, and the other, concerned with appetites and desires, in the belly. Humanity in *Timaeus* is a wonderful creation but not the one this book pursues, so I will leave the story here.

What, after all, was the point of this great literary flourish? No one was supposed to take it literally (though in later times, for lack of alternatives, many did), any more than Moby Dick was supposed to be a real whale. A few pages ago I mentioned two questions that have been at the edges of people's minds for a long time as they contemplated the cosmos: Why does the world exist and what are we doing in it? Timaeus answers that since it has always existed, no one can talk about who made it, but it is *as if* it had been made by a god who wanted it to be orderly, intelligent, and good. The alternative was chaotic nothingness. As to our place in it, he made our souls from the same stuff as the soul of the universe and decreed the law (Indians call it *dharma*) that governs their rebirth. And thus, to summarize this long story in two words, the world is Good, and we and it are One.

In discussing *Timaeus* I have paraphrased part of it and included a few quotations. The result gives about as accurate an impression of literary values as if I had done the same with *Moby Dick*. In each, artistry is inseparable from content, and the result is—a myth? Yes, if *Moby Dick* is a myth, if a story composed by a definite person at a definite time can be called that. Plato's story stands as he finished it, but the serious reader is left with a question. Behind the myth and the lessons it taught, what did Plato actually think about the structure of the world? We don't know for sure, but inside the Academy he delivered a famous set of lectures, "On the Good" in which, Aristotle reports, he said some of what he thought. Aristotle's version is so surprising that I quote it:

> The living universe[6] itself is compounded of the Idea of the One together with the primary length, breadth, and depth, everything else being similarly constituted. Again [Plato] puts his view in other terms: Mind is the monad [1]; science or knowledge the dyad [2] (because it goes undeviatingly from one point to another), opinion the number of the plane [3], because it is defined by three points], and sensation the number of the solid [4]; he identifies numbers with Ideas and forms them out of the elements of matter. Things are understood either by mind or science or opinion or sensation, and these same numbers are the Ideas of things.

No talk of souls here, no talk of the Good; this is not congruent with anything in *Timaeus*. It is as if Plato had one message for the reading public and another, purely Pythagorean, for the inner circle. Who knows what he really meant by these words, but if this was the way he talked to his students, one can understand the legend that over the gate to the Academy there was a sign: *Let no one ignorant of mathematics enter here.* Mathematics, remember, was the science of shapes and the nature and meaning of numbers. Aristotle was skeptical even so. According to a later commentator he said, "Thus if the Ideas are a different sort of number, not mathematical number, we can have no understanding of it, for . . . who understands any other [kind of] number?"

Plato's students had an agenda; it was to separate out what Plato "actually said," to study it, learn from it, criticize it, and go on from there. The most distinguished of them was Aristotle (384–322). Of all the intellects that have ever existed, Aristotle's, for better or for worse, is the one that had a millennial presence in Judaic, Christian, and Islamic thought. We will look carefully at what Aristotle says of the form and working of the cosmos, and glance briefly at the principles that govern movement and change. First, who was he and what kind of life did he lead?

[6] Literally, the living being.

4.4 ARISTOTLE'S OPTIMISM

In 367 BCE a Macedonian boy of seventeen named Aristoteles arrived in Athens to enter Plato's Academy. He came from Stagira, a Macedonian town on the north end of the Aegean, where his father was a well-known doctor with professional connections to the royal family. Aristotle stayed at the Academy for twenty years, studying, arguing, writing, probably involved in everything that the Academy's scholars were talking about.

Then Plato died, and as the unchallenged intellectual leader of the younger members, Aristotle waited for the directorship. Another man, Speucippus, was named. The Athenians had been intermittently at war with Macedon for ten years as King Philip tried to push his way into Greece, and they feared, correctly, that worse was to come. They were not about to entrust the Academy to a man whose father had been doctor to Philip's father. Aristotle left Athens, traveled for a while, and spent a few years in the Lydian town of Assos and on the nearby island of Lesbos. Here he may have begun the studies of marine biology for which he is still honored today; then Philip of Macedon invited—or ordered—him home to Stagira to tutor his fourteen-year-old son. Alexander turned out to be more interested in conquering the world than in learning rhetoric and political theory, and when he went off about his business, Aristotle stayed in Stagira. In 338 Philip's army overwhelmed a Greek alliance and three years later, when Athens had been cleaned up, Aristotle, aged about fifty, was able to return there and open his own research center known as the Lyceum. Like the Academy it was on the grounds of a former gymnasium, and here Aristotle walked and talked with his students for a decade while he wrote or dictated book after book describing his own findings and what he had learned from the scholars around him. (Life was, however, a little less rosy now for the intelligentsia than it had been in Plato's time because one of Philip's first acts had been to free most of the slaves on whom their leisure depended.) In 323, Alexander died, the Macedonian grip loosened for a moment and anti-Macedonian riots broke out. Aristotle fled to the old city of Chalcis, farther north, and there, in the next year, he too died.

His works in the standard English edition fill about 2,500 pages, and a similar amount has been lost. Though he loved and revered Plato, he followed a different path. Plato (in the works that remain) puts us in touch with the mysteries of existence; Aristotle is more interested in how things work: how reasoning works, how arguments work, how plants and animals grow and change. He wrote about physics, cosmology, meteorology, psychology, sense perception, memory, sleep, dreams, how animals function, on logic, biology, metaphysics, ethics, politics, economics, poetics, on and on. His readers looked up from their books and saw the world around them not as a vast enigma governed by supernatural forces but as a system made and functioning in a way they could at least partially understand, with the promise that further study would tell them more. The world, for Aristotle, was a project. Except in studies of nature that described things clearly observed, he is more the searcher than the finder; the phrase "the science we are seeking" occurs again and again. Some of his conclusions have lasted and some have not, but in every case he started with the idea that explanations are possible, and he got his students to think that way. We are all his students.

Aristotle wrote dialogues in his youth, but only pieces of them have survived. At his death what remained in Athens was a pile of draft manuscripts and lecture notes written by him and others, waiting to be put into order. It is said (nobody really knows) that most of the notes were inherited by relatives who in those uncertain times did not know what to do with them. They hid them, some say in a well, to keep them safe from a Macedonian general who wanted them for his own library. The story says that they stayed in the well for three centuries; then the last head of the Lyceum hauled them out and began the long task of putting what remained of them in order. They were finally published during the great burst of intellectual activity, centered in Alexandria, that was released by the period of peace and prosperity that ended the Hellenistic Age, and scholars hurried to write commentaries on them.

In Rome, Aristotle was respected and a few intellectuals like Cicero studied him carefully, but his greatest influence on Western culture came long afterward. It began in the eleventh century, when scholars from northern Europe visited Spain and found to their amazement that Arab and Judaic scholars were writing commentaries on Greek works of which northerners had only dimly heard. Most of the manuscripts were in Arabic, produced in a translation workshop called the House of Wisdom that the caliph al-Ma'mun had established in Baghdad four hundred years earlier; a few were still in Greek. Excited by these discoveries, educated Latins living in Constantinople searched the libraries there, found manuscripts of Aristotle, and began putting them into Latin. The first results, including *Physics, Metaphysics,* and *On the Soul,* reached Europe in about 1150. A century later, a Dominican monk named William of Moerbeke learned Greek while attached to Catholic parishes near Constantinople. In twenty years he translated a mass of texts and commentaries which, when copied, flooded the European universities.

The early translations of Aristotle were welcomed at Oxford and some provincial French universities, but at first sight of them the older professors in Paris drew their wagons into a circle; in 1210 and again in 1215 they banned the books from the curriculum. They had some reason to do this. Aristotle mentioned gods as if there were more than one; he thought the universe extends to infinity in past and future; he doubted that the soul is immortal. After these and other errors had been corrected he began to appear on the syllabus, and then something remarkable happened. It was noticed that on the whole, men who had studied his logical and rhetorical works and been trained to argue questions pro and con in public made better bankers and businessmen, better administrators of civil and ecclesiastical property than those who had not, while still others were better able to analyze and explain the mysteries of Christian doctrine. Aristotle became known as the Master of those who know, or, more simply, the Philosopher.

• • •

As we look around us we see things, and if we watch for a while they change. Aristotle seeks to understand the first principles of things, why they are as they are, and what makes them change. "The first principles and the causes are most knowable," he says, "and from these, all other things come to be known." What Aristotle calls *epistēmē* and we, lacking a better word, translate as *science* comprises what we actually know, what has to be true and could not be otherwise. That kind of knowledge may be suggested by observing some particular occurrence, but because we see it once, or even many times, that does not mean it could never happen otherwise. To really know it will always happen one must know how it happens and, just as important, what causes it to happen. Astronomy and medicine were not sciences in this sense; they were arts (the word is *technē*) along with architecture, farming, medicine, and war.

An example of a statement that is scientific in Aristotle's sense is "All things that are in motion must be moved by something." There were many kinds of motion. One was natural motion. The natural motion of Earth and Water and things in which they predominate is toward the center of the universe; that of Fire is away from it, and Air is neutral. The science Aristotle sought would consist of statements as solid as these, together with their necessary consequences. In the following pages I will tell first what Aristotle thought about the cosmos and the Earth's situation in it, then how he explained what happens in it, how it changes, and what it does. Most of the first is from the book called *On the Heavens;* the second is in *Physics,* but these works were never prepared for publication, and the arrangement we have is the best his ancient editors could do with the mass of papyrus that came to them. Look first at the picture of the cosmos that these texts reveal.

• • •

In the days of Plato and the ages before him, it was proper to explain the universe with myths like Love and Hate, which were supposed to contain truth even if they did not represent an actual state of affairs. A generation later Aristotle and his students, digging after facts, stored myth on the shelf labeled Uplifting Literary Entertainment and looked for explanations connecting what is seen and verifiable with fundamental principles of nature. Aristotle's first assumption in *On the Heavens* is that the Earth is round. We do not see this directly but the evidence is strong, for when it eclipses the Moon the shadow it casts is always an arc of a circle. Further, he says, the Earth is "of no great size," and he quotes an estimate that its circumference is 400,000 stades. A stade is the length of the straight running track in the city stadium. What city? If we choose Olympia, Greece's spiritual center, the stade is 630 feet and this makes the circumference 48,000 miles, almost twice the actual figure but not bad for a first guess.

Aristotle's second assumption is that the Earth does not move. The two assumptions are carefully argued in book 2 of *On the Heavens*: if the natural motion of heavy matter is toward the center of the universe, "the jostling of parts" would tend to ease the Earth into a motionless spherical form centered on that point. That the Earth does not turn follows from the fact that the natural motion of its parts is downward and not sidewise.

The central sphere in Aristotle's cosmos is named Earth because that's what it mostly is, and it is enveloped by successive spherical layers of Water, Air, and Fire. (Note that if the element Fire gave off light the sky would be light all the time, so Fire must be transparent.) Above the sphere of Fire is only a fifth element, Ether. Now Aristotle lays out the rest of the universe.

First, consider the nightly movement of the fixed stars. They wheel around a point in the northern sky with no change in their relative positions. Something must be rigidly holding them in place: they must all be attached to a vast surface that rotates daily around an axis through the North and South Poles. The stars seen in every direction are about equally bright, which suggests that all are about equally distant: the surface must be spherical. This is what Plato and everyone else had already supposed, but Aristotle gives the reason. "The shape of the heaven," he writes, "is of necessity spherical; for that shape is most appropriate to its substance and is also by nature primary." Aristotle is grasping for a fundamental principle that would require the existence of a sphere, but he had a tough, skeptical mind, and he doesn't let it go at that. He points out that his model is the simplest and most obvious way of understanding the turning sky, and of course the Earth itself, in the middle of it, is also a sphere. I think that if one accepts his assumption that the Earth does not move, the celestial sphere is the only reasonable model that explains the facts. We need a name for structures of this kind: Earth in the center, stars fixed to a great rotating sphere, planets in the space between them. I will use Aristotle's word, *cosmos.*[7]

Even though Aristotle's Earth stood still, there was a well-known Athenian intellectual named Heracleides, a member of Plato's circle, whom Aristotle must have known and who thought otherwise. He argued that it is much more reasonable to set the Earth turning once a day than to assume a gigantic sphere that carries the stars around with it at unimaginable speeds. He also proposed that at least two of the planets, Mercury and Venus, orbit around the Sun. This was a shrewd guess, but it died with him. After all, if the Earth revolved once a day and if its circumference was 48,000 miles, a point on the equator would be moving at 2,000 miles an hour, and the solid soil of Greece only a little less. One would expect that there would be a gigantic wind, and if you dropped something it would seem not to fall straight down, but nobody notices anything.

Against the starry background of Aristotle's heavenly sphere move seven planets: Sun, Moon, Mercury, Venus, Mars, Jupiter, and Saturn. The last five, the ones we still call planets today, travel across the sky in ways that are hard to predict, even reversing their motions from time to time. The first astronomer to offer an explanation of these motions was one of Plato's students, a mathematical genius named Eudoxus. He succeeded in inventing visualizable models—separate ones for the Sun, the Moon, and each planet—which account for the motions we observe. Each model consisted of a series of concentric spheres that rotate at constant speed. Why spheres? We have just seen that the outer surface of the cosmos is spherical and so is the Earth, but there was a more fundamental reason. The stars were gods, the whole heavenly display was divine, and a sphere was the shape that most perfectly expressed that divinity. And divinity does not speed up and slow down; hence the uniform motion.

Eudoxus's spheres occupied the space between Earth and the fixed stars. To describe the motion of the Sun, which travels in an inclined path around the zodiac at a rate that seems not quite constant, mount a smaller transparent sphere (fig. 4.2) so that it turns inside the sphere of fixed stars on an axis pinned to two opposite points of that sphere. Now mount the Sun on the equator of the inner sphere. By choosing the size of this sphere and the points where its axes are pinned and making the inner one rotate with respect to the outer at one revolution per year, we can model the Sun's apparent motion as seen from the Earth. And of course, if the Earth is assumed stationary, the whole construction spins once a day around it. To account for small effects Eudoxus added a third sphere,

---

[7] The word refers to a beautiful arrangement and survives in our word "cosmetic."

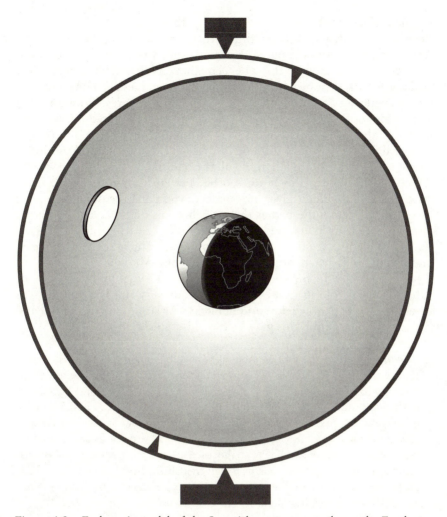

Figure 4.2   Eudoxus's model of the Sun, 4th century BCE, shows the Earth stationary at the center of two rotating spheres. The outer sphere carries the stars;
the inner one carries the Sun around the zodiac. The offset pivots provide the inclination of the zodiacal path. The Sun is drawn in perspective to suggest that for
most people it was a circle in the sky, not a sphere.

and three were also enough to account for the motions of the Moon. The
other planets required four to model the stops and starts as they moved
across the sky. These models did not account for all the observations, and
later workers improved them with more spheres.

Eudoxus's spheres were mathematical; they were like the lines that one
draws to prove a theorem of geometry, and they represented the planets

one at a time. There was no question of their being physical, actually there, because they could not all fit around the Earth at the same time. Aristotle, twenty years younger than Eudoxus, was not much of a mathematician, but he thought the spheres might be real, and with great effort he designed a system by which the whole Eudoxan model could be packed into a single set of fifty-five concentric material spheres. Why does the system move? Because the spheres and the planets mounted on them are made of Ether, whose property is to be continually in motion. If it moved in a straight line it would soon be lost to sight, but its natural motion is circular, and this motion inheres in all the celestial bodies. Ether pervades the cosmos, down to the Moon's orbit, perhaps down to Earth, for it gives us heat and life, it gives motion to the Sun and the other stars; it is "something beyond the bodies that are about us on this Earth, different and separate from them, and the superior glory of its nature is proportionate to its distance from this world of ours." The other four elements interchange with each other—Water becomes Air when it boils—but Ether is eternal, which means that like stars and other gods, it is divine. When Timaeus told us that every planet and every star is a god, he was repeating ancient knowledge, and in fact, as Aristotle says in a grand and memorable paragraph, this may be the oldest knowledge of all:

> Our forefathers in the most remote ages have handed down to us their posterity a tradition, in the form of a myth, that the stars are gods. The rest of the tradition has been added later in mythical form with a view to the persuasion of the multitude and to its legal and utilitarian expediency. . . . But if we were to separate the first point from these additions and take it alone—that they thought the first substances to be gods—we must regard this as an inspired utterance, and reflect that, while probably each art and science has often been developed as far as possible and has again perished, these opinions have been preserved like relics until the present. Only thus far, then, is the opinion of our ancestors and our earliest predecessors clear to us.

But do educated people still believe this? Aristotle fears not: "We think of the stars as mere bodies, and as units with a serial order indeed but entirely inanimate; but we should rather conceive them as enjoying life and action." This statement is memorable for what it says but also for what it denies. Only a century earlier Anaxagoras had gotten into trouble for suggesting that the cosmos consisted of random accumulations of matter. Now it seems that some educated people were thinking the same way.

Do stars have souls? Timaeus says yes. For Aristotle, soul belongs to life, but exactly what is a soul? In his book *On the Soul* he struggles with words: "Soul is the actualization of a natural body that has life potentially within it." Then stars, being alive, have souls, and since stars are immortal their souls are too, though generally he is skeptical about immortal souls. "If there is any way of acting or being acted upon proper to soul, soul will be capable of separate existence; if there is none, its separate existence is impossible." He argues the matter carefully, and at the end he seems to conclude that souls are mortal, though in *Metaphysics* he concedes that the part of the soul that is pure reason may survive death.

So much for Aristotle's model of the world. But the world is not just a collection of things; it changes, more or less, all the time. Now we will look at some principles of change.

• • •

Like Aristotle's idea of science, his idea of motion is not the same as ours, for it includes every kind of change. The simplest and most obvious kind is what we call motion. When something goes from place to place, he calls it local motion, but another kind might be the ripening of a peach. No special purpose is served by a rolling stone, but if peaches did not ripen the world would be a poorer place. There would be no more peaches or peach trees, birds and insects would be disappointed, and we would not have the pleasure of eating the ripe fruit. Since almost everything Aristotle writes about motion is supposed to be applicable generally, it is good to keep this example in mind. A green peach ripens, but a ripe peach never turns green. The nature of peaches is such that the green one tends toward an actualization that is ripeness.

Just as a fruit ripens, a stone released from the hand falls toward the ground. Both these occurrences, says Aristotle, are natural motions, but there is a difference. The peach's ripening, and most of an animal's movements, are natural and self-caused. The stone, on the other hand, falls only because it has first been lifted. Such motion is called violent, and the fundamental principle governing it is that it lasts only as long as the force causing it: "Everything that is in motion must be moved by something. For if it has not the source of motion in itself, it is evident that it must be moved by something other than itself."

Aristotle's principle, even among those who never heard his name, ruled people's thoughts about change, particularly local motion, for two thousand years. Consider, for example, the motion of a cannon ball. Exploding gunpowder gives it violent motion. It leaves the barrel in a straight line and continues, pushed by the air behind it. The push weakens and the ball slows. Then, after an interval of mixed motion, its natural motion takes over and it drops straight onto its target. Three-part trajectories, as well as some that omitted the mixed section entirely, were part of artillery theory into the seventeenth century. Figure 4.3 comes from a manual of applied mathematics published in 1561.

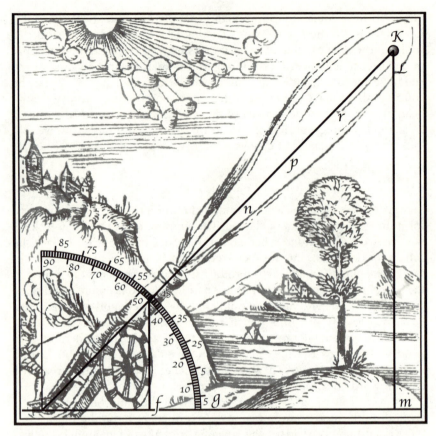

Figure 4.3   Artillery theory in the sixteenth century. The projectile's violent motion is abruptly replaced by natural motion. (From Santbech 1561.)

• • •

In Aristotle's view, the cosmos is full of violent motion. The spheres turn naturally and steadily but otherwise nature is marked by starts and stops and sudden changes. We have been told that everything that is in motion must be moved by something. But if there is some moving principle that energizes the world, what moves that? Aristotle escapes an infinite regress in the only possible way: "There must necessarily be something eternal, whether one or many, that first imparts motion, and this first mover must be unmoved." There are many examples of unmoved movers: objects of love or desire or fear exist, in our minds and in the world outside, which without moving inspire us to act. To keep the world going, says Aristotle, the first mover (or *Prime Mover,* or God), "produces motion by being loved, and it moves the other moving things."

The other moving things are the heavenly spheres. They turn, slowly and exactly, for the love of divine goodness, and they do this because they have intelligent souls. As to whether the prime mover is one or many, Aristotle seems not to have made up his mind, but his fifty-five spheres move, and they move for love. The prime mover (if there is only one) exists outside the cosmos; "There is neither place, nor void, nor time outside the cosmos. Hence whatever is there is of such a nature as not to occupy any place, nor does time age it."

The celestial machinery turns around us, once a day. Slowly? Even the most conservative estimate of its size reveals that its equatorial region travels thousands of miles every minute, a gigantic rush of motion. The spheres turn, the planets execute their dances, and two of them especially, the Sun and the Moon, act via Ether to produce motion in the Earth's air and water and the creatures that live in them. Humans and animals may seem to move only when they choose, but while they live their bodies are always in motion, and the internal organs from which they derive their strength are moved by what they receive from the world outside them. Without this contribution they cannot live. So step by step, the motion of the spheres reaches the ground, refined and attenuated, until it makes the smallest flower spread its petals toward the Sun. The stars and planets do this? "We must," Aristotle says, "assign causality in the sense of the originating principle of motion to the power of the eternally moving bodies." So saying, he opens a door that leads to astrology, but he does not walk through it.

Of course, causality involves more than a source of energy, and Aristotle, for purposes of discussion, distinguishes four kinds of cause. Consider how a sheep is made. It starts with the coupling of a ram and a ewe; Aristotle calls this an *efficient cause.* Then as a young lamb grows into a sheep, other causes operate. There is the grass it eats, a *material cause.* Inherent in the lamb is some principle which determines that it will grow into a sheep rather than a fish or a camel: the *formal cause.* There is a fourth cause that goes deeper, the reason *why* the lamb matures: so that it can reproduce its own kind, and perhaps, if one feels that way, so that we can have it for dinner. The lamb doesn't know this, there is no perceptible influence that pushes it toward maturity, but if it did not move that way there would be no sheep. This is called a *final cause,* the purpose behind motion. In our time, scientists try to explain the world in terms of substances and mechanisms, tacitly rejecting formal and final causes. They assume that what happens at each moment is determined by preexisting conditions and not by anything in the future, but for Aristotle the final cause operates everywhere: "God and nature create nothing that does not fulfill a purpose." More than that, "The good and the beautiful are the beginning both of the knowledge and of the movement of many things." And as to nature itself, "In all things, as we affirm, nature always strikes after the better." This can be compared with a remark by the American physicist Steven Weinberg, toward the end of his book titled *The First Three Minutes: A Modern View of the Origin of the Universe*: "The more the universe seems comprehensible, the more it also seems pointless." One writer finds nature aiming toward some kind of good; the other finds no sign of an aim. I think that most scientists today would say that the question is not a scientific one and cannot be settled that way.

A final remark: One sometimes reads that Aristotle divides all causes into four piles, but that was only by way of illustration. Cicero writes that Aristotle himself realized the situation is more complicated, "for thought, foresight, learning, teaching, discovery, remembering many things, love and hate, desire and fear, distress and joy, these and their like he thinks cannot be included in any of those four kinds [and so] he adds a fifth kind, . . . the mind itself."

• • •

Goodness and love rule the world; Plato invented the idea and Aristotle worked it out in his own terms. This is his great inheritance from Plato. Timaeus's world is a myth, explained in a long and complicated story that adds dignity to life by showing that mankind has a special place in the cosmos. Aristotle was a biologist, perhaps the greatest that has ever lived. Most of his surviving work is biological and shows patient and accurate observation. He has studied the habits of birds, insects, animals, and fish. When he theorizes in biology, it is because he knows. In cosmology the method was different because the known facts were so few. He states his assumptions—spherical heavens and motionless spherical Earth—and shows that they agree with the available evidence. Then, starting with these and a few other assumptions, for example that a thing only changes as long as something is pushing it to change, he draws conclusions, following the method that Greek mathematics had been developing since the time of Thales. This is no cosmos described in terms of myth; Aristotle is trying to establish what it actually is and how it actually works. It must have taken an immense effort to combine Eudoxus's seven planetary mobiles so that they could all move at once without running into one another. The fact that he did it shows not that this is the way it necessarily is, but that the thing is possible, if not this way then some other. Has Aristotle given us another myth? It is not a myth like Plato's, for he digs into the actual working of things. Still, there is the prime mover. At present we just don't know enough about the world to explain it with axioms, and people no longer try to do so.

FIVE

Beginnings and Endings

> From Harmony, from heav'nly Harmony
> This universal frame began:
> When Nature underneath a heap
> Of jarring Atomes lay,
> And could not heave her Head,
> The tuneful voice was heard from high:
> Arise, ye more than dead.
> Then cold and hot and moist and dry
> In order to their stations leap,
> And MUSICK's power obey.
>
> —John Dryden, 1687

THE BIBLICAL CREATION IS a short diary of an eventful week, but it has always raised more questions than answers, and we must take time to look at a few of each. A typical question: What was God doing before he created the world? Another: Why did he locate the center of the Earth where he did and not one foot to the right or left of there? These questions, though frivolous in tone, raise issues that shouldn't be brushed aside. It is easy to invent glib answers, but a careful one requires some historical background. Those two sample questions invite us to think about time and space and what, if anything, God has to do with them.

5.1 TIME AND SPACE

Augustine of Hippo (354–430), whom we will meet formally a little further on, ponders the question, "What is time?" in the autobiography that he calls his *Confessions,* adding an unhelpful remark that is much quoted because it represents the state of mind of people facing the question for the first time: "If nobody asks me I know, but if someone asks me, I do not know." The remark is unhelpful because the problem of "what" is much larger than the example suggests. Socrates asked what is justice, what is love; everybody knew until he asked them. Plato invented the theory of Ideas because he could not answer them in the terms of ordinary discourse. But then, what is the taste of an apple?

Time, in Plato's *Timaeus,* is not an Idea but part of the machinery of the created world. Ideas (or ideas—lower case—for that matter) hang uselessly in the sky unless they are put into action, and action involves time, "a moving image of eternity," for every act has a before and a during; and the world is changed, much or little, forever after. Hence we derive the world of experience, explained as an imperfect model of the real world of Ideas (though I think most people think that experience is what is really real). But it is possible to look a little deeper.

• • •

Time, says Aristotle, is often said to be motion or change, but these are properties of individual things, whereas time is the same everywhere. Furthermore, if we try to judge whether change is fast or slow, we are measuring it with respect to time, and this is futile because, as he reminds us, "time is not defined by time." "Time is not described as fast or slow but as many or few, and as long or short." Nevertheless, we know time by experience: we look at a thing, we look again and it is not the same. Time, though it is not change, is a word that pertains to change. In fact, says Aristotle, "Not only do we measure change by time but also time by change, because they define each other." Time is a measure of change. It is "the number of change with respect to before and after," and it is continuous because change itself is continuous. The last remark is necessary because numbers, to Aristotle and his friends, are integers and ill-adapted to describing a continuous process. We count whole days and estimate whole hours, but even when cloth is sold by the yard, everyone understands that the cloth itself is continuous.

This all looks quite simple when one reads it, but the lesson is not easily absorbed, for the tendency to think of time as some invisible thing that moves alongside the stream of perceptible events is very strong. Of course something is moving: in the world around us one motion causes another just as in our own minds one thought leads to another. Thinking of how the world moves, the emperor Marcus Aurelius writes, "The universal cause is like a torrent: it carries everything along with it." Here he is talking about actual events, but when Thoreau says, "Time is but the stream I go a-fishing in," that is metaphor.

Aristotle's discussion is long; it raises and answers many questions, but in the end it teaches a simple lesson that, if properly absorbed, would have defeated an army of later philosophical arguments before they began: contrary to the message of a hundred clichés, time is not something that exists and causes change. It is a measurement, an attribute of change, including change that takes place in the mind. In this way it is exactly like length or distance, which also do not exist by themselves. As to the world's age, Aristotle argues long and forcefully in *On the Universe* that, contrary to the opinion of many of the first philosophers, it has always existed. Tides of humanity may rise and fall like the wash of the sea, but the heavens move exactly as they always have and always will.

This condensation of Aristotle's attempt to establish once and for all the meaning of the slippery word *time* has no immediate bearing on discussions of the Creation. For the next two thousand years, however, almost everyone who wrote about Creation had studied Aristotle and tried to learn his way of thinking, and we can best follow them with Aristotle in mind. Now what about space?

• • •

The first thinkers to tackle the question of space were the Pythagoreans. Most of what we know about them comes from Aristotle as he pushes the views of his predecessors aside. He writes, "The Pythagoreans . . . construct the whole universe out of numbers, but not numbers considered as abstract units, for they suppose the units to have spatial magnitude. But how the first 1 was constructed so as to have magnitude, they seem unable to say." We have already seen an example of this in figure 3.1, which shows dots laid out on a regular grid with a unit spacing between them. In this sense (extended to three dimensions), we can see how it could be said that numbers occupy space, but it was hard to talk about the obvious fact that space is continuous and the numbers they talked about are not.

In another place Aristotle tells us more about these numbers. The Pythagoreans, it appears, thought that the space between the Earth's surface and the starry sphere is full of *pneuma*, a kind of breath, essentially air, and that outside the sphere is an infinite void. But void is not the same as nothing, for "the Pythagoreans held that void exists and that it enters the world from the infinite *pneuma* outside, as though the world were inhaling the void which marks off the natures of things, as if the void were what separates and distinguishes things situated next to each other. And this applies first of all to the numbers, for it is the void that distinguishes their natures." This strange passage suggests that it is not numbers but the continuous intervals between them that occupy and measure space, but it does not say what space is.

The atomists assumed that atoms and the void were equally real partners in existence, and when Plato assembled triangles into the fundamental particles that correspond to the four elements, he had no trouble in putting them into space. It is, he says, "everlasting, not admitting destruction; providing a situation for all things that come into being, but itself apprehended without the senses by a sort of bastard reasoning, and hardly an object of belief," and he went on to build a world in it. But even though Plato's void is emptiness that you can measure and talk about, it had an opponent who was able to exclude it from learned discourse for a long time.

• • •

Aristotle argues that the atomists' void is impossible. It is not just Parmenides' argument that what does not exist does not exist; that's too simple. He has a deeper objection, that what does not exist could have no properties at all: no size, no location, not even dimensionality. There could be no *there* in it, so that one could not say that anything is located or moving in it. Nothing could happen in it, since no cause could operate.

For Aristotle, space has no physical or logical existence independent of matter. If a room is empty except for the air in it and a cat walks in, how do we define the cat's place? It is, according to Aristotle, the cat-shaped boundary between the air and the cat. But there is a difficulty. Suppose there is wind in the room. The air flows past but the cat does not move. Is the cat's place changing? Aristotle tells us to ignore the moving air and look at the walls of the room: "The place of a thing is the innermost *motionless* boundary of what contains it." But now of course the boundary no longer fits the cat, and the definition no longer fits our notion of place. Having reached that nervous definition of place, Aristotle can now define space as the totality of places a thing can occupy.

The inability to separate place from matter was always awkward, and several writers produced other explanations. Probably the clearest of these was by John Philoponus, one of the last classical philosophers, who lived in Alexandria about 530 CE. He writes, "Space is not the limiting surface of the surrounding body . . . it is a certain interval, measurable in three dimensions, incorporeal in its very nature and different from the bodies it contains, for its nature is incorporeal. It is pure dimensionality, and indeed as far as matter is concerned, space and the void are the same." I think that this, almost nine hundred years after Plato, is an example of Plato's "bastard reasoning." We are about to consider the Creation, and questions naturally arise: Did space already exist? If so, was there anything in it, and why did God choose one spot, rather than another, to plant the Earth? Similarly with time: Why not earlier or later? Jews and Christians shared the book of Genesis, but it seems to have been mostly Christians educated in the Greek tradition who labored over questions like these. Let us see how they answered them.

5.2 CREATION

The earliest Sumerian and Egyptian creation stories start with two motionless divine bodies coupled in a watery chaos. The Hebrew version starts with a few words spoken by God, but the Buddhist version (sec. 1.3) and those of Greek thinkers all start from the experience of sleeping outside and occasionally waking to see that the northern stars have revolved around a certain stationary point in the sky. We seem to live at the center of a great whirl of stars, and it was natural to imagine that our world consists of heavy matter that gathered at this point.[1]

Anaximander, of whom we know so little, seems to have thought so. He imagines a universe permeated by the Boundless, that "indefinite thing from which come into being all the heavens and the worlds in them." It contains every quality: heat and cold, light and dark, and all the rest, and it is always in motion, otherwise nothing would ever happen. In the sixth century CE, Simplicius of Cilicia says "he explains the creation of the world not by any change of the Boundless but by the separation of opposites as a result of its eternal movement." Later, "A kind of sphere of flame was formed round the air surrounding the Earth, like the bark of a tree. When this was broken away and became shut off in certain circles, the Sun and the Moon and the stars were formed." We have already seen how the drum-shaped Earth sat in the center of rings of fire and smoke. The universe we know was formed from this one vortex, but there are others in the Boundless: "From the Boundless arise the creation of the world and its destruction. From it the heavens were separated off and all the innumerable worlds." Since the Boundless is infinite in time as well as space, it suggests the kind of cosmos that Anaxagoras and the atomists proclaimed a little later.

The beginning that most of us know in one version or another as the Creation is the one sketched in Genesis. It is not clear whether a watery waste was present before God began his work. A Jewish commentator points out that if one reads the words with a microscope, the word "created" comes before any mention of the waters; therefore the Creation came first. But consider: after God created light he saw "that it was good"; when he ordered the firmament "it was so"; the mention of each creative act is finished off in the same way except for the one that would have produced the opening scene. Does that mean that the opening scene was already there? A speaker in the Book of the Wisdom of Solomon, first century BCE, refers to "thy almighty hand, that made the world of

---

[1] Put some sugar into a cup of tea, stir it a couple of times, and see what the sugar does.

matter without form," but a phrase from a book written nine centuries after Solomon does not settle the question. I mention it only to introduce discussions by Jews and Christians who wanted to extract the last drop of truth from the trickle of words that begin the Bible. Jews, possessing the texts in their original language, were expert at teasing out implications by careful study of the words. Most early Christian scholars read them in a Greek translation called the Septuagint, one step removed from the original, but they were educated in Greek philosophy that produced bold inventions where knowledge was lacking. Because the curious inhabitants of any model of the world have a right to know whether it has been there forever or, if not, how it began, the question of beginnings is a vital part of our history and deserves to be discussed.

• • •

In the fourth century, the Roman province of Cappadocia in what is now central Turkey was the intellectual center of the Christian movement that Emperor Constantine had legitimized in 312. There, a few years later, a man known afterward as Basil the Great was born in the town of Caesarea, now Kayseri. After studies in Constantinople and Athens he returned to Caesarea, and for a time before his early death he was the bishop.

In his sermons on the Creation Basil fills out the bare narrative of Genesis. The curtain rises on an "older state," unformed, chaotic, and timeless; in the opening scene Spirit hovers over it. Then, "Let there be light: and there was light." The Gospel says nothing more about this light. The Sun had not been created; what was its source? At the end of the day did it just go out, leaving the windy ocean in darkness as before? Basil explains:

> The air was lighted up, or rather it made the light circulate mixed with its substance, and, distributing its splendor rapidly in every direction, dispersed itself to its extreme limits. . . . For the Ether is such a subtle substance and so transparent that it needs not the space of a moment for light to pass through it. Just as it carries our sight instantaneously to the object of vision, so without the least interval, with a rapidity that thought cannot conceive, it receives these rays of light in its uttermost limits. . . . Since the birth of the Sun, the light it diffuses in the air when it shines on our hemisphere is day; and the shadow produced by its disappearance is night. But at that time it was not by the movement of the Sun, but by this primitive light spread abroad in the air, and withdrawn in a measure determined by God, that day came and was followed by night.

In this way God established the daily rhythm of time that Sun and Moon would continue when he had made them. Before the beginning was darkness; now it was night.

On the fourth day "God made the two great lights: the greater light to rule the day, and the lesser light to rule the night; and he made the stars also . . . to divide the day from the night; and let them be for signs and for seasons, and for days and years." The signs are to show farmers when to plant and when to reap—Basil is firm on that point. Many people said that "for signs" justifies astrology. That, he claims, is impious and absurd.

Even after Basil has explained it, the scriptural story raises questions: What about the heavenly spheres, vastly larger than the Earth, whose existence had been demonstrated by Eudoxus and Aristotle? This huge machine, involving not only gigantic size but also unimaginable precision of construction and operation, is nowhere mentioned in the Bible. Was it there already? And was it made so that once started it could run by itself, or does God keep it going? The life of the natural world suggests that it can keep itself going, but what of humanity? Man's will is free, but does this mean that since Creation, except for miraculous interventions such as a virgin who gives birth or a man who rises from the dead, the whole world has run without God? This question is still asked. Those who answer yes are called deists; the contrary minded are theists. Basil was named a Father of Fathers of the Greek church; several of the other Fathers, including his own younger brother Gregory of Nyssa,[2] wrote their own reconstructions of what happened in the Six Days. They tend to be intelligent, persuasive, and quite different; one stands out especially.

• • •

Aurelius Augustinus, known to Catholics and Eastern Orthodox Christians as Saint Augustine (354–430) was born and lived much of his life in what is now Algeria, and his name shows he was a Roman citizen. After a youth that his autobiography, the *Confessions,* represents as rather wild, he joined the Manichean sect and later, studying philosophy in Milan, he became a Christian. Back in Africa, he planned to spend his life thinking and writing but was drawn into parochial affairs and ended his life as bishop of Hippo, a small city on the coast west of Carthage. He wrote hundreds of letters and sermons and dozens of treatises, and he was certainly the most influential teacher in the first thousand years of Christianity. He expounded the Creation in several works, including chapter 12 of the *Confessions* and a two-volume book called *The Literal Interpretation of Genesis,* from which most of the following account is taken.

Augustine probably spoke Greek but did not read it easily. The Bible he used was what is known as the Old Latin version, and it says that in the

---

[2] See Park 1997, 29f.

beginning the world was invisible and formless. "Before [God] formed this unformed matter and fashioned it into kinds there was no separate being, no color, no shape, no body, no spirit. Yet there was not absolutely nothing: there was a certain formlessness devoid of any specific character." Then, on command, the Creation began. Augustine is not sure of any of this, but wants to help believers understand God's immense intellectual achievement in conceiving without error the interlocking complexity of the natural and moral worlds. The whole cosmos began to function the instant it was created and has continued without change, but things are always being created. There are changes in the landscape, trees grow, children are born. What is God's role as the world goes on? In a remarkable paragraph of his book *On the Trinity,* Augustine distinguishes two kinds of creation. He begins:

> It is one thing to make and rule over some kind of being in a free act of creation; only God does this. It is something quite different for external forces to operate so that what has already been created may later, in one way or another, come into being. For all such things have already been created in a sort of texture of the elements, and they come forth when they can. As mothers are pregnant with their children, so the Earth is pregnant with the causes of things that are always being born.

Note "when they can." The world was not created with its future locked in, but with potentialities that become actual when time and opportunity arrive.

This discussion of beginnings started with two apparently frivolous questions: What was God doing before he created the world? and, Why didn't he center the world one foot to the left or right of where he did? Augustine answers both in book 11 of his *Confessions.* God, he says, lives outside of time. His Word has existed since eternity ("In the beginning was the Word," says St. John) and all things were created by it, and yet they were not all made at once. How does it happen, then, that God intervened at specific moments, as when he ordained the Flood? The answer is that the Word, in its temporal sequence, is spoken through the physical body of Christ, who lives eternally in time. What was God doing before the Creation? The question implies a temporal process that denies the eternality of God. There was no time before the Creation. Why was the universe not centered one foot away from where it is? This question, like a few others, is left as an exercise for the reader.

• • •

Before the Church Fathers, Judaic scholars were studying and interpreting sacred texts word by word, a method known as *midrash,* meaning a search or an investigation. These studies are rarely as detailed as those of the Fathers, but they are full of acute and sometimes unexpected insights. In section 1.1 we have seen a midrash on the separation of the waters. On the question, what was the source of light during the first three days, Rabbi Eliezer (c. 40–c. 120) and some others suggest that God made the "two great lights" and the stars at the same time as he created Heaven and Earth, but only hung them up during Day Four after the firmament was installed. The explanation is awkward, however, for the lights move across the sky. Greek mathematicians had studied these motions but Eliezer does not mention them. Is there a better explanation? Yes, a good one, but only if we jump forward several centuries.

• • •

Moses Maimonides (1135–1204) was a rabbi in Cairo, known familiarly to later Jews as Rambam, who took Scripture and Aristotle seriously and did what he could to harmonize them. He was born into a learned and prosperous family in Córdoba, the largest and most cultivated city in Europe and the capital of Muslim Spain. Famous scholars taught in schools joined to its mosques, and a talented student could not have had a better education anywhere else. When he was about thirteen this city of mosques and palaces was conquered by an army of Muslim fanatics and life became increasingly dangerous for its Jews. The Maimon family stayed under worsening conditions for a few years while their son studied religion and medicine and Greek philosophy; then they moved to Fez and finally to Cairo where, since they were now in poverty, Maimonides supported them as a doctor while he studied and taught. His religious writings fill volumes, the most famous of which is the *Guide of the Perplexed*.

One of the perplexing questions was this: a philosophically educated person examines the version of religion that serves simple people and concludes that it is one-dimensional, logically incoherent, and unlikely to be correct. Clear thinking is based on a knowledge of philosophy (which of course includes science), but philosophy and religion often reach different conclusions. How to combine religious and philosophical thought so that one does not overcome the other? Maimonides' aim was not to subordinate either but to present the conflicts to properly prepared minds in language that cannot be used by the ignorant to promote an unbalanced view. For this purpose some points must be expressed discreetly. "Not everything mentioned in the Torah concerning the Account of the Beginning is as the vulgar imagine," he says, "for if the matter were such, . . . the Sages would not have expatiated on its being kept secret. . . . The correct thing to do is to refrain, if one lacks all knowledge of the sciences, from considering those texts merely with the imagination." He goes on to explain that God requires those who have the necessary knowledge to share the secrets it opens to them, but in flashes hidden among mundane remarks, and not all at once. He exemplifies this in his own writing, for one of the challenges of the Guide is to decide which, if any, of several contradictory statements on the same subject is the one directed to the person who is reading. Luckily, his discussion of the Fourth Day seems fairly straightforward.

The first point is that since, as Aristotle said, a star or planet is a glowing concentration of Ether on an Ethereal sphere, they must both have been created at the same time, and this explains why Scripture makes no special mention of the spheres. There are at least eighteen spheres, and perhaps many more. "As for the bodies that move in circles, they are animate, endowed with a soul that makes them move. . . . As regards the question whether they have an intellect by means of which they make mental representations: this does not become clear except after subtle speculation." The heavenly motions produce change on Earth: weather; growth of plants, animals, and minerals. "And just as an individual would die and his motions and forces would be abolished if his heart were to come to rest even for an instant, so the death of the world as a whole and the abolition of everything within it would result if the heavens were to come to rest." Here he perhaps takes too literally Aristotle's remark that assigns causality to the power of the planets.

As to the subtle speculations, Maimonides argues in another place that when Psalm 19 says "The heavens tell out the glory of God," it shows that the heavens have intellect; he continues, "and this without speech or language, or the sound of any voice." And if Genesis says that the Sun and Moon are "to govern day and night and separate light from darkness," Maimonides says, "it is absurd to assume that he who governs should not know that which he governs." In fact, he tells us, the spheres are angels, "For the spheres and the intellects apprehend their acts, choose freely, and govern, but in a way that is not like free choice and our governance, which deal solely with things which are produced anew." This is as close as we will ever come to understanding how planets can have perfect freedom and yet keep perfect time, and thus Maimonides arrives by a different route at exactly what Aristotle seems to have thought.

5.3. THE UNIVERSE RECYCLED

In section 1.5 I told how an Egyptian priest told Solon, "There have been, and will be again, many destructions of mankind arising out of many causes." He specially mentions fires, which destroy hill dwellers, and floods, which eradicate cities, so that the civilization as well as inhabitants of every region must periodically be renewed. This is the first report of such a speculation in Greece, but it must have existed earlier.

Anaximander, born around the time Solon returned from Egypt, talked of such historical cycles, and a few years later Pythagoras went further: each cycle is an exact repetition of the one before it. I have not found any claim that the gods also repeat their quarrels and adulteries, but since they occasionally interact with humanity this seems to be implied. If general cataclysms occur at exactly equal intervals, any information from an earlier cycle would enable us to tell when our own will end, but it seems the human race barely survives each time, and the destruction is so nearly complete that all information is lost. Later a student of Aristotle's named Eudemus of Rhodes argued that if the cycles of existence are really indistinguishable they must be identical, so that there is in fact only one existence, and one time. That is, the scale along which time is measured has the character of a circle, not an infinite line, thus answering the otherwise difficult question of how the whole thing got started. We are told that Heraclitus believed in exact repetitions and so did Empedocles, for whom it was a sign of the alternate mastery of Love and Strife.

In *Laws* and in the fragment called *Critias,* Plato again takes up the Egyptian priest's story and tells what happens when there is a deluge: cities on the coast, where the arts of civilization have flourished, are washed away, and the only people left are simple shepherds living high in the hills. For a long time they live in peace and in fear of another flood; slowly some escape the idiocy of rural life by moving down to the plains and seacoast. Cities arise, and then, one by one, the arts and graces are reinvented and civilization flowers again.

Aristotle distinguishes two levels of catastrophe. At definite intervals there is a destructive regional flood like Deucalion's, which occurred when Zeus was angry at a local despot and produced rain. But not everything is lost in such a flood—Aristotle (above, in sec. 4.4) mentions ancient wisdom that survives as relics from our ancestors and earliest predecessors. He called the interval between floods a Great Year, but there is also a Greatest Year. At very long intervals, Sun, Moon, and the rest of the planets all arrive in the same part of the sky; then a more general destruction takes place: a flood if it happens in winter, a fire in summer. The date of a planetary alignment can in principle be calculated, but alignment must be clearly defined: Must the planets all be in an exact straight line or is it enough that they be in the same constellation? Some writers attached the figure of 36,000 years to the Greatest Year, but nobody ever said how it was calculated, and in the second century CE the astronomer Ptolemy, who really could calculate planetary motions, wrote that an exact alignment would "take place not at all, or at least not in the period of time that falls within the experience of man."

• • •

Why did so many Greek thinkers whose thoughts otherwise differed widely believe in a periodic destruction of mankind? Fragments of evidence suggest that the idea came from Mesopotamia. Bishop Aëtius tells us that Heraclitus posited periodic destructions every 10,800 years. This looks at first like an arbitrary figure, but written in the Babylonian system it is simple. The number is equal to $3 \times 60 \times 60$, written as 3;0;0, the sort of number a Babylonian savant might choose. Let us follow this clue a little further.

In India, in the seventh century CE, there lived a mathematician named Brahmagupta. He was the first person we know of who calculated with negative numbers and the first to use zero as a number, as in $6 + 0 = 6$. He lists four successive ages of mankind, called *yugas,* with their durations in years. I will write them in decimal notation and then in sexagesimal numbers (sec. 2.3):

Yuga	Duration, Years	Sexagesimal
Krita	1,728,000	8;0;0;0
Treta	1,296,000	6;0;0;0
Dvapara	864,000	4;0;0;0
Kali	432,000	2;0;0;0

The Krita Yuga was an age of peace and fertility. People lived long lives and their deaths were like going to sleep; there were many Brahmins. The Treta was only three-quarters as peaceful as before, with more Kshatryas (warriors) than Brahmins. In the Dvapara, every good is balanced by an evil, and murder is common. In the Kali Yuga, our own time, society turns upside down; Sudras, the lowest caste, will rebel against Brahmins; caste will finally be abolished, there will be no justice, temples, or schools, and many books will be written.[3] At the end of the Kali Yuga the world will suffer catastrophic floods, but just as Plato told us, a few people will survive on mountains and in caves, and the story will begin again. It is said that we are now only about five thousand years into the Kali Yuga, at the head of a long downward path.

The cycle of yugas lasting $4,320,000 = 20;0;0;0$ years is called a *caturyuga,* and it corresponds to Aristotle's Great Year. A thousand repetitions of this make a *kalpa,* a Greatest Year; then the world dissolves into its

---

[3] Thirteen centuries earlier Hesiod, in *Works and Days,* made a similar division of the ages of mankind into Gold, Silver, Bronze, Heroic (the warriors of Troy), and Iron.

atoms, and after a long interval the entire process slowly begins again. The kalpa betrays a mixed origin. Indian mathematicians were probably aware of the Babylonian notation but they themselves invented the decimal system and used no other. The caturyuga looks like a Babylonian interval; then the kalpa was added by Indian astrologers, who enjoyed big numbers.

• • •

Not everyone believed that events in successive Great (or Greatest) Years repeated each other exactly. For one thing, until astrology became generally accepted, no one thought the planets cause any particular event to happen; they only mark the time, and Aristotle claimed no more than that they are an originating principle of motion on Earth. There was no reason to think that if at some time they repeat their motion they will cause an exact repetition of events on Earth. But still, the same people and circumstances might show up again, even if inessentials like the colors of their clothes were different. (I remember a conversation with university students in India in which their consensus amounted to exactly this, and the clothes were their example.)

• • •

The Bible does not mention earlier creations, but a fourth-century commentary on Genesis by Rabbi Tanhuma finds a trace of them. When Genesis 1:31 says that "God saw all that he had made, and it was very good," this implies a comparison. "We deduce therefrom that God created previous worlds and destroyed them, until he built the present one, declaring 'This pleases me, the others did not.'" The unconvincing nature of this argument shows, I think, that the conclusion was already in the commentator's mind and that he was looking for words from which he could hang it.

In the third century CE, Origen, one of the earliest Fathers of the Church, answered the question as to what God was doing before the Creation, two hundred years before Augustine. "It is," he writes, "both impious and absurd to say that God's nature is to be at ease and never to move, or to suppose that there was a time when goodness did not do good and omnipotence did not exercise its power." Therefore, he continues, it is logical to suppose that "God did not begin to work for the first time when he made this visible world, but just as after the dissolution of this world there will be another one, so also we believe that there were others before this one existed." But of course, says Origen, successive worlds do not repeat exactly, for then there would be no such thing as free will. Later, St.

Jerome supports Origen with scripture: What else does Isaiah mean when he says "There shall be a new Heaven and a new Earth," and surely Origen has also explained the words of Ecclesiastes, "What is it that hath been? Even that which shall be. . . . And there is nothing new under the Sun, nothing which can speak and say, Behold this is new. For it hath been already, in ages of old which were before us." Augustine has a ready answer: read the words in their context. It is clear that the prophet did not mean to be taken literally, for a little further on he says, "I have seen everything that has been done under the Sun; it is all futility and a chasing of the wind." He surely doesn't mean that literally. But even so, the idea of transitory worlds in the hands of eternal God reappears from time to time in Scripture, as when the Psalmist sings: "Of old hast thou laid the foundations of the Earth: and the heavens are the work of thy hands. They shall perish, but thou shalt endure: yea, all of them shall wax old like a garment; as a vesture shalt thou change them, and they shall be changed; but thou art the same, and thy years shall have no end."

None of these claims has ever touched the general consensus among Jews, Christians, or Muslims. Their teachings shared the same story of the beginning of everything, and for all three, when the end comes it will be the end. But what will actually happen?

5.4 The End of Everything

To the average person living in the world it seems clear that it had a beginning, for nature tends to run down. Mountains wear away, mines become exhausted, the Earth can't be terribly old now, and many legends have arisen to tell how it started. But what starts does not necessarily end, and if you think of the Earth as sinking slowly toward weariness and exhaustion, it is not surprising that more has been said about its dawn than its sunset. For those taking an impersonal view, the two are equally worthy of study. Heraclitus and others after him imagined worlds that form from random accumulations of matter and dissolve after a while, but these are only worlds. When ours dissolves, there will be others, and the universe remains. Will it always remain?

In section 1.4, Seneca cheerfully predicted how whole regions of the Earth will some day be flooded out, but these are just preludes to the end of everything, and one of his letters tells how that will actually happen. It is a long essay addressed to a brilliant and strong-minded woman named Marcia who had seen her father starve himself to death in Tiberius's prison and was now desolated by the death of her last remaining son. At the letter's end, Seneca reminds her that not only mankind but the universe itself is mortal:

Fate will flatten mountains and raise new and lofty summits; it will dry seas, divert rivers, and by suppressing communication between nations it will abolish the society and unity of the human race. Elsewhere it will open gulfs that swallow cities; other cities will be destroyed by earthquakes, and pestilential vapors will rise from the Earth. Fate will cover the inhabited world with floods that kill every animal; it will spread devastating flames to consume everything else that breathes. And when the hour arrives for the universe to be destroyed and recreated, every substance will destroy itself, stars will crash together, and fire will envelop the universe and consume the ordered lights of heaven. Your son is happy, Marcia! Now he understands these mysteries.

Even though the word "recreated" offers a new world, the ending Seneca imagines is not the mere separation of particles previously bound together while the rest of the universe goes on as before. There is purpose in this destruction, as if something or someone is really out to get us and the Earth and the whole universe of stars. Why should this happen? Seneca doesn't say, but it might reflect the lesson learned from a good farmer who plows up his field and lets it lie fallow for a year or so before he plants it again.

• • •

For the Peoples of the Book there was one Creation, but the end is less certain. Isaiah, in a strange passage, imagines an ending (fig. 5.1), when

> the Lord's anger is against all nations
> and his wrath against all their hordes;
> He gives them over to slaughter and destruction. . . .
> All the hosts of heaven will crumble into nothing,
> the heavens will be rolled up like a scroll,
> and all their host fade away,
> as the foliage withers from the vine
> and the ripened fruit from the fig tree.

A few verses later it turns out that this terrible wrath is directed only against the neighboring land of Edom: "Age after age will it lie waste and no one will ever again pass through it. Horned owl and bustard will make it their home; it will be the haunt of screech-owl and raven." Critics say that the first of these quotations is part of an older poem, inserted by later editors. Clearly it refers not to Edom but to the world, and that's why I print it.

The Book of Revelations tells that there will be a Last Judgment and, for those who are saved, a new Heaven and a new Earth: "The first Earth had vanished and there was no longer any sea. I saw the holy city, new

Figure 5.1 The heavens will be rolled up like a scroll. Fresco in the Kariye Camii, Church of the Chora, Istanbul, early fourteenth century. (Dumbarton Oaks, Byzantine Photograph and Fieldwork Archives, Washington, D.C.)

Jerusalem, coming down out of Heaven from God, made ready like a bride adorned for her husband. . . . It shone with the glory of God: it had the radiance of some priceless jewel, like a jasper, clear as crystal." This city is 150 miles on a side and just as high, but like the Lamb and the Beast and the Great Whore of Babylon, it is made only of allegory.

More specific prophecies are available in late Jewish and Christian addenda to the Bible. Writing in the first century CE, the Jewish author of the book *IV Ezra* tells how we shall know when the Last Days have arrived: "One-year-old children shall speak with their voices; pregnant women shall bring forth untimely births at three or four months, and these shall live and dance. And suddenly shall the sown places appear unsown, and the full storehouses shall suddenly be found empty. And the trumpet shall sound aloud, at which all men, when they hear it, shall be struck with sudden fear."

The Sibylline Oracles, mostly Jewish in origin but worked over by Christian editors and dating from about the third century CE, are (or pretend to be) utterances of a dozen oracles in the east Mediterranean area, collected and studied because they were thought to be divinely inspired. Here is what will happen before the Judgment begins:

And then a great river of blazing fire
will flow from heaven and will consume every place,
land and great ocean and gleaming sea,
lakes and rivers, springs and implacable Hades
and the heavenly vault. But the heavenly lights
will crash together into an utterly desolate form.
For all the stars will fall together from heaven onto the sea.

This destruction is prelude to punishment of the wicked and settlement of the just in a new Earth with "springs of wine, honey, and milk. The Earth will belong equally to all, undivided by walls or fences. . . . Lives will be in common and wealth will have no divisions. For there will be no poor man there, no rich, and no tyrant, no slave."

Man builds his version of the world from the materials that surround him, but experience is the architect. Most people in antiquity lived sad, short lives, and the chorus in Sophocles' *Oedipus at Colonus* agrees that the best fate for anyone would be never to have been born, with early death the second best. All the accounts of the end of the world quoted above have a certain similarity, and perhaps the message common to all of them is, "Destroy it, sweep away what is left, and perhaps, just perhaps, try again."

The Koranic version differs mostly in details. The trumpet will sound, and then:

Frail and tottering, the sky will be rent asunder on that day, and the angels will stand on all its sides with eight of them carrying the throne of your Lord above their heads. On that day you shall be utterly exposed, and all your secrets will be brought to light. (69:15)

When the Sun ceases to shine, when the stars fall and the mountains are blown away; when camels big with young are left untended, and the wild beasts are brought together; when the seas are set alight and men's souls are reunited; when the infant girl, buried alive, is asked for what crime she was slain;[4] when the records of men's deeds are laid open, and heaven is stripped bare; when Hell burns fiercely and Paradise is brought near; then each soul shall learn what it has done. (88:1)

Hell is fire and smoke, and "as for the righteous, they shall be lodged in peace together amid gardens and fountains, arrayed in rich silk and fine brocades. Even thus: and we shall wed them to dark-eyed houris" (44:54).

• • •

---

[4] This refers to the pre-Islamic custom of burying unwanted infant girls.

While the Koran doesn't tell its readers when to expect the end, early Christians thought it was coming soon. Jesus had warned them, "There are some standing here which shall not taste of death till they see the Son of Man coming in his kingdom" (Matthew 16:28). That short interval had already passed, but it was still possible that the Last Days were near. Scenarios for those days originate in Jewish interpretations of scriptural prophecies such as the seventh chapter of Daniel. The last two millennia have produced many versions; the one that in various forms is now current in American fundamentalist thought dates from the early nineteenth century in England, when the cruelty of industrial society caused many despairing people to hope for an end to it all. It goes something like this. There will be a Great Tribulation, seven years of agony under the domination of a king whom originally both Christians and Jews (for different reasons) called Antichrist, during which the whole world will experience war, pestilence, famine, floods, and a collapse of moral standards. At the end of this time, Christ and his saints will come at the head of a big army and in the Battle of Armageddon they will defeat Antichrist but not kill him.[5] Christ will then rule the world for a thousand years, imposing justice and righteousness everywhere. At the end, Antichrist will return, but he will be vanquished and thrown into a Lake of Fire. Then comes the Last Judgment, when the living and the dead will be judged and consigned to heaven or hell.

What good Christians feared was the Great Tribulation. If their salvation was assured, why did they have to go through all that? In the mid-nineteenth century, John Darby, an Irish ex-priest, discovered an old doctrine dating from the third century that spares them those seven years of suffering. It is anchored in the words of St. Paul, "For the Lord himself shall descend from heaven with a shout, with the voice of the archangel, and with the trump of God: and the dead in Christ shall rise first: then we which are alive and remain shall be caught up together with them in the clouds, to meet the Lord in the air: and so we shall ever be with the Lord" (1 Thessalonians 16,17). The faithful, then, whatever they may be doing at that moment, will hear a great sound. A moment later comes the Rapture (from Latin *rapio,* to seize): they will all be caught up together. Those left behind will see friends and neighbors vanish and will know that the Great Tribulation has begun. Like the other prophecies, this one says nothing about what will happen to the Earth when mankind has finished with it. I will discuss this from another angle in chapter 11.

John Dryden began this chapter; let him end it.

---

[5] The name may mean Hill of Megiddo, a former city southeast of Haifa.

As from the power of Sacred Lays
    The spheres began to move,
And Sung the great Creator's Praise
    To all the bless'd above;
So, when the last and dreadful Hour
This crumbling Pageant shall devour,
The TRUMPET shall be heard on high,
The dead shall live, the living die,
And MUSICK shall untune the sky.

SIX

Philosophy Continued

> Such appears to be the truth about the generation of bees,
> judging from theory and from what are believed to be the facts
> about them; the facts, however, have not yet been sufficiently
> grasped; if ever they are, then credit must be given rather to
> observation than to theories, and to theories only if what they
> affirm agrees with the observed facts.
>
> —Aristotle

IN THE OLD DAYS, copies of books were heavily used. Papyrus fell to pieces in a hundred years or so, destroying books and documents. Parchment lasted longer, but not for centuries. In Europe courts and universities had libraries, and monasteries had workshops where monks copied books of literary and religious interest and kept them alive. Very few scholars in the West before the Italian Renaissance learned Greek, and books in that language almost disappeared, but they survived in the Eastern Empire because wealthy nobility preserved them as a mark of distinction and as also as a duty. Without this tradition, we would have almost nothing of the Greek poets and dramatists; as it is, we have little enough. Greek works of science, philosophy, and mathematics fared better because they were selected for translation into Arabic in the caliphs' workshops. Books on subjects like economics and engineering disappeared entirely, but a few authors left works on astronomy that have survived.

6.1 THE STARS IN MOTION

Plato's friend Heracleides suggested that fixed stars and a turning Earth would be more economical than the conventional arrangement. This notion did not fly because it disputed the obvious fact that the element Earth moves down and not sidewise, but a century later Aristarchus of Samos (c. 310–c. 230) carried sidewise motion even farther. We know his idea only incompletely, from a single short book and some anecdotal mentions. The book is called *On the Sizes and Distances of the Sun and Moon* and it comes in two versions: one in a collection of short astronomical texts translated into Arabic in the tenth century and into Latin five hundred years later; the other in a Greek manuscript that surfaced briefly after the fall of Constantinople. It was copied and then vanished.

Aristarchus assumed that the Sun is the only source of light in the universe. Here he would have lost most of his audience right away, for stars were gods. On his assumption, if the Earth were larger than the Sun it would cast a shadow on the sphere of fixed stars and a star would occasionally go out, but this never happens; it follows that the Earth is much smaller than the Sun. From this and other observations he deduces by a clever geometrical argument that the Sun is between 6.3 and 7.2 times as large as the Earth, while the Earth is between 2.5 and 3.2 times as large as the Moon. He also shows that the Sun is 19 times as far from the Earth as the Moon. The numbers are much too small, but the reasoning is exact. The errors stem from a very small angle that he could not measure and had to estimate by eye, and he got it 20 times too large.

Still, following Aristarchus, the Sun's diameter is at least 6.3 times the Earth's, but how much does the Sun weigh? He proves that the volume of a sphere varies as the cube of the diameter, and so the Sun has at least 250 times the Earth's volume. If the Sun, like the Earth, is made mostly of rock, then it is at least 250 times as heavy as the Earth. We do not know what Aristarchus concluded; his commentators imagined the Sun whirling around a stationary Earth the way a stone is whirled in a sling, but if the stone weighs 250 times more than the slinger, the mind gives up. Perhaps Aristarchus reasoned that way, for the only other thing we know about his astronomy is that he proposed that the Sun stands still and the Earth travels around it once a year, and at the same time it spins, one revolution per day.

Plutarch is one of those who discuss Aristarchus's idea. He mentions that Cleanthes the Stoic thought he, like Anaxagoras, should be indicted for impiety, but in another place he says that Aristarchus was speaking as a mathematician and should not be indicted for making hypotheses. In those days there was a great difference between a mathematician and a philosopher. A mathematician was somebody who used shopkeepers' numerical tricks to calculate future events. He used Eudoxus's spheres or whatever other hypotheses were needed to get his answer, without worrying about philosophic truth. A philosopher explained what was happening and why it had to happen by relating it to fundamental principles that could not be otherwise. Aristarchus's book disappeared. Nobody was much interested when a mathematician started talking philosophy.

Aristarchus was the first to calculate the planets' sizes and distances, but Eudoxus had already shown how one could use an imaginary system of rotating crystalline spheres to calculate their changing positions in the sky. The results were not exact, but they accounted for the various kinds of celestial motion. There were other reasons besides inaccuracy why his model could not be taken literally. For example, in any arrangement where turning spheres are centered on the Earth, the planets are always at the same distance from us, whereas even simple measurements show that the apparent diameter of a full Moon is appreciably larger at one time than at another. Also, the brightness of planets, especially Mars and Venus, can vary widely from month to month. If they are self-luminous, this suggests that their distances must vary. Eudoxus's model, though it accounted for planetary positions, could not be correct, and if not for its mention by Aristotle it would have been quickly forgotten.

In the following centuries several first-class mathematicians worked to produce models in which steadily rotating spheres replicate the changing look of the heavens. These models became very complicated and there is no point in following their development, but they culminated in a standard version that, with minor adjustments, endured for about thirteen centuries.

• • •

The compiler of this theory was Claudius Ptolemaeus (c. 100–c. 175 CE). Everyone calls him Ptolemy even though it makes him sound like a pharaoh, and some old pictures show him wearing a crown. He seems to have spent his life in or near Alexandria; his surname shows that his family was either Greek or Hellenized Egyptian, and his first name shows he was a Roman citizen. Figure 6.1, taken from an old map, dresses him up as a desert Arab with a Jacob's staff, used to measure the angular elevation of a star. Ptolemy lived during the era known as *pax romana,* the Roman peace, during which the barbarians, fenced out at the limits of the empire, waited for their time, and Roman citizens were able to call the Mediterranean "our sea." Nothing more is known about Ptolemy except for his work. This work is important to our story in several ways. I will sketch it briefly and then draw conclusions.

Ptolemy's most famous book is his treatise on astronomy, so let us start with that. Its modern name, *Almagest,* reflects its history, for it was among the works translated into Arabic. Ptolemy had originally called his treatise *Mathematike Syntaxis,* The Mathematical Compilation. Later it seems to have been known as *Megiste Syntaxis,* The Greatest Compilation. The Arabs adopted this as *al-Majesti,* which in Latin became *Almagestum.*

Almagest is a splendid book, more than six hundred pages in the translation. It includes the basic theory underlying the model, mathematical procedures for interpreting astronomical measurements, instrumental techniques, and a mass of observational data, some of which were new and others centuries old.

After a few philosophical remarks, Almagest begins with a physical description of the cosmos. Observation shows that the heavens move like a

Figure 6.1   Claudius Ptolemaeus (Ptolemy) as shown on an old map.
(From Bunbury's *Ancient Geography,* 1883.)

sphere, turning around the poles at an exactly constant rate. It shows also
that the Earth is round and located at the center of the sphere; if it were
off-center, the stars would seem to move unevenly. Since at any time of
night an observer in a flat place sees exactly half the celestial sphere, the
Earth must be negligibly small compared with the sphere. The Earth does
not rotate; he nails this down with powerful arguments. We know from

Aristotle that the natural motion of heavy objects is downward. They move in any other direction only when something pushes them; if the push stops, they stop. Nothing pushes the Earth to rotate. If it did rotate, it would have to carry the air with it or else there would be a tremendous wind. Now throw a ball upward. As it falls, its natural motion (according to Aristotle) carries it toward the center of the Earth, and unless the air dragged it violently along, it would land to the west of where it was thrown. But note that a dropped ball falls quickly toward the ground, so that air does not affect its motion very much and therefore cannot drag it anywhere. Ptolemy does not mention the philosophy of the heavens except for a brief remark on what drives its motions: "Now the first cause of the first motion of the universe, if one considers it simply, can be thought of as an invisible and motionless deity; . . . [but] this kind of activity, somewhere up in the highest reaches of the universe, can only be imagined, and is completely separated from perceptible reality" (Almagest I.1).

As to the motions of the planets, Ptolemy mounts them without comment on a system of huge spheres called epicycles. Each planet has a steadily rotating sphere with the Earth as center, but what is mounted on that sphere is not the planet but the center of another steadily rotating sphere. If the planet is mounted on that sphere, things can be arranged so that its apparent motion as seen from the Earth is roughly duplicated. For more accuracy, add more spheres. Ptolemy stopped at thirty-four (if one counts only the useful ones) and was able to calculate with reasonable accuracy where Sun, Moon, or any planet was at any reasonable date.

Were Ptolemy's spheres real spheres or only a moving diagram? If one thought these gigantic structures really existed, one might be tempted to talk about them. Ptolemy doesn't. There is another reason why I think his spheres were mathematical. His lunar theory gives the Moon's position in the sky quite accurately but requires that its distance from the Earth, and consequently its apparent size, should vary by a factor of 2. It actually varies by about 12 percent, but Ptolemy doesn't consider the discrepancy worth mentioning. A little adjustment would have given the Moon's size correctly, but then its position would have been wrong. For constructing a calendar or a horoscope, the Moon's size isn't important and theory doesn't have to get it right. From this point on, Almagest becomes technical, analyzing centuries of observations to establish the measurements and motions of the parts of the great celestial machine.

The book you are reading is intended as a study of the ways people have thought about the Earth and its spatial environment. Most people of Ptolemy's time and the following centuries knew nothing of his machinery and imagined the cosmos, if they imagined it at all, as a series of solid, transparent, concentric spheres. Figure 6.2 shows a fifteenth-century version. In the center is the Earth (the little landscape it shows is

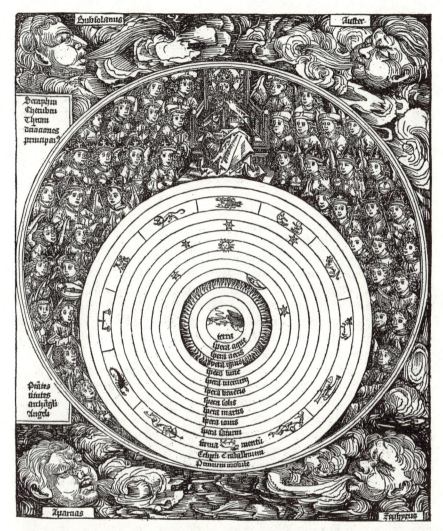

Figure 6.2   The universe as a system of concentric spheres, as shown in the *Nürnberg Chronicle* (Schedel 1493). Gutenberg's bible, the first printed book, had appeared about forty years earlier. This volume of six hundred pages compiled by Hartmann Schedel, a learned doctor, came out in Latin and German editions at a time when publishers were turning from religious and utilitarian texts to books for a literate public that was hungry for knowledge. It is an encyclopedia of the Western world presented in chronological order, from Creation to Last Judgment, covering popes, kings, emperors, scholars, saints, martyrs, cities, comets, floods, earthquakes, epidemics, and two-headed babies. It is illustrated with hundreds of woodcuts, some of them probably sketched by Albrecht Dürer. More than a thousand copies still exist, some handsomely illuminated. (Courtesy of the Chapin Library of Williams College.)

upside down). Reading outward, the spheres are Water, Air, Fire, Moon, Mercury, Venus, Sun, Mars, Jupiter, Saturn, Firmament (which carries the fixed stars and is ornamented with signs of the zodiac), Crystalline (the waters above the firmament, which, if they had not crystallized, would not have stayed in place), and *Primum Mobile,* the Prime Mover that motivates all the lower spheres. The region outside, where there is neither space nor time, is the Empyreum, the domain of God and the Elect. On the left are the nine orders of the celestial hierarchy: Seraphim, Cherubim, Thrones, Dominations, Principalities, Powers, Virtues, Archangels, and Angels. This was the popular cosmology, and until the seventeenth century it served well enough for most people.

6.2 Stars, Earth, and Numbers

Why did so many work so hard over these models of the cosmos? Apart from philosophic curiosity, there were good reasons. Because eclipses warned of changes to come, it was important to know when they might happen. Lunar eclipses were comparatively easy; Babylonians, Chinese, Mayas, and the astronomers we have been talking about could get them right. Solar eclipses were harder to predict, but from early times astronomers could say when there was danger of one.

The most important reason for knowing planetary positions was to forecast floods, droughts, and political changes and, for those who believed, the fortunes of an individual. When a child was born one could know the time, say 3½ hours after sunrise, but where in that bright sky were the stars and planets at the moment of birth? Each year their positions were calculated and published in almanacs. Once the planets were located, they could be read, and an important reason, perhaps the main reason why Ptolemy worked so hard to collect observations and refine his theory was to make him a better astrologer. There is no point in pretending that astronomy and fortune-telling came out of different hemispheres in Ptolemy's brain; his world was One and his astronomy included both.

• • •

But how are planetary alignments connected with what happens to weather and human fortunes? The nature of this linkage was a problem long before Ptolemy's time. Aristotle had attributed causality to the planets and suggested Ether as the link that connects Earth and cosmos. But how does this actually work? Ptolemy's *Apotelesmatica,* On Astrology, also known as *Tetrabiblos,* The Work in Four Rolls, is a companion volume to Almagest. Almagest doesn't mention influences at all; *Tetrabiblos* says nothing about planetary theory, but it is the bible of astrology. It was written about 150 CE, entering Europe via Arabic versions in 1138, and as late as 1553 Martin Luther's collaborator Philipp Melanchthon was making a fresh Latin translation.

Like Almagest, Tetrabiblos is a masterpiece of patient explanation as it sets out a vast and ancient doctrine. Some of the examples go back to the Babylonian *Enuma Anu Enlil*. Ptolemy begins by telling how astronomy helps us know the future. First, it locates the planets at any past or future moment; this is the way of Almagest. And second, it studies the changes planets produce in the world as they come and go overhead. Almagest's calculations are reasonably precise, but astrology cannot be an exact science on account of "the weakness and unpredictability of material qualities found in individual things." The heavens move in accordance with principles that we may not know but they are mathematical in nature, whereas the Earth and the space around it, everything below the sphere of the Moon, are subject to the changes and uncertainties of the material world. Tetrabiblos says, "A certain power emanating from the eternal ethereal substance permeates the whole region around the Earth, of which all parts are subject to change. The spheres of Fire and Air are changed by Ether, and in turn they change the Earth and Water of the lower spheres together with the plants and animals that live in them." Later writers refer to this influence, whatever it may be, as radiation. Ptolemy then lists the effects of the Sun and Moon. Those of the Sun are well known, he says, but also the Moon's influence is everywhere, for things of our world, animate and inanimate, "are sympathetic to her and change in company with her: rivers increase and diminish their streams with her light, the seas turn their own tides with her rising and setting, and plants and animals in whole or in some part wax and wane with her." In another place Ptolemy tells us that the bodies of mice are in tune with the heavenly bodies, "for the number of their liver filaments becomes greater or less with the light of the Moon." And of course the cosmos wrapped around the Earth is connected with wind and weather.

The techniques of astrology do not interest us here; what is important is the way the art (Aristotle would not have called it a science) is built into Ptolemy's picture of the world, and what it means for humanity. There are two kinds of astronomic indicators. The first is the regular and predictable motions of stars and planets that are the basis of a horoscope and of predictions of the success or failure of an undertaking. Nothing can be done about the stars except to profit when they are favorable and be warned when they are not. As to undertakings, there were then, and are now in many places, good and bad moments to open a meeting, celebrate a marriage, or set the first stone of a building. The other kind of indicator is an omen, something unusual and unexpected that happens and needs to be interpreted. Eclipses and meteoric showers can be predicted with varying success, but a comet is always a surprise. "Comets generally foretell droughts or winds, and the larger the number of parts that are found in their heads and the greater their size, the more severe the winds." Their shape is important. "The so-called beams, trumpets, jars, and the like naturally produce effects peculiar to Mars and to Mercury—wars, hot weather, disturbed conditions; and they show, by the parts of the zodiac in which their heads appear and through the directions in which their tails point, the regions that will be affected."

These predictions, of course, are not absolute. They indicate the direction in which events are tending as well as dangers that should be looked out for, but the sky is a vast picture with constellations rising and descending and planets alone or in groups. Sometimes all the tendencies work in the same direction, sometimes they oppose one another; even clouds and weather play their parts to strengthen or weaken a forecast. But whatever happens in the sky, its influence on earthly affairs is a real physical influence. Fate, for Ptolemy, is pure causality acting on the Earth, on the cloudy vapor of the material world, and on what Kant called the crooked timber of humanity.

By the time Ptolemy has finished explaining astrology as a study of the balance of opposing tendencies (just as the real world is), we are not surprised that astrological forecasts do not seem to come out any better than if they were made at random. Why, then, in the absence of any convincing evidence, does this great scientist believe that we can learn anything by studying the stars? I think the reason lies in philosophy. Suppose we ask: Why do stars and planets exist at all? We run through the four kinds of causation: the material cause is Ether; the efficient cause, I suppose, is the mechanism of celestial spheres; the formal cause is the design of this mechanism, which Ptolemy has done more than anyone else to make clear; and what is the final cause, the reason for the whole show? Obviously the Sun has to be there or there would be no life. The Moon is responsible for the fruitfulness of the Earth and its creatures. And the celestial spheres? They are the link through which divine energy and purpose reach the Earth to direct a million modes of action and keep them going. This machinery would be invisible and unfathomable except that the great celestial sphere carries visible constellations, while inside it some spheres carry a glowing planet that tells us, like a lamp at the masthead of a boat, how that sphere is oriented and how it is moving. This much evidence is given to our eyes; the rest is up to us.

• • •

So much, for the moment, for the way in which order in the cosmos is reflected in the order of human affairs. But things on Earth, too, have kinds of order that can be read by a trained mind. Consider numbers, for example. I have mentioned that Plato identifies them with Ideas, and he writes in the *Republic:* "The qualities of number appear to lead to the apprehension of truth." Even Aristotle, who usually avoids this kind of talk, writes: "Nature is everywhere the cause of order . . . [and] order always means ratio." But what are the qualities of number? How, other than in the size of the quantity represented, does one number differ from another? There are a few fragments of Greek answers to this question, but most of what we know comes, directly or indirectly, from an *Introduction to Arithmetic* written in about 100 CE by Nicomachus of Gerasa, a city in Palestine.

First, Nicomachus explains why his subject is important: the pattern of the universe "was fixed, like a preliminary sketch, by the domination of number preexistent in the mind of the world-creating God, number conceptual only and immaterial in every way, but at the same time the true and the eternal essence, so that with reference to it, as to an artistic plan, should be created all these things, time, motion, the heavens, the stars, all sorts of revolutions." Numbers, for those who think like Nicomachus, define the forms that determine the world's qualities as well as the active principle—we might call it energy—that drives it. A few simple examples will have to suggest the richness of neo-Pythagorean arithmetic.

It starts by connecting numbers with letters, words, and concepts so that a person, situation, or proposed course of action can be summed up in a number; then the qualities of the number are examined. Consider the number 12. Its factors are 6, 4, 3, 2, and 1, and the sum of these is 16, greater than 12. Thus 12 is called a *superabundant number.* On the other hand, the factors of 14 are 7, 2, and 1; their sum is 10, which is less than 14, and so 14 is called deficient. "The superabundant and deficient [numbers] are distinguished from one another in the relation of inequality in the directions of the greater and the less; for apart from these no other form of inequality could be conceived, nor could evil, disease, disproportion, unseemliness, or any such thing, save in terms of excess or deficiency." Between these are the perfect numbers already mentioned; an example is 6, whose factors add up to 6. These are perfect because they are neither superabundant nor deficient, and we have seen that, like perfection itself, they don't often occur. They measure what is best and most desirable. Themes of love and harmony are expressed in pairs of amicable numbers; 220 and 284 are an example: each is the sum of the factors of the other. There is much more in Nicomachus, including the relations between number and geometrical shape—of which figure 3.1

shows a simple example—and the Pythagorean theory of musical harmony. The book survived into the Middle Ages in a Greek manuscript and a Latin translation by Boethius, the standard reference in the centuries that followed.

A question occurs to many modern readers: How can the mere presence of a number or a numerical relation cause anything to happen? Aristotle, in book 14 of his *Metaphysics,* concludes that there may be a kind of parallelism between the qualities of a thing and those of a number, but in no way can either of them cause the other. The only effect of this judgment has been to convince believers that God's plan for the universe involves numbers, and that by discovering a little bit about them we gain some hint of occult connections, unknown to Aristotle, that permeate the world. They are strong but hidden from view, and must be considered when we plan our own actions or interpret those of others. Boethius will show up again in section 6.4, but first a few pages that describe some wider regions of the occult.

6.3 OMENS AND DEMONS

The stars and their spheres transmit the influence of the Prime Mover down to Earth and its inhabitants. Being visible, the stars can also be studied for news of future events, but this involves understanding the language of their motions and groupings, and as to that there were many opinions. How much better if there were someone, or something, whom one could just ask, who could provide information and even, perhaps, practical help. Such beings existed, the good or bad *daimones* who lived, invisible, in the air. Daimones have always been present in folk religion; in one form or another they can still be found, but until Plato opened the door for them they were rarely mentioned by philosophers. In *Timaeus* the fashioning of us humans was left to "lesser gods," and in *Laws*, Cronus "gave our communities as their kings and magistrates, not men but spirits, beings of a diviner and superior kind. . . . We do not set oxen to manage oxen, or goats to manage goats; we, as their betters in kind, act as their masters ourselves." This is metaphorical talk, just as *Timaeus* is, but about a generation after Aristotle a president of the Academy, Xenocrates of Chalcedon, seems to have taken it literally. He imagined that space is filled with daimones, and that those in the region between Earth and Moon affect human lives in many ways. One of his intentions was to shield the gods of the Greek pantheon from the libelous stories told about them—it was daimones that had done these things, not gods. When Christians came, they taught that all the pagan gods were daimones. These beings are much like us; their nature is partly corporeal, they experience pleasure and pain, some are good and some are bad. Since they had always been so much a part of common belief and practice, none of this came as a surprise. In Greece you could not walk more than a little way along a city street or a country road without seeing a pile of stones, an image, or a few flowers honoring some spirit that lived nearby.

A strange thing: Aristotle never mentions daimones, and having provided the theoretical picture of Prime Mover and the spheres' descending powers, he says nothing about the supernatural. Pliny, reporting on the vast real world and its peoples, saw no more spirits than Aristotle did. The medical treatments he reports (whether or not he thinks they work) involve a hundred ingredients and procedures but they share one feature: no words are spoken.[1] If it is true that the ashes of a weasel relieve pains in the shoulders, there must be a link between weasel-ash and shoulders. Even if it was occult, Pliny assumed it was natural. It would have been senseless to produce words if no one was listening.

Under the Roman Empire people gave presents to their familiar daimones, but when serious questions arose they went to someone who knew about omens and stars. The emperors found themselves facing powers they could not control. Suetonius says that Augustus collected more than two thousand books on divination and burned all but a few old and honored Sibylline Oracles. What was so bad about a little fortune-telling if everyone believed that signs were there to be read? Tacitus, writing of the time after Nero died, explains: as long as there were astrologers and enough people believed them, any politician could stir up trouble by circulating a rumor that one of them had predicted a death in the imperial family or a commotion in the streets. They were, Tacitus said, "a class of men who betray princes and deceive the ambitious, and in our state will be forever both forbidden and retained."

After Augustus, smoke of the burned prophetic books hung in the Roman air but it didn't stop divination. *Genii,* after all, existed, and why not consult them? A dinner guest of Plutarch's, at the end of the first century CE, echoes Xenocrates: Air, which transmits celestial influences down to the human world, contains daimones that mediate between gods and men. Some are guardians of sacred rites; others avenge acts of injustice. The Olympians in Homer and the dramatists were daimones and not the true gods. Yes, there are rambunctious daimones, even bad ones, but basically, just like ordinary human beings, most of them are good.

• • •

[1] I know one exception. In Pliny 1938, 27.75, eczema is touched with a specially prepared stone and told to go away. Later, in 28.3, is a discussion as to whether spoken words do any good.

The Old Testament is full of references to people who consulted spirits. Probably most of these people did not intend to harm anyone, but it was always a possibility, and that may be why it was forbidden in the laws of Moses (e.g., Deuteronomy 18:10) and afterward by prophet after prophet (e.g., Isaiah 8:19), but of course it continued. In the New Testament the only reference to astrology I can find is Jesus' prediction that as the end of the world approaches, "portents will appear in Sun and Moon and stars" (Luke 21:25). There is a mild reference to "subversive spirits and demon-inspired doctrines" (1 Timothy 4:1); beyond that the only demons mentioned are those cast out of demented people. This however happens often, and one demon, asked his name, answers "My name is Legion, there are so many of us" (Mark 5:9).

Christianity, originating in the Middle East, was burdened with the assumption, common in most of that region's cults, that the world is the scene of an endless battle between forces of good and evil. In Christian minds daimones became an army of demons, all of them wicked and commanded by Satan whose kingdom was Earth, the only imperfect part of the cosmos, an ugly, dirty little sector overflowing with passion and evil deeds. Because of Adam's sin and fall, Satan had been crowned prince of this world, and his devils were everywhere. They were ready to seize on any weakness that would give them control over a human life, ready to enter a human body at a moment when it was unguarded: that is why people started saying "God bless you" at the gulp of air before a sneeze. Christians easily understood why the Bible (Matthew 8) says that after Jesus expelled devils from two madmen into a herd of swine, the reaction of townspeople who heard about it was to ask him to go away. They didn't want anyone in their neighborhood who dealt with devils.

Emperors could burn books and make laws, but no policeman could prevent a private citizen from talking with a learned friend once they were indoors; therefore learned friends had to go. An edict of Constantine in about the year 319 decrees, "No soothsayer shall approach the threshold of another person, not even for some other reason, and the friendship of such men shall be rejected. That soothsayer shall be burned alive, . . . and the person who summoned him by persuasion or rewards shall be exiled to an island after confiscation of his property."

But it was not only, or even primarily, divination that worried the emperor. Another Constantinian edict, later copied into Visigothic law, refers to "the science of those men who are revealed to have worked against the safety of men." Clearly, though the edicts nowhere mention demons, more is involved here than fortune-telling. The fortune-teller may (or may not) be a simple reader of stars and omens, but workers against mankind are men and women allied with evil powers that can commit crimes against people and institutions with little danger of being caught.

In spite of all this, in the popular mind astrologers were as respectable as doctors (not all of whom were respectable), and at length the emperors decided that even if they were political irritants, lumping them with magicians was counterproductive. The stars, after all, are up there, and ordinary people trying to make plans in an uncertain world would always be consulting them; the ban could never be enforced, and magicians would continue to act. Therefore, fifty years later, emperor Valentinian opens a distinction: "I judge that divination has no connection with cases of magic, and I do not consider this superstition, or any other that was allowed by our elders, to be a kind of crime."

• • •

Augustine, the great synthesizer of early Christianity, studied astrology but finally renounced it. Not that the stars have no influence on human life. Of course they have, for reasons I have explained. It is just that there is no practical way to find out what the influence is going to be. Augustine repeats Cicero's observation that twins born a few minutes apart often lead very different lives. Perhaps those few minutes make all the difference? No; what finally made him lose interest in astrology was a friend's report of a rich and cultivated man who happened to have been born at exactly the same moment as a slave child on his family's estate. The rich child grew up in luxury; the slave remained a slave. Clearly, the moment of their birth had little, perhaps nothing at all, to do with the way their lives ran. Augustine cautions: this does not, of course, mean that prophecy is impossible; that would contradict Scripture. God's foreknowledge is absolute, and he communicates it to anyone he chooses.

Augustine has no doubt about the existence of demons. He treats the Greek myths of gods and heroes as if they were factual except that the great gods were really great demons with faults and virtues like the rest of us, and he understood that feats of Egyptian and Babylonian magic described in the Bible were performed with demonic help. There is a minor treatise called *The Divination of Demons* in which he says that as inhabitants of the region between Earth and Moon, demons are made differently from us and have special abilities. Their aerial and therefore invisible bodies can enter a room without going through the door. They can enter our own bodies to diagnose and treat an illness; they can also cause one, but they cannot touch the mind because that is incorporeal. Their senses are keener than ours and they move more quickly. With these advantages, and because they live so long that they accumulate much more experience than we do, they can read people's intentions, predict events long before they happen, and perform miraculous acts. But all this, says Augustine, is no reason to worship these beings, any more than we worship a vulture because it goes higher and sees better than we do. Demons are no better than we are, and they belong to a different order of nature. We should leave them alone.

There remains, however, a nagging question: Why did God make demons in the first place, or how did they come to exist? Nine centuries later Thomas Aquinas, in his immense *Summa theologica,* says that God did not make them, for at the end of the Six Days he was able to look at all that he had made, "and behold, it was very good." He had, however, made them possible when he created angels in great number and variety to be his messengers and agents, for he gave them the same free will that he gave Adam and Eve, and some, driven by envy and pride (the only sins possible for angels), chose "to hate men and fight against God's justice." Demons are such fallen angels. Envy and pride are not wicked, and angels who cannot resist them are not evil by nature. "It is not evil in the fox to be sly, since it is natural to him; just as it is not evil in the dog to be fierce." But when an angel joins the "concord of demons," he becomes evil. And why did God make a world in which this could happen? Because he knew that agents with special qualities would sometimes be useful, as when it was necessary for evil men to be punished or the faith of good ones to be tried. In our own world, not all the best policemen are of saintly disposition.

Demons had a long life in Europe and perhaps a few are still at work. Figure 6.3, for example, shows the reputed inventor of gunpowder and his assistants. They were not necessarily visible, but this is how they might have looked if they were visible.

• • •

A sixteenth-century manual of magic recently showed up in Munich. Most of the instructions in it are easy to follow.

On a clear and calm Sunday morning in the summer of 1454, a merchant from Hamburg walked through one of the gates of Munich toward a low hill in the countryside. He had chosen the place because no one lived nearby. The sack on his back, containing three clay pots, several sheets of paper, and a few other supplies, was cumbersome to carry, and occasionally he shifted it from one shoulder to the other. In his mind were grief and fear: grief because a letter from Hamburg had just told him that his wife was ill and not expected to live much longer; fear because of what he had decided to do. Hamburg was far off, the time was short, and he needed the help of invisible powers. He carried the instructions for

Figure 6.3 Berthold the Black was a German monk who was also an alchemist, active c. 1320. He is said to have invented gunpowder and made a brass cannon. This sixteenth-century German woodcut shows him doing both, helped by demons.

what he was about to do, copied from a book that a client had kindly lent him. But how could a man who had no knowledge or experience in these matters command spirits he could not see? And the book itself, was it correct? Was it genuine? Or was the procedure he was going to try only a trick designed to involve him with the powers of darkness?

The merchant from Hamburg reached the hilltop and looked around. No one was in sight. The pots clinked together as he lowered his sack and spread its contents on the ground. He gathered grass and sticks and with his flint and steel struck a little fire that burned almost invisibly in the morning sunlight. Then, with the papers trembling in his hand, he began to read in a voice as strong and confident as he could make it: *Holy Mary, pray for us according to thy great mercy*, finishing with the rest of Psalm 51. Then, *Glory to the Father and to the Son . . .* , the familiar words bringing some comfort to him. *Into thy hands O Lord I commend all my spirit and body, Lord God, Father almighty.* Finally he read the 43rd Psalm, *Judge me, O God.* After these words and another scan of the horizon, he drew a large circle on the ground with an equilateral triangle inside it that had one vertex toward the north, and placed a pot on each vertex. Into the northern one he poured ashes and flour, into the eastern one salt and a few burning twigs, into the western one some water and a shoe. For the dozenth time he reviewed the paper, though by now he knew what it told him to do: he should step into the middle of the triangle and tell Bartha the cloud-king to send three spirits—Saltim, Balthim, and Gehim—who had power over clouds and winds, commanding them to take him to Hamburg without harm or danger to body or soul. At once, it said, he would see a little cloud, and from the pots would issue three more little clouds that would speak the words *Ascend, ascend, ascend.* Then, when he had said three times, bravely and undaunted, *Make the chair ready*, he would see a throne in the middle of the sky in which he should sit and declaim:

> King Bartha and Generals Saltim, Balthim, and Gehim, carry me peacefully, without fear or impediment or any danger to my body and soul to Hamburg; lift me gently and gently set me down. . . . I powerfully call on you in the name of those high angels whose power is in the air: Mastiesel, Emedel, Emethel . . . [there are 29 more names]. By these names I invoke and conjure and exorcise and compel you to take me from this place and carry me kindly and gently to Hamburg without harm or danger to my body or soul. Let this be done: *Fiat, fiat, Amen.*

Once more the merchant looked around him. Far away on the road someone walked toward the hill, but there was still time. A little smoke still came out of the eastern pot. There was no sound but he went on, softly reciting *Our Father.* He crossed himself, stepped into the center of the triangle, filled his lungs, and bellowed into the quiet air, "Make the chair ready!"

But it is time for us to leave the demons alone. Even if supernatural beings are an important part of many people's vision of the world, they belong to a different order of nature and should be allowed some privacy.

6.4 REMEMBRANCE OF THINGS PAST

It was only toward the end of the next thousand years that people once again took up the study of the natural world. The reasons for this neglect are complex, but two stand out: upheavals of society drove much of the old leisured and intellectual class into obscurity, and for the educated few who remained there was more to be gained from studying a Book than a star or a flower. The next few pages will show how, during centuries of change, some fragments of the old culture survived until once more a society arose whose members had time and inclination to think about the world.

Ptolemy worked peacefully in Alexandria. Soon afterward, faraway peoples began to move. The Greeks called them barbarians, and they ended up transforming the empire. They carried destruction westward. Book by book, or in the burning of whole libraries, thus destroyed rolls containing the literature and philosophy of the Roman Peace. But not all of them disappeared, nor did all the knowledge they contained. A few educated men who watched their culture being trampled by warring tribes thought it was their duty to preserve what they could.

Anicius Manlius Severinus Boethius (c. 480–524) was born in Rome to a patrician family that had been Christian for several generations. He studied, probably in Alexandria, and became a high official of the court of the Ostrogothic king Theodoric in Ravenna. In moments of leisure he wrote books on music and arithmetic and translated some of Aristotle's logical works into Latin; then he decided to translate as much of Plato's and Aristotle's works as he could find. At some point he got into trouble and was imprisoned on a charge of treason. While waiting for the executioner he wrote *The Consolation of Philosophy,* poetry mixed with prose, in which a woman personifying philosophy appears to him and expounds the Platonic doctrine that the Good is the source of all reason and order in the universe. Evil, she says, is not a force, and whatever trouble befalls Boethius now is a step toward a greater good hereafter, but Boethius does not rise to this bait and continues to pray humbly for his salvation. With the possible exception of Jerome's Latin Bible, this was the most widely read book in medieval Europe. Boethius had finished only Aristotle's logical works when the executioner came, but these translations survived, and over the next thousand years they taught people in authority to make exact distinctions and draw necessary conclusions.

In southern Italy, a few years after Boethius, a patrician named Cassiodorus (c. 485–c. 580), after a long career of public service under Theodoric and his successors, established a monastery called Vivarium near Squillace. It sheltered a library and a copying workshop in which he hoped to preserve some of what still remained of classical culture. As a guide to the library, he compiled an *Introduction to Divine and Human Readings.* The divine part is a miscellany of Christian history and doctrine. The second part, information scooped out of Boethius and a few earlier writers, surveys the seven liberal arts: grammar, rhetoric, dialectic, arithmetic, music, geometry, and astronomy, a few pages for each. It is probably well that the Ostrogothic kingdom had little contact with the Eastern empire. Schools and libraries were still open in Athens and Pergamon, but Boethius and Cassiodorus probably knew little about them, and they may have fancied that the darkness that surrounded them and encouraged their labors was more universal than it really was.

Greeks came and settled in the toe of Italy and there was peace for a while. Then in the ninth century came Saracen raids, one after another. In 1060 Normans plundered the region and finally made it their own. No one knows what happened to Vivarium but the *Introduction* survived, and as late as the sixteenth century the mathematical and astronomical parts were reprinted as a convenient synopsis of the liberal arts. Just as Boethius's collection is narrow in scope, so is Cassiodorus's short and simple. It was time for a last effort.

• • •

In the middle of what are quite reasonably called the Dark Ages of Western Europe arose a man who attempted to make a wide survey of human knowledge. A bit of background first.

In 376 Hun horsemen had arrived in Romania from somewhere beyond the Volga, sweeping westward toward the Germanic tribes of eastern Europe. As Gibbon writes, "The number, the strength, and the rapid motions, and the implacable cruelty of the Huns were felt, and dreaded, and magnified by the astonished Goths; who beheld their fields and villages consumed with flames, and deluged with indiscriminate slaughter." When Huns attacked, the only way to survive was to run, and the surviving inhabitants of what is now Romania, notably a tribe called the Visigoths, set out to find a place where they could be safe. After defeating a Roman force at Adrianople (modern Edirne), they settled for a while in the Balkans, where they were converted to the Arian version of Christianity, in which Jesus comes from God but is of a different nature and God reigns solely supreme.

A leader named Alaric arose and led several invasions of Italy in search of land and, of course, plunder. In Rome the emperor Honorius, whose character combined weakness in action with foolish stubbornness in matters of principle, refused to deal with Alaric, who responded by laying siege to the city. In August of 410, on the third try and aided by a revolt of the city's 40,000 slaves, he succeeded.[2] Gibbon again: "At the hour of

---

[2] Historians disagree concerning this revolt. Procopius, writing about a century later, says that Alaric presented some wealthy citizens of Rome with young soldiers who, until a moment set in advance, pretended to be slaves. Gibbon, *Decline and Fall of the Roman Empire,* ch. 31, thinks it was an internal revolt.

midnight the Salarian gate was silently opened, and the inhabitants were awakened by the tremendous sound of the Gothic trumpet. Eleven hundred and sixty years after the foundation of Rome, the Imperial city, which had subdued and civilized so considerable a part of mankind, was delivered to the licentious fury of the tribes of Germany and Scythia." Soldiers plundered the city during three days of rape and murder, but on Alaric's orders the Vatican was not entered and many found shelter there.

The Visigoths were a small tribe, Rome was a single city, it was occupied for only three days, other cities were suffering the same or worse fate all the time, the burning of Rome's wooden houses left most of its monuments undamaged, and its fall did not greatly affect the daily life of the empire. Yet Rome was Rome, and its sack in 410 registered as a huge and portentous event in the mind of the Western world. For those with a sense of history it was the end of an era as old as the city, and nothing would ever be the same again.

Eventually the Visigoths settled in Spain. There this small tribe, rough and savage compared to the polished Spanish-Roman population around them, seized power and chose a king. At first the faraway Roman emperor stipulated that this new king ruled only the Visigoths, but they converted to Catholicism and learned Latin and became residents. Thereafter, in tumult, murder, and near-anarchy, they ruled Spain until 711, when the Arabs arrived.

Isidore (c. 560–636) was born in Seville into a Spanish-Roman family. He entered the Church and in about 600 was named archbishop of Seville. He became Spain's leading ecclesiastical statesman, the friend and counselor of its Visigothic kings, and the author of statutes decreeing the union of church and state and the toleration of Jews. Educated two hundred years after Rome's fall, he had read the principal Latin authors and knew something of the Greeks as well. The compass in his head pointed toward Rome, but here he was in Spain, surrounded by men with blond hair and skins whiter than his, some still speaking a guttural tongue that had no literature whatever. And outside the range of their own tribal culture these people knew absolutely nothing about the world, its languages, history, its arts and sciences and philosophy. Isidore, whose education honored the past, understood that what he saw around him was the future, and with the old culture's rolls and books dusty and disintegrating in the episcopal library, he resolved to save what could be saved so that the new world could begin to educate itself.

Isidore wrote several books, but we will look at only two of them. The first is an encyclopedia known as *Etymologies* because it is organized according to the imagined origin of each term he discusses. Most of these etymologies are based on accidental resemblances and are not easy to accept. An example taken at random: "*Mundus,* meaning the entire universe, is so called because it is always *in motu,* in motion, for the elements composing it never rest." I conjecture that his method was inspired by the story of the Tower of Babel. When God saw the huge tower going up, he feared that all mankind in a single nation would be too powerful, so he created and gave to each family a different language, and the building project ended in confusion. He had to put dozens of languages together immediately and he may have constructed them around similarities of sound. But why hang his encyclopedia on these frail resemblances? Isidore had been brought up in Latin; most of his intended readers had not. Perhaps he used similarities of sound as aids to memory. The Russian word for a dog is *sobaka.* When I say it I hear a dog barking. Don't we all learn a new language that way?

Etymologies is a compendium of the liberal and practical arts, including chapters on grammar, music, theology, astronomy, medicine, architecture, agriculture, and warfare, and it is a treasury of information on life in that remote time. Isidore's other book is *De rerum natura,* echoing Lucretius, and it summarizes in short chapters the current information on Earth, Heaven, and phenomena such as weather. The excerpts to follow are not fairly chosen because they will emphasize points at which his understanding doesn't agree with ours.

Isidore's Earth is Scripture wrapped up in Ptolemy: a flat disk fitted into a spherical Heaven, but the disk is not level; we can think of it as stuck part way down the side of a globe. "The world is formed this way: just as the world rises toward the north it also slopes downward toward the south. Its head, one might almost say its face, is the eastern region; the other end is the western region, for the world has four divisions. The first is the east, the second the south, the third the west, and last is the north."

Having thus laid out his world with vague reference to the human body (the head being toward Jerusalem), Isidore inserts five climatic zones as Ptolemy defined them: arctic, north temperate, equatorial, south temperate, and antarctic. They fit nicely onto a sphere, but where do you locate them on a disc? Isidore arranges them in a circle, as in figure 6.4. "As to the equatorial zone, it is uninhabitable because the Sun, rushing through the sky, produces an unbearable heat. . . . On the other hand, the northern and southern circles, touching each other, are not inhabited because they are so far from the Sun."

The heavens enclose the slanted disk:

The sphere of Heaven is a certain form, spherical in shape. Its center is the Earth and it is shut in equally on all sides. They say that the sphere has neither beginning nor end; since it is round like a circle its beginning and end cannot readily be seen. . . . Heaven has two gates, east and west, for the Sun issues

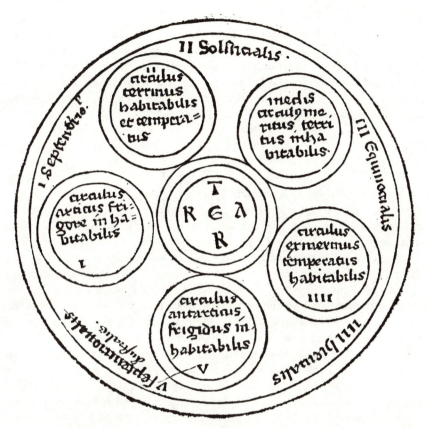

Figure 6.4   Isidore of Seville arranges Ptolemy's five climatic zones (shown in fig. I.7) so as to cover a flat Earth. Starting with no. 1: Arctic circle, uninhabitable because of cold. 2: Terrestrial circle, habitable and temperate. 3. Middle circle, uninhabitable. *Territus* should read *torridus, meritus* should read *[iso]emerinus,* meaning equinoctial or equatorial. 4: Temperate, inhabitable. 5. Antarctic circle, cold, uninhabitable. (From Isidore's *Etymologiae,* c. 1473; courtesy of the Chapin Library, Williams College.)

from one and retires into the other. . . . The rising Sun follows a southerly path, and after it comes to the west and has dipped into the ocean it passes by unknown ways beneath the earth and again returns to the east.

The voice that dealt firmly with diplomatic issues and spoke into the ear of kings is here puzzled and uncertain, producing words in the absence of any mental picture. For centuries, those who wanted to know how the vast world actually is drank from this trickle of knowledge, but later, as Ptolemy's astronomy and his geographical writings became available and people traveled more, ideas of the Earth became more accurate. It was not so with other branches of knowledge that rested less firmly on experience. Perhaps the best example is provided by Isidore's answers to the question, What is everything actually made of? The atomists and Epicurus had done their best to answer this question, and Lucretius had written a poem about atoms, but it was Isidore's writings that kept the idea alive during the long dark ages.

6.5 MOTES OF DUST

People who theorized about matter were never popular with the Church. Atoms, particularly, were associated with Epicurus, who by this time was cataloged as the philosopher who praised sensual excess, and he served as a stick to beat the other Greek materialists. But there was worse: the atomists, from Leucippus on, had mechanized nature by teaching that most kinds of change involve the law-governed but otherwise random motion of atoms, and this allows no room for Providence and leaves Creation high and dry. Further, as Christian doctrine evolved, there was the question of the Eucharist. The Church maintained the doctrine—and sharpened it during a series of councils—that at the moment in the Mass when the priest consecrates bread, the bread's qualities remain those of bread but its substance turns into the body of Christ, and likewise wine turns into blood. This was possible in Aristotelian philosophy for it distinguished substance from quality; but in the materialistic philosophy the qualities of a thing emerge from its substance and cannot be separated from it. Even so, there were atomists among the churchmen of the Middle Ages. I can think of two reasons why they existed. First, much of what they knew about the world came from the big volumes of Isidore of Seville; and second, there was the enduring popularity of Lucretius. His *On the Nature of Things*, which like *Etymologies* explored the material reality beyond the world's appearances, was admired and must have been copied and recopied.

Here is Isidore's introduction to atoms:

> Philosophers call by the name of atoms certain parts of bodies in the universe so very minute that they do not appear to the sight, nor admit of *tomein*,[3] that is, of being cut; therefore they are called *atomoi*, atoms. They are said to flit through the void of the whole universe with restless motions, and to move hither and thither like the finest dust that is seen when the Sun's rays pour through a window. From these certain philosophers of the heathen have thought

---

[3] For once the etymology is correct, though the verb form isn't.

that trees are produced, and herbs and all fruits, and fire and water, and all things are made out of them. Atoms exist in a body, or in time, or in number, or in the letters.

Atoms "in a body" would remain, he says, if the body were ground finer and finer until it could not be reduced any further. They are made of *hylē,* a Greek word meaning timber or building material that here denotes the fundamental substance of which all things are made; hylē has no form of its own but can adopt a form that is given to it. An atom of each of the four elements has its own particular form, but there is no such thing as a pure sample of any element; everything in the world, he says, is a mix.

There are other atoms: the atom of number is 1, the letters of the alphabet cannot be further divided, and there is a shortest possible time interval. The simplicity and universality of his atomic concept made it attractive, and later writers of those dark centuries often mention it.

• • •

In the Arab world, the same centuries were bright and full of promise. Islam had planted its banners over the Middle East and Spain and in huge territories in North Africa while the arts of peace flourished at home. Muslim thinkers studied Greek philosophy as it emerged from translators' workshops and tried to fit it into the context of the Koran. I do not know whether there was an Arabic version of Isidore's work, but it was studied in Spain. At any rate, in the tenth century or perhaps earlier, Arabic scholars and theologians became interested in the idea of atoms and sought to incorporate it in the *Kalam.* This word can be translated as "Islamic philosophy," or more accurately as "the Word of God," but still more is caught in a definition given by the eleventh-century teacher al-Bakillani: "The *Kalam* is an entity subsisting within the soul which sometimes expresses itself in audible sounds."

The audible part of the *Kalam* was the special concern of a diverse group of savants called the Mutakallimun whose theological explanation of the universe was justified by its derivation from the Koran and not by anything directly experienced. A clear account of this doctrine is given by Maimonides, who doesn't believe in it, in chapters 73ff of his *Guide for the Perplexed.*[4]

The world, for the Mutakallimun, consists of emptiness and two kinds of atoms: atoms of material substance, which account for its look and its structure, and atoms of time, called *instants,* which account for its changes and its preservation. For Aristotle, the qualities of a thing inhere

[4] See also Pullman 1998, ch. 11.

in the thing as a whole; here they reside in each individual atom, and an atom without qualities is impossible. Some thinkers went so far as to claim that a man's soul or his intelligence are qualities of one particular atom in his body. But what is radically new in this Arab atomism is that no quality lasts more than one instant, and if God does not continually renew it, the atom to which it pertains ceases to exist. In this way, from instant to instant, God governs every change and every constancy, and any idea that one motion directly causes another is an illusion. It is as if its history were recorded, instant by instant, on motion-picture film, with each successive frame designed by God. A favorite metaphor in the poetry of the period compares God to a gardener without whose continual care, in that desert climate, the garden of the world would perish. Above the courtyard of the great eighth-century mosque of Damascus runs a mosaic frieze of garden motifs that beautifully represents God's nature as Islam perceives it: omnipotence and compassion.

Atoms did not resurface in Europe until the twelfth century, after Europe had largely recovered from two centuries—the ninth and tenth—of anarchy and foreign invasions.

• • •

The invasions started after Charlemagne died in 814 and his empire was split among his sons. From North Africa came Saracens who took Sicily and established bases in Provence and Italy from which they raided and plundered, almost at will, the weakly defended places of the Mediterranean world. Magyar horsemen invaded from the east, and this was also the time when the Norsemen, isolated for thousands of years in Scandinavia, discovered Europe. They sailed to England and Ireland in their brightly painted boats and southward along the European coast, up the rivers, and finally into the Mediterranean. Wherever they went, they found treasure in churches, monasteries, and in noble households. There were books, which meant nothing to them, but more silver and gold than they had ever imagined. They took what they wanted, including thousands of young men and women whom they sold as slaves in North Africa, and increasingly, instead of loading their ships and returning home, they settled down. The descendants of Emperor Charlemagne reigned over smaller and smaller territories, and finally, for most of the tenth century, there was no emperor at all.

Finally, Byzantine forces drove the Saracens out of Italy. The Scandinavians were Christianized and began to feel they belonged to Europe, while Norsemen who had settled in the south married into the local populations. King Otto of Germany defeated the Magyars and drove them back into what is now Hungary. The establishment of something like peace was slow and took several lifetimes; but when Otto was crowned emperor in 962, Europeans sensed that perhaps, just perhaps, the worst was over.

In about 990 a school was set up at Chartres to teach the liberal arts. Traditionally these were divided into two levels: the *trivium*—grammar, logic, and rhetoric; and the *quadrivium*—arithmetic, music, geometry, and astronomy. These were considered to be the skills appropriate for free men who did not have to work with their hands. At that time and for more than a century afterward, Plato reigned supreme among philosophers as the author of *Timaeus,* while Aristotle was known as a writer of difficult logical texts. In this atmosphere, new ideas sprang up overnight. A teacher named Thierry interpreted the biblical account of Creation in purely physical terms. He explained that Moses, generally considered to be the author of Genesis and other books, constructed the six-day story to encourage us to know God by studying his works. The theme was further developed by one of his colleagues, William of Conches (c. 1100–1154), whose philosophy was a Christian Platonism that owed more to Plato than to Scripture. For him the world is a work planned after an ideal model in the mind of its Creator, but when he discusses the fine details of the Creation, the thoughts are new. His elements are not the classical ones, for he defines an element as "the smallest and simplest part of any thing." Elements in this sense are particles. Individually they are too small to see, but if enough are clumped together they form what he calls the world-elements, the usual four: "An element is what is found first in a body and last in its breaking-down. . . . Now reason requires that just as every body can be divided into two largest parts, so it can be broken down into infinite [for William, infinite means too many to know] smallest parts, for every body has a limit and a boundary."

But why does the process of breaking down reach a point where it cannot be continued? Any three-dimensional object, however small, can be cut in two. William answers that an element is a point, defined by Euclid as that which has no parts, so that the idea of cutting a point doesn't even make sense. Dimension is created only when there is a large enough mass. It is the same in geometry. Draw a square on a piece of paper. The area inside consists of an infinite number of dimensionless points, so many that they fill the square with no gaps between them. One point has neither dimension nor area. All together, they create both. In this way, William avoids having to decide whether or not God could cut an element. The world-elements never appear in pure form; perhaps he takes this idea from Isidore. He says, "True elements are things which do not exist by themselves, but they are understood by themselves."

The Creation, for William as for Thierry, is a process in two parts: first the work of the Creator, then the work of nature, which brings God's work to completion and continues today. Adam and Eve came to exist by natural processes; the Creation story is only a myth intended to make people remember that God is the Creator.

The authorities of the university did not agree: "We condemn and ex-communicate the absurdities of these atoms and philosophical theories as well as everything that follows from them." What William had written was offensive not only because it disputed the scriptural account but because this kind of speculation was a waste of time. Whatever happens in nature was what God wills to happen, and the only reason for studying nature was to bring the mind closer to God. To wonder about it for its own sake was to succumb to curiosity, which Augustine had nailed long ago in his *Confessions*: "This disease of curiosity is responsible for the strange sights that are shown in the theatre. It urges men to search out the hidden powers of nature, which does them no good, simply because they desire to know." Faced with charges of heresy, William retracted some of his claims. For the compilers of *Kalam,* theology demanded a universe of atoms. In the eyes of the Church, the same doctrine was heretical. The City of God has many gates.

Shortly after William's surrender, translations of Aristotle's *Physics* from Greek and Arabic manuscripts reached France. A carefully reasoned physics featuring inner urges up, down, and around in circles began to replace one based on atoms which—since Lucretius is the only ancient writer who supports the atomic theory with sensory evidence—were largely mythical anyhow. The atomic version of things had few followers, and we will hear little more of it before the seventeenth century.

The next two centuries belonged to Aristotle. Students learned him, debated him, spotted weak points, and tried to perfect his science. Take for example Aristotle's theory that a ball continues to move after it leaves the hand because air pushes it. In about 1340 Jean Buridan of Paris easily disposed of Aristotle's arguments and put forward an idea that John Philoponus had already proposed: when a thing is thrown, the mover gives it an *impetus,* something like a force, which acts in the direction of the motion, straight ahead or in other cases circular, and which tends to keep the thing going. The faster the motion, the more the impetus. But also, why can we throw a stone farther than a feather? The stone must have a greater impetus; therefore, Buridan concludes, the amount of impetus must be given by multiplying quantity of matter by speed, and if you want to know the quantity of matter in a stone or a feather you can just take it to the nearest shop and have it weighed.

The theory of impetus had an important application: "Since the Bible does not state that appropriate intelligences move the celestial bodies, it could be said that it does not appear necessary to posit intelligences of this kind, because it could be answered that God, when he created the world, moved each of the celestial orbs as He pleased, and in moving them He impressed in them impetuses which moved them without his having to move them any more."

Finding that the celestial bodies could be moved without the help of angels did not, of course, imply that they actually were, but it is not hard to guess that Buridan was pleased to find that impetus, having solved one problem, had led him to a possible solution of another. Finding some of the science Aristotle had sought and not found did not refute the great system but helped to perfect it; this was cutting-edge stuff.

The last few pages have dealt with the world's microscopic structure. Before finishing with the Middle Ages, a few more on the world's overall plan, about which Aristotle, except for one momentous passage, had little to say.

6.6 THE GREAT DESIGN

What Isidore showed his Visigothic overlords were fragments of knowledge concerning the world's overall structure and function but there was no hint as to why it had been made that way. The Bible does not tell us what was in God's mind when he created the world. He acted; we are told none of the particulars, and we learn no more. Thus far, the book you are reading is the story of humanity dissatisfied with gaps in its knowledge and filling them with products of inexhaustible imagination. We have paid less attention to discerning an overall plan. Plato says that in the beginning there was a timeless chaos and a god who acted to make it better. Here is a motive that Scripture doesn't mention, but as it goes on, Plato's story is hard for people of the Book to accept. The god did not create mankind; he created daimones (as one might call them) who seem to have worked on the project together. Soul was what is important; it came first, and bodies were made afterward to house bits of it. Souls are reincarnated into other species. Clearly, Plato had got the main idea correct and gone astray in the details. In Timaeus he gives a scheme in which the ecliptic and zodiacal planes seem to cross at an angle but otherwise he avoids questions of layout. Genesis says that stars and planets serve as signs for festivals and seasons and years, but the question that dominates all others is the one asked in Psalm 8:

> What is man, that thou art mindful of him? and the son of man, that thou visitest him? For thou hast made him a little lower than the angels, and hast crowned him with glory and honor. Thou madest him to have dominion over the works of thy hands; thou hast put all things under his feet. All sheep and oxen, yea, and the beasts of the field; the fowl of the air, and the fish of the sea, and whatsoever passeth through the paths of the seas.

Man is of course the central concern: his place among living creatures, the reason for his creation, and ultimately, his position in the universe of created beings. The psalmist has already suggested part of the answer: there is a scale of being, with mankind somewhere near the top of it. When Aristotle arrived in Europe it turned out that he too had the idea of a scale, which he explains carefully:

> Nature proceeds little by little from lifeless things to animal life in such a way that it is impossible to determine the exact line of demarcation, nor on which side thereof an intermediate form should lie. Thus, next after lifeless things comes the plant, and of plants one will differ from another as to the amount of its apparent vitality. . . . Indeed . . . there is observed in plants a continuous scale towards the animal. So in the sea, there are certain objects concerning which one would be at a loss to determine whether they are animal or vegetable. . . . For instance, several of these objects are firmly rooted, and perish if detached.

Continuing this theme, in about 400 CE the essayist Ambrosius Theodosius Macrobius wrote a treatise on the world in the form of a commentary on a passage in Cicero. He says: "Since Mind emanates from the supreme God and Soul from Mind, and Mind, indeed, forms and suffuses all below with life, . . . and since all follow on in continuous succession, degenerating step by step in their downward course, the close observer will find that from the supreme God even to the bottommost dregs of the universe there is one tie, binding at every link and never broken."

Macrobius compares his sequence of life forms to a golden chain that the Iliad mentions as hanging from heaven to Earth. During the European Middle Ages the chain became a dominating metaphor, and it remained so into the eighteenth century. In Europe the chain's sections were divided according to habitat: the lion (some said the elephant) was the king of beasts. The dolphin was the top fish; lacking this knowledge it would be hard to guess why the heir to the French throne was known as the dauphin. The eagle ruled over the other birds and is pictured on every dollar bill. Alexander Pope, in the *Essay on Man* (1734), refers to the "vast chain of being," and the American Arthur Lovejoy, who explored this philosophic principle during the 1920s and 1930s, wrote an excellent book about it called *The Great Chain of Being*.

The chain explains why God did not make various creatures that would have been useful to humanity, and it also explains why the world contains things that seem to be useless or harmful. To choose an obvious example, why mosquitoes? The conventional answer was that, far down the chain between fleas and bedbugs, they are a necessary link. But why, then, did God create a world in which, along with these minor scourges, there are so many human forms of evil? In the thirteenth century, Thomas

Figure 6.5   Whale and calves. (From Gesner's *Historia animalium,* 1604.)

Aquinas writes that if existence is a good and if God always chooses the good, then "a universe in which there was no evil would not be so good as the actual universe." And this principle refines the answer to the question I asked before: Why did God create the universe? The universe, and each thing in it, is its own reason for existing. Look around you: everything you see is holy, and the world is better, even if we do not understand how, than it would be without it.

In these centuries, a new pattern was discovered in the Great Design: in the beginning, God had planned the world so that from everything in nature, if one looked hard enough, a lesson could be drawn that is useful to humanity. An example: you never find oak saplings growing in the shadow of a great oak. Lesson: All power in a kingdom must belong to the king, and until his death the heirs to the throne must have none of it. And long before, Augustine had taken pages to show how exactly Noah and the ark prefigure Christ and the Church. Finding interpretations of this kind became a popular sport, and books were published explaining how the (real and imagined) habits of animals were designed to instruct us. Many of these books, known as bestiaries, were well illustrated; figure 6.5 shows a whale and her calves.

• • •

During the thirteenth century, European philosophers began to study translations of Greek and Arabic texts. These explained some of God's purposes and acts but did not do much to answer a question that every thinking person asked: What of our own species—where do we fit in? The 8th Psalm seems to situate mankind in the chain of being, but there is a tradition claimed to be just as venerable that treats him as independent of God's other creations. A ninth-century Byzantine bishop named Photius reports,

> Pythagoras said that man is a little universe, meaning a compendium of the universe. This is not because, like other animals, even the least, he is made of the four elements, but because he contains all the powers of the cosmos. For the universe contains gods, the four elements, animals, and plants. All of these powers are contained in man. He has reason, which is a divine power; he has the nature of the elements, and the powers of moving, growing, and reproduction.

From hints like this an idea grew during the Middle Ages: the universe is perfect, no one ever criticizes the design of a constellation, flower, or an animal, and mankind is in some sense a replica of that perfect structure and shares some of its perfection.

The physical relation between the human body and the cosmos begins with the four elements of which both are made, but since every part of the body involves all the elements, this doesn't say much. It was Empedocles who identified the four elements responsible for the properties of matter, and it was also he, or someone else at about the same time, who identified four fluids in the human body that determine each person's ruling mental and physical characteristics and whose proper balance is necessary and almost sufficient for good health. The four humors, as they were called, and the individual types associated with them, are blood (sanguine), yellow bile (choleric), phlegm (phlegmatic), and black bile (melancholic). Sanguine people are strong, heavily built, cheerful, and active. Choleric people are hairy, brave, wrathful, and generous. Phlegmatics are cold, lazy, and tend to be stupid, while melancholics are sad, envious, and covetous. The human world is composed of such types and their mixtures just as elements make up the physical cosmos.

By Isidore's time, this idea was banal enough to be written into his encyclopedia. "The world," he says, "is actually the symbol of man. For just as the world is made of four elements, so man is composed of four humors which combine to form a single being. The ancients concluded that man is intimately united with the world."

The body's physical nature is also tied to the stars. Figure 6.6, from a twelfth-century manuscript, shows what is called a zodiacal man. He is depicted as standing, but we must think of him as wrapped around the Earth and turning with the stars, his head in Aries, the constellation of the spring equinox, and his feet next door in Pisces. Table 6.1 shows how each part of the body is placed under the dominion of one of the signs (the list varies from one author to another); furthermore, the seven openings of the head belonged to the seven planets. Therefore, as we have learned from Chaucer and many others, a doctor ignorant of astrology

TABLE 6.1
The Zodiacal Man

Aries	head	*Libra*	kidneys
Taurus	neck and shoulders	*Scorpio*	genitals
Gemini	arms and hands	*Sagittarius*	thighs
Cancer	chest	*Capricorn*	knees
Leo	heart	*Aquarius*	lower legs
Virgo	stomach and intestines	*Pisces*	feet

was no doctor at all. He not only consulted the patient's horoscope, but, because each hour of the day was ruled by a planet and different planets favored different treatments, in his treatment he had to take into account both hour and sign. If pills were prescribed, an odd number of them was considered more effective than an even number. Table 6.2 shows the system, dating from the time of Constantine, that connects hours with the cosmos. Suppose it is Saturday and you have forgotten what day comes next. The reason today is called Saturday is that its first hour is ruled by Saturn. Saturn (fig. 6.2) is the outermost planet. Counting downward, the next hour belongs to Jupiter and so on. When you reach the bottom, start again at the top. The table, in which some of the planets are also given their names in Norse mythology, assigns a planet to each hour of Saturday. The twenty-fourth hour of Saturday belongs to Mars; the next belongs to Sun, and therefore the day after Saturday is called Sunday. Start again from Sun, count to 25, and you will land on the Moon. Combine the demands of the body with astrological conditions

TABLE 6.2
Hours and Planets (Beginning on Saturday)

Saturn	1	8	15	22
Jupiter (Thor)	2	9	16	23
Mars (Tiu)	3	10	17	24
Sun	4	11	18	1
Venus (Frigg)	5	12	19	
Mercury (Woden)	6	13	20	
Moon	7	14	21	

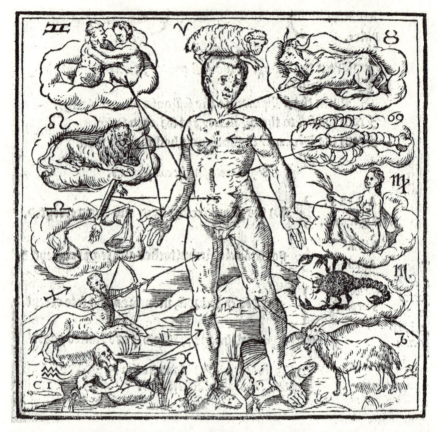

Figure 6.6   Zodiacal man, showing how parts of the body are governed by zodiacal signs. (From Digges 1556; courtesy of J. M. Pasachoff and the Chapin Library of Williams College.)

imposed by humors, numbers, the rising and setting of planets, the patient's horoscope, the positions of the zodiacal signs, the planetary rulers of the hours, and the astrological complications of the various medicines. Clearly the intellectual demands on a well-trained and conscientious doctor in the Middle Ages were at least comparable to what they would be on a good doctor today, and perhaps greater.

The reason why our bodies were thought to be so much governed by the stars is that they, like stars, are physical things. Will and intellect are faculties of the mind that have no physical connection and are therefore not governed, but every object you see is physical. Many of them are of a nature that can affect your life, and all have stellar and planetary connections, even if we don't know what they are. This view accounts for the absolute domination of astrology over all medieval thinking about the material world. Ptolemy's book on geography, carefully studied by scholars of the Middle Ages, lists the latitudes of the countries and cities he mentions and tells how they were estimated. This information was needed because latitude determines what stars and planets are visible at each season, and therefore it affects the qualities of life and action. Anyone living or transacting business in Cairo had to know its latitude. And longitude was also factored in, since conditions in eastern countries, where the Sun rises, differ in many ways from those in the West. Every country, every city had its own unique relation with planets, stars, and constellations. Why was it all made so complicated? That is not for us to know. God's plan is mathematically exact and there is a reason for every detail.

• • •

In 1440 appeared a book that closed the link between God and human in a new and unexpected way: just as a human is in some sense an imperfect copy of the universe, it claimed, so the universe is an imperfect copy of God. The author was born as Nikolaus Chrypffs, or Kryfts (the modern form is Krebs) in Cues on the Moselle River, where his father was a boatman. In the Church he became Nicholas Cusanus and wrote many books, but he is known mainly for the first one, called *On Learned Ignorance*. It is short, plainly written, and confusing to read since it claims that the path toward understanding climbs over the wreckage of generally accepted ideas. The following sketch jumps over formal arguments and stops at some conclusions.

God is boundless. If the nature and form of the universe reflect his reality, then it too is boundless. A boundless region has no center; therefore, "the Earth is not the center of the eighth sphere or of any other sphere." In fact, he says, the only reason we think the Earth is motionless is because we happen to live on it. "[If a man] were on the Sun he would fix a set of poles[5] in relation to himself; if on Earth, another set; on the Moon, another; on Mars, another, and so on. Hence the world machine will have its center everywhere and its circumference nowhere, so to speak; for God, who is everywhere and nowhere, is its circumference and center." Here Nicholas arrives at a formula already known through the writings of the German mystic and theologian Meister Eckhart: God is a circle whose center is everywhere and whose circumference is nowhere.

One consequence of what Nicholas has just said is so obvious that he does not even mention it: the distinction between the regions above and

[5] I.e., the axes of his celestial spheres.

below the sphere of our Moon—the unchanging ethereal cosmos above and the world of four elements below—can no longer be made, and if we believe Nicholas we are obliged to admit what he claims, that thinkers had assumed much too much about that upper world.

Having proposed that each planet has an equally good claim on the central position, Nicholas sees no reason why they should not all be inhabited. He plays with the idea: "We surmise that in the solar region there are inhabitants which are more solar, brilliant, industrious, intellectual—being even more spiritlike than those on the Moon where the inhabitants are more Moonlike, and than those on the Earth, where they are more material and more solidified."

Learned Ignorance is fun to read, often puzzling, a mind at the end of its tether. It was not much noticed. Fifteen years later the first printed books appeared, but by then it was no longer new. It resurfaced only in the seventeenth century when people began to enjoy it for its originality and its occasional anticipation of future developments, but there is little evidence that it ever changed any minds.

The next chapter will begin nine years after Nicholas died. It will show how people started to explain the universe in a new way—not by telling why it was made and what it means, but by telling what it is made of, how it is laid out, and how it works. But first there will be a fifteen-minute intermission during which you are invited to get up and join a tour of the world.

The World Map

> And the captain [Vasco da Gama] told him [the King of
> Calicut] that he was the ambassador of the King of Portugal,
> who was lord of many countries and the possessor of great
> wealth of every description, exceeding that of any king of
> these parts; that for a period of sixty years his ancestors
> had annually sent out vessels to make discoveries in the
> direction of India, as they knew there were Christian
> kings there like themselves. This, he said, is the reason
> which induced them to order this country to be discovered,
> not because they sought gold or silver, for of this they had
> such abundance that they needed not what was to be found
> in this country.
>
> —Journal of an anonymous companion of
> Vasco da Gama on his arrival in India in 1498

IN OUR AGE OF INFORMATION, with the stuff shoved at us from every side, it is hard to imagine what it felt like to be starved for it, and yet for most of human history that has been the normal condition of people who want to think. In Greek and Roman times, books were not plentiful. Later, monasteries had libraries, and one of the duties of a monk was to maintain or enlarge it by making copies. Courts and some towns had libraries, but few families could afford one of their own. In the 1450s, printing was invented in Germany, and the speed with which it spread shows what a hunger for information there was. In 1500 there were 150 presses in Venice alone. There Aldus Manutius invented the cheap edition, and now a man coming from work could buy a copy of Herodotus and carry it home in the pocket of his coat. At first most books were in Latin, but that is not as exclusive as perhaps it sounds because there was not yet a standard form of any European language. A book printed in Venetian dialect would have annoyed a Florentine.

Given the sudden availability of information, what was it that ordinary bourgeois readers specially wanted to know? They had questions about points in law and religion, but equally they wanted to know about the great world, of which news had long been drifting back to Europe. The

Interlude that follows tells of explorers who made their way into unknown regions, what they found there, how news of their discoveries was spread across Europe, and what happened afterward.

I.1 EARTH AND COSMOS

Long ago, over much of the world, people grew up learning about two maps. One was rich in tradition and associations that everyone shared; the other was sketchy, hazy in details, and in most ways less important. The first was the sky, ornamented with constellations that told familiar stories and together with wandering planets exerted pressures on human destiny. The other map referred to places: over there somewhere is a city called Eridu, near it is a big river, and they say that in the south, very far away, is bitter water.

Since the sky is plainly a hemisphere and a hemisphere sits neatly on a circle, the Earth was usually depicted as round. Homer and Hesiod lived on that Earth. It is flat, with the inhabited lands in the center encircled by a great river called Oceanus, which both Homer and Hesiod said "flows back on itself." Whatever they may actually have meant, later writers concluded that the river flows round and round. Since the constellations also flow round and round, it may have seemed reasonable that water would move this way where Earth touches sky. Oceanus was also a god. Hesiod writes that he is of the second generation, born of Uranus, the sky, and Gaia, the Earth, and he calls him father of gods and of all things. For the ancient Greeks, Oceanus was not very far away. The Argonauts found it in the east somewhere past the Black Sea, and to the west it was a little beyond the Pillars of Hercules. In fact the Pillars, framing the Strait of Gibraltar, were said to touch the vault of the sky as it curved downward to meet the ground at the other side of the river. When Gilgamesh went to consult the deathless Utnapishtim he crossed a river. To find the deathless Tiresias, Odysseus headed north to the land of the Cimmerians, so far from the Sun that they lived in darkness. There he crossed a series of streams that marked the edge of the world of life: Styx, a sort of marsh, is the most famous, but there were also Lethe, forgetfulness; Cocytus, sorrow; and Phlegathon, the burning river. The ever-flowing Oceanus has a useful function in the scheme of the world: according to a poem by the sixth-century Greek poet Mimnermus, after the Sun descends into the western water, "he sleeps peacefully in a beautiful bed made from honored gold by Hephaistos' hands. It skims over the water from the Hisperides to the Ethiopians' land where horses and the swift chariot stand waiting for early-born Dawn."

Homer never mentions Babylonia or Assyria. He speaks of the tribe of Sicels that inhabits Sicily, but not of any Greek settlements there or in southern Italy. He reports that in the far southern part of Africa, near where Oceanus sweeps past, are little people no bigger than your fist (the Greek word for fist is *pygmē*). They live in fear of cranes, which "flee from the coming of winter and sudden rains and fly with clamor toward the streams of Ocean, bearing slaughter and fate to the pygmy men."

In this ancient world of sky and water, the sky is a vault across which the Sun moves by day and the stars by night; there is a quantity of water beyond the Pillars; the reason why the Aethiopes of Africa have dark skins is because the Sun rises and sets so close to them. (The Greek word aethiopes means bright-eyes, and Aethiopes supposedly lived in both east and west.)

But that simple simple old picture could not survive for long. Though it set the stage for wonderful stories, there was too much that it did not explain: How are the stars synchronized so perfectly that as they move across the sky, the shapes of the constellations never change? Why do the seven planets move according to seven different rules? How is it that a week's journey southward brings stars into view along the southern horizon that one has never seen before? Everyone knew that Canopus, invisible in Athens and occasionally visible from the top of a tower in Rhodes, was easy to see from the Nile. Classical astronomers answered these questions by replacing the old disk with a spherical Earth surrounded by stars and planets mounted on rotating concentric spheres. Aristotle's estimate for the Earth's size was about twice the actual value, but a century later Eratosthenes of Alexandria made a careful measurement. The answer was 252,000 stades, the second 2 being tacked on so that there would be just 700 stades in one degree of latitude; again, we do not know what stadium he had in mind. The best estimate is probably Pliny's. A stade is one-tenth of a Roman mile, and with that the result was 25,200 Roman miles. A Roman mile is about .92 statute mile; that would give 23,200 statute miles. At any rate, Eratosthenes' number is not far from 24,900, the one we use today.

Later, when navigators explored the world, it was important to know how big the Earth is, but in Aristotle's time it was more important to understand that it is round. This brought problems of its own, for now one had to understand why everything not located at the top of the sphere doesn't slide off. Of course, any point on a sphere can be taken as the top, and as long as the Earth was stationary and it was the sky that turned, words such as "north" were defined by winds and celestial motions and not by anything fastened to the Earth. The weather was cold under the pole and hot under the equinox, but temperature is determined celestially, by the path of the Sun. The idea of the Earth oriented so that north is at the top would have been absurd, for then all of Greece would have been tilted more than 45 degrees away from the horizontal.

As long as everyone stayed close to home it was easy and natural for them to think they lived on top of the world, but Greeks traveled between colonies in Asia Minor and Sicily and as far away as Spain. Not all these places could be on top. Still, it was hard to think in any other way. A theory of gravity was needed, or at least a theory of down. People educated in Aristotle knew that the natural motion of a heavy substance was toward the center of the cosmos. That is why the Earth is located there and why it is a sphere, for heavy material tries to get as close to the center as possible, but few of those who made their living on the sea had ever heard of Aristotle. Tacitus was born four centuries after Aristotle, and a remark in his *Life of Julius Agricola* shows that this highly intelligent man didn't even know the world is round. According to his mumbling explanation of the midnight Sun, in lower latitudes the Sun sets behind mountains, but since "the flat extremities of the Earth are all water except for a few islands, it does not set in the north because there is nothing up there for it to hide behind."

So much for the attempt to make an intellectual model of the Heaven and the Earth. Greek astronomers could observe about 90 percent of the entire sky, all except a cone toward the south. They mapped and measured this domain with naked-eye accuracy, a few minutes of arc, whereas anyone wishing to survey the Earth had a much harder problem. Thus in the classical age, and for 1,500 years thereafter, astronomy was far ahead of geography, and knowledge of the Earth accumulated only piecemeal. But there were numerous travelers, driven by curiosity, love of adventure, or hope of trade—and a few of them left accounts of where they had been.

I.2 Explorers and Traders

As early as the eighth century, Greeks established colonies on the shores of Anatolia. They also moved westward, with big settlements in Sicily and southern Italy as well as trading communities in North Africa and southernern France and Spain. But in what is now Tunisia there was the Phoenician colony of Carthage, and as it grew into a great city it took control of the seas around it, so that except for a few small towns and the city of Massilia, now Marseille, it effectively excluded Greece from the western Mediterranean and prevented any curious Greek from finding what lay beyond the Pillars.

There were good reasons to be curious. For centuries, traders had walked trails leading north and west to bring home goods from distant lands. There were amber beads from the North Sea in the royal graves in Mycenae, about 1500 BCE. A piece of Chinese jade was found far down in an early archeological level of Troy. In Hallstatt, Austria, an Iron Age chieftain, roughly contemporary with Anaximander, was laid to rest in clothes embroidered with silk from China. The swords and spear points that Homer celebrates would not have been possible without bronze, copper alloyed with tin that came in Phoenician ships from distant islands called the Cassiterides. These lay somewhere beyond the Pillars, but Phoenician traders jingled the money in their pockets and would not say where.

In about 300 BCE a Massilian sailor named Pytheas sailed his ship along the Spanish coast and sneaked into the Atlantic when no one was looking. He was a careful navigator who recorded winds, currents, tides, and latitudes. He sailed up the Spanish and French coasts and then across to Cornwall, where he found the tin mines. It was summer, and he went farther north until the nights were only two or three hours long. Statements like this, which made the old historians doubt him, now confirm his story. He was rarely out of sight of land, but his 7,000-mile voyage was twice as long as Columbus's.

• • •

There is no space here for a thorough account of early exploration, but it may be interesting to follow a few ships to India and China.

In about 320 BCE, Alexander led his army overland as far as the Indus River in what is now Pakistan. There he established kingdoms which, before they gradually reverted to native rule, provided trade routes to India that were safe from the usual bandits and demands for customs duties. But leading a train of pack animals over a journey of 3,000 miles that included the Kara Kum Desert and the mountains of the Hindu Kush was not easy or fun; much better to go by water. One could ship goods by land from the Nile to the shore of the Red Sea, then by water along the north coast of the Indian Ocean and down the west of India to cities where goods were plentiful and rich princes had their courts, but the route was long and coastal waters were full of pirates. Tradition tells that about 150 BCE an Indian living in Alexandria told King Ptolemy VII about the monsoons, winds that blow fairly steadily from the southwest in spring and summer and from the northeast in autumn and winter. Compasses were still unknown, but guided by the wind a ship could sail straight from the mouth of the Red Sea to an Indian port in about forty days with minimal risk from pirates. The geographer Strabo writes that by the time of Augustus, 120 ships a year, some as large as 500 tons, sailed to India or down the east coast of Africa. A few continued farther east, around Cape Comorin, up to the mouths of the Ganges and even beyond. Hoards of Roman silver and gold coins found in several ports, notably Pondicherry, indicate that Greek and Roman traders imported more than they sent out, making up the difference in cash.

In Rome one could find Indian merchants as well as Indian fortune-tellers, conjurors, dancing-girls, and mahouts with their elephants. Brahmins taught philosophy and some families had an Indian cook. Roman consumers demanded perfumes, spices, jewels, textiles, rice, sugar, and ivory (the museum at Herculaneum shows an ivory statuette from India). Then as now, what Indians wanted was gold, and there were times when the evaporation of Roman coinage rocked the finances of the empire.

• • •

What Romans, specially Roman ladies, wanted most was silk, and that came from China, where its production was a very ancient art. During the first millennium BCE silken textiles came overland to India and Persia, but where they came from and how they were produced was a secret. In the fifth century Herodotus, born in Halicarnassus (now Bodrum) on the western edge of the Persian Empire, wrote that in the farthest reaches of India (he knew nothing of China) "there are trees which grow wild there, the fruit whereof is a wool exceeding in beauty and goodness that of sheep. The natives make their clothes of this tree-wool." Two hundred years later women in Athens were wearing clothes at home that were nearly transparent. In Pliny's time the Chinese secret was still intact:

> The Chinese [he calls them the Seres] are famous for the woollen substance obtained from their forests; after soaking in water they comb off the white down of the leaves, and so supply the women with the double task of unravelling the threads and weaving them together again, so manifold is the labor employed and so distant is the region of the globe drawn upon, to enable the Roman matron to wear transparent clothes in public. The Chinese, though mild in character, yet resemble wild animals, in that they shun the company of the remainder of mankind and wait for trade to come to them.

Herodotus may have been conflating stories of cotton and silk, but Pliny seems to be talking about silk. Shortly afterward the secret leaked to India, but until then anyone who wanted silk had to go to the traders with money in his hand, plenty of it, for after a journey of 7,000 miles by land, further by sea, it was worth more than its weight in gold. To keep up with Greek and Roman styles, dyers and weavers in the cities opened shops where they got most of the yarn they used by unweaving oriental fabrics.

During the two centuries around the year 1, the caravan routes to China were secure, crossing empires that maintained them with entrepôts where duties were paid and goods and money transferred from one caravan to another (nobody went the whole route). Later, the roads were no longer safe and most goods moved by sea, transferred in ports along the route from Chinese to Indian to Arabian bottoms.

I made this commercial digression to show how European people began to gain a knowledge of the world. By the time of the Roman Empire the regions mentioned here, together with parts of Africa, were covered with an organized network of commerce much as they are today. An enormous amount of information was available from men who had made these land and sea voyages, especially from sailors who recorded estimates of latitude, longitude, and distance along the route so that they could find their way home. In the beginning, this knowledge was recorded as instructions for getting from one place to another, but imagination needs a map of the whole.

• • •

Ptolemy wrote a book on geography that has come down without the maps. What remains is a list of places, mostly cities, whose latitude and longitude had been reported to him, combined with brief notes on topography and a discussion of the mathematical theory that tells how to project an image of part of the spherical Earth onto a flat piece of paper. He explains how his atlas is arranged with the world as he knew it stretched over exactly 180 degrees, the zero of longitude being taken as the Fortunate Isles, the westernmost point of the known world. These are probably the Canary Islands. But longitude is notoriously hard to measure. Simple astronomical sightings, for example the height of the Sun at noon on the equinox, give the latitude of a place, but longitude was usually figured from sailing time and an estimate of a vessel's speed. To find the speed a sailor threw a log of wood off the bow of a boat and measured the time that elapsed before it came opposite the stern. This was not a very accurate procedure (it ignores currents), and boats rarely sail in a straight line.

Latitude was important for astrology, but errors in longitude could have mortal consequences. Using the information available to him, Ptolemy judged that the Mediterranean, from the Pillars of Hercules to the Palestinian coast, covered 68 degrees of longitude. The correct value is 41. This stretch of the longitudinal scale, followed more or less faithfully by geographers of the Middle Ages and Renaissance, resulted in a vast overestimate of the size of Asia and a corresponding underestimate of the Pacific Ocean that persisted even after Magellan's ships crossed it in 1521.

I.3 THE CHRISTIAN EARTH

As the Christian religion spread over the Western lands, scholars discovered how little of them had been anticipated by earlier philosophers. Plato's *Timaeus* was mostly fable, while Aristotle thought the world was eternal, misrepresented the Lord of Hosts, and knew nothing of the Hosts themselves. Realizing these shortcomings, Christians were skeptical of everything the ancients had written about the natural world. Bishop Lactantius Firmianus, born about the year 250 when Christians were a persecuted minority, lived to see his faith declared the official religion of the Roman empire. In his *Divinae institutiones,* Divine Precepts, Lactantius tells what he thinks of the speculations that were the specialty of Greek science: "To investigate or wish to know the causes of natural things—whether the Sun is as great as it appears to be, or is many times greater than the whole of this Earth [he continues the list] . . . is as though we should wish to discuss what we may suppose to be the character of a city in some very remote country which we have never seen, and of which we have heard nothing more than the name." There is one bit of philosophy that he considers especially ridiculous, the claim that the Earth is round. "Is there anyone so stupid as to believe that there are men whose footprints are higher than their heads? Or that things which lie straight out with us hang upside down there; that grains and trees grow downwards; that rain and snow and hail fall upwards upon the Earth?" Lactantius is aware that Aristotle explained all this with the principle of natural motions, but he doesn't believe that theory. Up is up, down is down, and he doesn't know whether to dismiss Aristotle's argument as a joke or as a lie.

As for the arguments based on observation that favor a round Earth, Lactantius does not mention them. A century later, Augustine calls the round Earth ridiculous—it cannot be proved by reasoning and there is not even any evidence. He lived in a seacoast town from which he could see ships drop below the horizon; he traveled to Milan, 600 miles north, where the evening sky looks different, but he didn't look. The senses give useful indications but everyone knew that they are often deceived. Truth, real truth, was issued from Scripture, from reason, and from the mouths of the Fathers. It was not sought in the evidence of the senses.

• • •

Slowly, geographical knowledge disappeared from the domain of scholars and churchmen, but it did not vanish altogether. In the sixth century there was still overland traffic with the Orient—the silk trade died only at the end of that century after two monks smuggled silkworm eggs and mulberry seeds to Constantinople in the hollows of their staffs. Trade with India for jewels and spices never stopped entirely, and a book, *Christian Topography*, by an Egyptian who called himself Cosmas Indicopleustes, Cosmas the Indian Navigator, shows firsthand knowledge of Africa, India, Taprobane (now Sri Lanka), and the routes leading there in about the year 550. But most of the topography derives from Scripture.

Cosmas's world is not a sphere; he draws figure I.1 to show the silliness of that idea. Instead, the twenty-seventh chapter of Exodus reveals to him that the civilized part of it is the cosmic prototype of the Tabernacle which the Lord commanded to be built in the desert. His main text is a single phrase from Isaiah 50:22, "He who established the Heaven as a vault." These words tell him that the civilized world, the world around the Mediterranean, has the form of a vaulted box, as shown in figure I.2. Its plan is rectangular—about 10,000 miles long and, as in Exodus, half that in width. Above is the vault of the sky, but between it and the ground is the firmament (fig. I.3), a flat ceiling that restrains the waters above. Sun, Moon, planets, and stars are all below the firmament and are moved by angels. In the north is a huge mountain behind which Sun and Moon pass in their daily motion; when the Sun disappears it is night. In the east is Eden from which flow four rivers (Genesis 2:10): the Tigris, Euphrates, and two others called Pishon and Gihon, which were later often identified with the Don and the Nile. Figure I.4 is Cosmas's map of the wider world, the world beyond the ocean. Pishon flows eastward to become the Indus or the Ganges and is not shown. Oceanus is rectangular of course, and mankind was created somewhere on its outer shore. After the Expulsion, Cosmas says, the population spread out along that shore. Then came the Flood; Noah's ark crossed Oceanus into the uninhabited area around the Mediterranean and landed on Mount Ararat. Since then, the four rivers have flowed under the sea whence they emerge in various places; a commentator remarks that their courses are hidden so that adventurous men cannot follow them back to Paradise. Cosmas also supplies drawings (fig. I.5) of animals that he has seen or almost seen. Everyone knows about the hippo's teeth, and almost everyone knows that a seal is a fish that barks like a dog; here they are. I suspect that in the sixth century this chapter of Cosmas's book was read more than any other.

Cosmas says he has sailed on oceans and stood on lands that are not in the oblong world he has just shown us. He must have seen, if not noticed, the evidence that the world is round. What is going on? I mentioned Emile Durkheim's division of the conceptual world into sacred and secular. Cosmas sailed the secular world and mapped the sacred one. In his mind the two must have been absolutely separate.

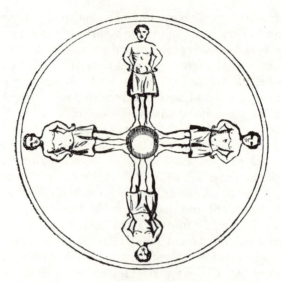

Figure I.1 Cosmas Indicopleustes points out how stupid one must be to suppose that the Earth is round.

Figure I.2 Cosmas's diagram of the civilized world as a vaulted box. The three inlets are (left to right) Mediterranean, Red Sea, and Persian Gulf. The lake is the Caspian Sea.

Figure I.3   Cross section of the box showing the firmament.

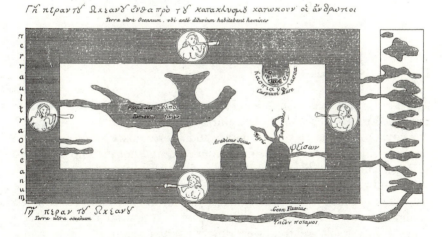

Figure I.4   Situation of the box surrounded by "the world beyond the ocean, . . . where mankind lived before the Flood."

Figure I.5 A seal and a hippopotamus according to Cosmas.

I.4 TRAVELERS' TALES

The maps in Cosmas's *Christian Topography* are said to be the first ever published in the Christian world, and how little they show us—not even one city. Ptolemy had given the names and approximate locations of hundreds of cities far away in the north and east, and not just those close to seacoasts and navigable rivers. Pilgrims who walked from Europe to Jerusalem and got home again described their experiences, and there were also reports of a wider world unimaginably far beyond the Holy Land. Their common theme was that the farther one went, the more different everything was. Animals and trees were different, and even the people were not like us. Herodotus writes of India, which he never visited: "In this desert there live amid the sand great ants, in size somewhat less than dogs but bigger than foxes. . . . These ants make their dwellings under ground and like the Greek ants, which they very much resemble in shape, throw up sand-heaps as they burrow. Now the sand which they throw up is full of gold," which accounts for much of the country's enormous wealth.

Pliny's fact-collectors dug up most of the information available up to his own time, and for fifteen hundred years his book was the main source of what Europeans knew about the rest of the world. Much of his information on trade, on minerals, and on cities and agricultural methods agrees with other reports, but beyond the range of normal commerce lay regions he knew only through travelers' tales. In Ethiopia there are "winged horses armed with horns, called *pegasi,*" and in the west "there are some people without necks, having their eyes in their shoulders." Fifteen hundred years later Desdemona's eyes filled with wonder when Othello described these creatures, and there are many more. In India there are "people with their feet turned backward and with eight toes on each foot, while on many of the mountains there is a tribe of human beings with dogs' heads" and "Ctesias describes a tribe of men called the Monocoli who have only one leg and who move in jumps with surprising speed; the same are called the Umbrella-foot tribe, because in the hotter weather they lie with their backs on the ground and protect themselves with the shadow of their feet." In the north, beyond the Alps, live the Arimaspi, "people remarkable for having one eye in the center of the forehead. These people wage continual war around their mines with the griffins, a kind of wild beast with wings, as commonly reported, that digs gold out of mines which the creatures guard and the Arimaspi try to take from them, both with remarkable covetousness."[1]

Though the size of Pliny's *Natural History* made the book expensive, there were dozens of copies in the libraries of medieval Europe. Most later encyclopedists did little more than select and rearrange from it, but a few added improvements of their own.

In about 240 CE, one Gaius Julius Solinus, not otherwise known, produced a *Collection of Memorable Matters* that survived to be one of the most popular books in the Middle Ages. His method of collection seems to have been simple. He read through Herodotus, Pliny, and other geographers. He then divided all the material into two piles: what is true, probably true, or at least reasonable, and what is purely fabulous. He threw away most of the first pile, sorted the rest, and wrote it down. Out went nine-tenths of Herodotus, three-quarters of Pliny, and all of Ptolemy. The gold-digging ants are now as big as mastiffs and have claws like a lion's, and in Germany he reports "a beast called Alce, much resembling a mule, with such a long upper lip that he cannot feed unless he walks backward." The hyena has a spine that is one continuous bone

---

[1] Adrienne Mayor (2000) suggests that Herodotus's griffins as well as huge reconstructions of the skeletons of Achilles and other heroes may have been produced when people tried to reassemble the fossil bones of prehistoric creatures such as *Protoceratops* which are occasionally found in that area.

from head to tail, and "many wondrous things are reported of it. First that it haunteth shepherds' cottages and by continual listening learneth some name, the which he expresseth by counterfeiting man's voice, with the intent to work his wrath upon men whom he calleth outside by this policy in the night time." Not all of Solinus is so bizarre and some is even correct, but it is a sobering thought that he was cited as an authority for a thousand years and that Isidore quotes him more than two hundred times. As brilliant and sophisticated an intellectual as Augustine cites the dog-headed and umbrella-footed peoples as examples of what the Lord can do when he chooses. He reminds us that if they really exist, and if they are human, then they, like us, are descended from Adam. There still remain over 150 manuscript copies of the *Collection* from the ninth and tenth centuries, and when printing was invented it was published many times. The modern editor of the 1587 English translation conjectures that it was used to teach schoolchildren about the wonders of the world they lived in.

• • •

I have already mentioned how Boethius and Isidore worked to preserve old knowledge as darkness closed in. Boethius does not mention the wider world. Isidore gives a rather disorganized sketch of geography as he knows it; he refers to Pliny a couple of times, mostly with reference to animals, but does not quote Strabo or any other classical geographer, nor does he try to describe the world's people, how they look or what they do.

In the Middle Ages, and as late as the fifteenth century, maps were usually drawn showing the inhabited world as a single landmass called Terra Firma, together with nearby islands, surrounded by a circle of water. Figure I.6 shows such a map, two temperate landmasses separated by a channel that the Sun's heat makes impassable, with frigid zones at top and bottom. The lower continent may or may not be there, but if there it is lifeless, and some said that everywhere below the equator primeval chaos still reigned. All mankind descends from Adam and Eve, and all animals, together with the plants to nourish them, must have originated at that same place and time or Adam could not have named them. For centuries, knowledge of the world beyond Europe and the Mediterranean had been accumulating among sailors and traders, but little was added to the literature of geography until that wider world arrived at the gates of Europe.

In 1241 the Golden Horde of Mongols and Turks burst open these gates, intent on plunder. They had sacked and burned Kiev; now a northern detachment defeated the Poles, and two days later the main force, commanded by Genghis Kahn's grandson Bathu, annihilated a

Figure I.6   World map attached to a 1483 edition of Macrobius's treatise on the world.

Hungarian army and devastated the country. For Mongols, the people of a conquered city were rubbish that got in the way as they took what they wanted, and by the time they left Hungary half the population was dead and most of the towns were in ruins. No European force could have stopped them from going farther, but at that moment Bathu learned that the Great Khan had died and he galloped off to the meeting that chose his successor. The Mongols' empire was already starting to break up. They never again came in full force, and when a smaller band returned a Russian army turned it back. If the Great Khan had not died then, the history of Europe might have turned out very differently.

Everyone expected the Horde to come back, and Pope Innocent IV decided that the only chance of safety was to civilize it by converting it to Christianity. There was another reason, too: the Crusades had set

Christian Europe at war with the Muslim world, and Christians were struggling with them for control of the eastern seas. The Horde's Turkish contingent was already Muslim, and if the whole mass of Turks embraced Islam the danger to Europe would be that much greater. In 1245 Innocent sent off several Franciscan friars to find the Great Khan and make friends with him. One of them, Giovanni da Pian del Carpini, wrote about his travels, as did Friar Willem van Ruysbroeck, dispatched a few years later by Louis IX of France.

These envoys were courteously received, listened to, given almost nothing to eat, and sent on their way. In time, internal politics weakened the Mongol threat, and Europeans read these travelers' accounts and a later one by Friar Odoric de Pordenone to learn about the countries they had seen. As far as the men actually went, their accounts of landscape and people reinforce each other, but though they tried hard to find reliable information about more distant regions, one could hardly expect them to get it right every time. Outside the city of Chanyl, says Carpini, for example, "wild men are certainly reported to inhabit, which cannot speak at all, and are destitute of joints in their legs, so that if they fall they cannot rise alone by their legs." Even as close to home as the shores of the Baltic there are "certain monsters, who in all things resembled the shape of men, saving that their feet were like the feet of an ox, and they have indeed men's heads but dogs' faces. They spake, as it were, two words like men, but at the third they barked like dogs."

Wherever the friars actually went, they found Europeans with whom they could speak. Some were traders; some were captives who, with no hope of going home, had made a life for themselves among Mongols; some were Nestorian Christians who had fled eastward to escape persecution. And always, all across Asia, the friars asked after Prester John.

• • •

First, something about Nestorius, who was bishop of Constantinople in the fifth century. He taught that the Jesus who was born of a woman in Palestine was in every sense a man and only a man, and that Mary, instead of being called Mother of God, should be called Mother of Christ. These claims were condemned as heretical at the Council of Ephesus in 431 but Persian Christians liked the Nestorian version, and it spread to the east and south for several hundred years. In the fourteenth century Nestorian bishoprics existed throughout the Middle East and as far away as Peking and India, and today a few ancient Nestorian churches still exist in the Middle East as well as in Kerala and on the Malabar Coast of South India. Europeans were aware of these other Christians, and in the course of time they came to imagine huge numbers of them ruled by a great and unimaginably wealthy king named John.

In the twelfth century some teller of tales wrote how Prester John (from Latin *presbyter*, priest), King of the Three Indias, had written to the Byzantine and Holy Roman emperors offering his help with the Crusades. To introduce himself he had described his kingdom, which extended from Mesopotamia to India. It contained marvels of every kind: Herodotus' gold-digging ants, a Fountain of Youth, a subterranean stream whose bed was nothing but jewels, a deposit of pebbles that make one invisible, and much more. In his court he was attended every day by seven kings and an army of lesser nobility, and on his castle's terrace was a mirror that showed him everything that went on in his dominions. Why did he call himself John the Priest? Because a ruler whose butler was an archbishop and whose head cook was a king had no need for fancy titles. He ruled a land of peace, justice, and plenty, and Europeans, who had never known any of these, thirsted for more news of it. The reports of Friar Giovanni and other travelers were awakening people to a world that was vast and full of wonders. The rumor of Prester John increased their interest and prepared them for a literary sensation, a book popularly known as *Il milione,* The Million.

Marco Polo, the youngest member of a family of Venetian merchant adventurers, left home in 1271 with his father and uncles to travel eastward. The older Polos had already visited Kublai Kahn and hoped to find him again in his summer capital of Shang-tu (Coleridge's Xanadu), northwest of Peking. After a long journey they arrived and were welcomed, and they ended up living in the emperor's entourage for seventeen years. If we believe Marco's account, he was a favorite of Kublai's and was sent several times as an unbiased observer to report on various parts of the empire. As Kublai aged the Polos grew uneasy; they got themselves attached to a mission to Persia from which, in 1295, they finally regained Venice. Shortly afterward, Marco was aboard a boat that the Genoese captured. Thrown into jail, he shared a cell with a man who had done some writing, to whom he dictated an account of his travels. It was full of huge numbers—the 12,000 stone bridges of Hangchow, the thousand cartloads of silk that arrived every day at the gates of Peking—nobody believed them and both Marco and his book were called "The Million."

Is Marco's story true? Church authorities accused him of lying when he said he had passed below the equator on the way home, since that central zone was uninhabitable. But was he ever really in China? There are doubters. Much of his information was not new, and he never mentions Chinese writing, the Great Wall, or the Chinese fondness for tea. Nevertheless, he gives more or less accurate descriptions of a dozen regions of which Europeans had known nothing except perhaps a name: Ceylon, Sumatra, Japan, Laos, Vietnam, and several more. What matters here is that the book is a wonderful story about a brave young man, that it was widely read, and that it made his readers hungry to know more.

• • •

According to Marco, Prester John had died in a battle with Genghis Khan, but the old king was not finished yet. About fifty years later, Europeans were at last able to read that John ruled not in India but in lands far beyond China. The book's author claims to be Sir John Mandeville, an English knight born in Saint Albans who went on pilgrimage to the Holy Land and didn't stop there. As far as Jerusalem his description agrees with other accounts, but beyond there he finds a world of marvels. Not far from Chaldea is the country of the Amazons, a land of women where men are admitted for only a few days at a time to keep the population up, and in Ethiopia (remember that there were *Aethiopes* in both east and west), where the Sun is very hot, he finds Pliny's Monocoli: "In that country be folk that have but one foot, and they go so blyve [quickly] that it is a marvel. And the foot is so large, that it shadoweth all the body against the Sun, when they will lie and rest them." Figure I.7, from an early printed edition, shows a few of these marvels.

Farther away than Ethiopia, the inhabited world is a series of islands. In one, "there be . . . a kind of snails that be so great, that many persons may lodge in their shells, as men would do in a little house." In Ceylon are the gold-digging ants, now as big as hounds, as well as some two-headed geese that Friar Odoric had already reported. Further along he meets more of Pliny's friends, "folk of foul stature and of cursed kind that have no heads. And their eyes be in their shoulders." Far along, even beyond China, is the domain of Prester John, who "hath under him seventy-two provinces, and in every province is a king." We are told much more, but of course the whole book is a spoof, compiled from Pliny, Odoric, and various accounts of the Holy Land, written in French and translated into Latin, then English (my quotations are from the fourteenth-century English version with spelling modernized). It was a great success in several languages, and it raises the suspicion that just as the French author must have smiled as he put in the two-headed geese, so whoever told Friar Odoric about them and Friar Giovanni about the men with no joints in their legs probably smiled too. Most Franciscans were simple men from modest backgrounds, and it must have been fun to tease them.

As for Prester John, when travelers to India and China did not find him, a rumor that he ruled in Africa began to spread. In 1513 the Por-

Figure I.7   Illustrations from a German translation of the tales of Sir John Mandeville, Strassburg, 1484. (Courtesy of the Chapin Library, Williams College.)

tuguese king Manuel told Pope Leo X that Alfonso de Albuquerque, the governor of Goa, had received an ambassador who purported to come from Ethiopia representing "Prester John, most powerful lord of Christians. He offered in the name of his monarch, as one Christian to another, all possible aid and everything necessary for a war against the enemies of the Catholic faith." As a token of sincerity, John sent with his letter a little piece of the True Cross. The letter was published, and because everybody had heard of Prester John there was great interest in it. But events unrolled at their usual pace, and it was seven years before Manuel sent a diplomatic mission to Ethiopia. The next year he reported to the pope, "With the favor of divine clemency we have at last found that most powerful bishop of the Indian and Ethiopian Christians, Prester John, Lord of the Province of Abyssinia." The mission spent six years in Abyssinia but nothing came of it. The king, whose real name was Lebna Dengel, acknowledged the pope's ecclesiastical supremacy, but at the same time it was clear from the ambassador's reports that the Ethiopian church was more Eastern than Catholic and that technically the king and his subjects were all heretics. Contrary to what tradition had led Europeans to expect, his realm offered neither peace nor justice nor plenty, and as for military aid, it turned out that for Lebna it was more blessed to receive than to give. All this happened at a time when Europe was convulsed by wars, and Abyssinia was very far away. After an exchange of letters between king and pope, the matter lapsed into silence.

I.5 THE AGE OF EXPLORATION

As Europeans woke from their dreams of a great Christian empire in the East, they were beginning to feel surrounded by Islam. Muslims controlled the south and east coasts of the Mediterranean as well as most of Spain; the Mongol empire was no longer unified under a single khan, but its pieces dominated much of Asia. Turks could advance into India any time they chose, and travelers even to the remote island of Java found Muslims there. Trading was dangerous. European ships trading in the Arabian Sea were threatened by pirates, and Arab captains, who probably paid for protection, charged whatever they liked. This meant that European bourgeois families had to do without Indian luxury goods and, more important, without spices from the Indies.

In about 1420, a century before the last spasm of interest in Prester John, a visionary Portuguese prince withdrew from the royal court and, on a rocky promontory at the country's southwestern tip, set up what we might call an institute for maritime studies: cartography, ship design, and navigation. He is called Henry the Navigator (1394–1460), though in fact he never sailed far from home. He was master of the wealthy Order of Christ and, seeing an opportunity to spread the faith while enriching his own small country, he combined the resources of the Order with his own fortune to sponsor exploration and develop trade. His first expedition headed south, claimed Madeira and the Azores for Portugal, and established colonies there. Meanwhile his designers were developing the caravel, a ship lighter and faster than the usual merchantman, and soon Portuguese sailors were sailing down the African coast.

With a larger power to the east and deep sea to the west, Portugal chose the sea. In 1488 a ship commanded by Bartolomeu Diaz rounded the Cape of Good Hope as far as what is now Mozambique. Rather than face Arabs farther north, Diaz turned back and informed his government that ships could now safely trade with India and the Spice Islands. He was followed in 1497 by Vasco da Gama, who finally reached the big commercial port of Calicut (now Kozhikode) in South India. Here, expecting to find Christians, he asked to be shown their church. His guides conducted him to what is clearly a Hindu temple but he took it for Christian. Here he and his men said their prayers and were anointed with "holy water" and a mixture of dust and cow-dung. That evening he was allowed to call on the rajah, but so many curious townsmen crowded around that his party was obliged to fight its way in, "giving many blows to the people. When at last we reached the door where the king was, there came forth from it a little old man who holds a position resembling that of a bishop, and whose advice the king acts upon in all affairs of the church. This man embraced the captain when he entered the door. Several men were wounded at this door, and we only got in by the use of much force." Once inside, da Gama treated the rajah to the parade of lies and arrogance printed at the head of this Interlude, and having heard him the Indians mopped up their blood and got rid of him as soon as they could. No doubt the court was amused, but were they also afraid?

Within a decade of this first voyage Portugal controlled the whole southeast coast of Africa and the sea routes to India and the islands beyond, and soon they were in India. It was not a peaceful penetration. In 1509 Egyptians joined with Arabs to oppose the Portuguese advance into the Indian Ocean. The Portuguese sailed a fleet around the Cape, and in an engagement known as the Battle of Diu they shattered the opposing forces and burned their ports. The next year, Albuquerque seized more than a thousand square miles of good land from the Sultan of Bijaipur and established the colony of Goa, which became the political and military headquarters of Portuguese possessions over a vast area. But this brilliant success could not last. Portugal was a small country with a small population; there were not enough people to work farms and industries at home, defend the homeland from the Spaniards, build ships, and man the forces of an empire. By the middle of the seventeenth century the empire was reduced to remnants and the Dutch were poised to take over, but now we have seen how the European penetration began, and the rest is another story.

• • •

How America was discovered—Norsemen, Christopher Columbus, Americus Vespucci—is so well known that we need not repeat much of it here. The discovery joined a new world to the old one; it opened the door for a new population and almost destroyed the one that was already there, but what is less familiar is how the enterprise was determined by the map Columbus held in his mind. Genoese by birth, Columbus petitioned Ferdinand and Isabella of Spain to finance an expedition aiming to reach the East by sailing westward. Success, he claimed, would open a new path for the advance of Christianity as well as give Spain access to silks, spices, and gold unhindered by pirates or Muslim sea power, while failure would mean only the loss of three unimportant vessels. Early in 1492, the royal couple decided to try their luck.

Although Columbus had sailed before, it does not appear that he knew much about navigation or had ever commanded a ship. He learned both as he went along, and he seems to have been born with a fine sense of place; in the Caribbean he learned to sail confidently from island to island in a way that amazed his subordinates. His program, the "Enterprise of the Indies," was based on several misconceptions. Figures in Ptolemy's *Geography* and its later continuations vastly overestimated the length of Asia. The Earth's circumference was known by the time Columbus and his consultants had spread the world onto a map, so they estimated Japan lay only 2,500 nautical miles west of the Canaries. Further encouraged by a strange figure for the nautical mile, only three-quarters of what it really was, Columbus cast off his hawsers. He sailed to the Canaries, took on supplies, and on September 6 he headed into the unknown. Thirty-five days later he found land about where he expected to find it and came ashore on a small island. Seeing no trace of civilization, he concluded that he must be somewhere off the main island of Japan. He sailed southwest to Cuba where he found many people but no gold or silk; now he decided he had reached a southern province of the Chinese Empire that Marco Polo had called Mangi. He anchored and "sent two men inland to find out if there were a king or great cities. They traveled for three days and found an infinity of small villages and uncountable people, but no sign of centralized government, and so they returned." Columbus turned back and came to a large island, and naming it Hispaniola he established a settlement. Here he found modest deposits of gold which suggested to him that perhaps he had found the land where the Queen of Sheba lived; at any rate these little chunks would encourage his supporters in Spain. Columbus, imagining himself so near the Orient, must have been surprised to see no shipping and no great seaports. He carried with him a letter from their Spanish Majesties to the Great Khan but there was no one to give it to, so after exploring the Caribbean for three months he headed home.

There were three more voyages. On the last one, Columbus sailed farther south and landed on a new coast where for the first time he tasted something like reality, for near an island he named Trinidad was a shore from which flowed an immense river, and he understood at once that so much fresh water could only come from a region of continental size. He had discovered South America and the river was a branch of the Orinoco, though he thought he might be looking at one of the Four Rivers of Paradise.

Columbus's real intentions may have been different from those he told the king and queen. If he had planned to bring home samples of the merchandise available in China and Japan he would have taken trading goods and he would have armed his ships to protect them. As it was, he took only glass beads and bits of lace, but even these saved him on several occasions when the going got tough. The contract he signed with Ferdinand and Isabella specified that he was to be viceroy and governor general of any lands he might discover or gain for the Spanish crown. Had he landed in Japan or on the coast of China, this promise would have done him little good; instead, it was only in his last voyage that he pushed westward as far as he could go and sailed along the coast of Panama and Honduras looking for the route that he thought Marco Polo must have taken when he sailed from Mangi to India. Was he just feathering his own nest?

To the end of his life, Columbus believed, against all appearances, that he had come within three weeks' sail of the mouths of the Ganges. Not all experts were so optimistic, but Figure I.8 shows part of an immense world map published in Rome in 1508 that represents some experts' ideas of the globe's far side after Columbus's second voyage. On the right side, the scroll on the large island says "The ships of Ferdinand King of Spain came this far," which identifies it as the east end of Cuba. Java Major is an old name for Sumatra; Java Minor and Candy are anyone's guess. A scan from north to south reveals, first, at a latitude somewhat north of Iceland's, an enclosed settlement of Jews (*Judei inclusi*), bordered by a great lake on the north and on the south by a chain of mountains. This protects them from the hostile powers of Gog and Magog (Ezekiel 38f; Revelations 20:8; Koran 18) who live nearby. And what is this settlement of Jews? Some old maps place the earthly Paradise, "in Eden away to the east," on the coast of China. This is not because Eden was identified with China but because older maps had east where we have north. It was natural to put Eden at the top, and this explains Columbus's hypothesis concerning the Orinoco.

A little south of Gog and Magog are the Karakoram Mountains along with Tibet, Bengal, and a city called Cathay. Japan and Formosa are not there. Loac may represent Laos, and on the islands south of it, "men have the heads of dogs, cows are worshipped, and the forests are full of spices." More than 30 degrees south of the equator is part of the island of Ceylon. But for Ferdinand and Isabella, if they saw this map, the main interest was that Cuba is almost inside what may be the Gulf of Tonkin. Next to Cuba is Hispaniola where Columbus had his settlement, north of it is Greenland, and in the upper corner are the Azores, not far from Spain. Even as late as 1540, a popular atlas (fig I.9) shows how hard it

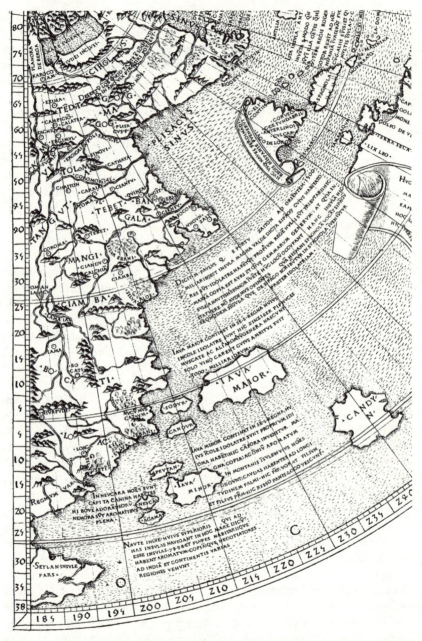

Figure I.8 Part of a world map (Ruysch, *Universalior cogniti orbis Tabula*, 1508) showing the seas and lands between Ceylon (now Sri Lanka, lower left) and some Caribbean islands (upper right).

Figure I.9   Map of America from an edition of Ptolemy's *Geography*, Basel, 1540. Zipangri is Japan.

• • •

was to get longitudes right. It shows a ship sailing the Pacific but puts Zipangri (Japan) just beyond where Catalina Island is.

Marco Polo had written of an Ocean of Asia. In Columbus's time explorers of the Spice Islands had reported a great sea to the east of them, and in 1513 Vasco Nùñez de Balboa, one of the Spanish conquistadores, stood on this sea's eastern shore and claimed it all for Spain. His right to do this had been established in 1494 in the Treaty of Tordesillas by which Spain and Portugal agreed to divide the newly discovered and imperfectly charted world. New lands east of about the 46th meridian west were assigned to Portugal; all lands farther west went to Spain. Measured from there, each got half the world. In theory their realms touched at meridian 134 east, but that didn't mean much because nobody knew what was there. Portugal got India and Africa but in the Atlantic it only got Brazil, while Spain got the rest of the New World. Needless to say, no European power except the signatories ever paid any attention to what the treaty said. Balboa's claim, even if taken seriously, referred only to a coastal strip of uncertain length. Spain's rights extended to land far west of there, but what land? Where? How far? If the Spaniards really wanted to own the immense new territories they had been given, they had better go and claim them.

Fernão de Magalhães (1480–1521)—I will call him Magellan—grew up in the Portuguese court and served with the navy in the Indian Ocean where he fought in the Battle of Diu. Something went wrong in Lisbon, and following Columbus and a number of other sailors, he transferred his allegiance to Spain. There he worked out a great plan to secure both edges of the Pacific Ocean for his masters by sailing west to claim the Spice Islands, now called the Moluccas. For years, sailors had been trying and failing to reach the Orient through the icebound waters north of Canada, and if there was no western outlet from the Gulf of Mexico, what remained was to find a way either around the southern continent or through it.

In 1518 Magellan was given command of five ships that set out the following year, and after another year he was sailing down the South American coast, probing bays and estuaries for a passage westward. For many weeks, farther and farther south, he found none until at latitude 53 degrees south (about as far south as London is north) he arrived at an opening that looked as if ships could pass through. On October 21, 1520 he decided to try. The passage was delicate, 350 miles of dangerous rocks and high winds. Three ships emerged into calm water on the other side. Of the five that started out, one had been wrecked and another, carrying most of the provisions, had deserted. Looking at the long gentle waves, Magellan called the ocean Pacific and the name stuck.

He could have taken his ships up the coast to pick up water and supplies. Instead, like someone in the dark who didn't know the level of water in a swimming pool and dove in anyway, he set out at once to cross the ocean. Why? I think he had absorbed the idea that the Pacific Ocean is not very wide. And why did he attempt the dangerous strait when, by sailing a few days farther south, he could have rounded Cape Horn in open water? Nobody knew there was a cape, and it is not shown even on Mercator's excellent map (fig. I.10). There were rumors of a great southern continent and there was no reason to expect another waterway.

After ninety-nine days at sea, almost three times the length of Columbus's run, Magellan and his men dragged themselves ashore, tortured by scurvy and in the last extremes of hunger and thirst, onto a tropical island that was probably one of the Marianas. When they had recovered their strength, they set off again and reached what must have been the Philippines. One of Magellan's crew was a slave named Henrique who came from the Moluccas. The captain told him to hail some men in a boat that came alongside and they answered in the same language. Everyone on board realized what this meant. Near there, while interven-

Figure I.10   From Mercator's world map, 1587. New Guinea is identified as an island but Australia has no south coast, and almost seventy years after Magellan's passage, Cape Horn is still not shown.

ing in a tribal dispute, Magellan lost his life. The ships continued under another commander who somewhere made the ritual claim of sovereignty; then they took on spices and continued west. In September 1522, three years after their departure, a single ship, leaking badly from woodworms and manned by eighteen sailors in dying condition but loaded with spices, tied up at a pier in Seville. It had put a girdle around the Earth. Three ships and many lives had been lost. When the spices were sold at market they returned to the Crown the whole cost of the expedition, but the Magellan family received nothing. Eighteen years later the map in figure I.9 was published. Evidently the cartographers, adding up longitudes, couldn't believe the Pacific was ninety-nine days wide.

It was just a century since Henry the Navigator had first dreamed of exploring the world by water. In that time, ships and navigation had improved and nations were contending for domination of distant seas. The look of the world had also changed. Before Prince Henry, it was something that might have been designed by Sir John Mandeville. Your own part of it—Catalonia, Burgundy, Yorkshire, whatever it was—contained people like yourself who spoke your language. On a larger scale were distant kings and wars and strange animals. Somewhere beyond that, Terra Firma ended and the biblical chaos began, cold to the north and hot to the south, stretching off into a haze of water and mystery. But now the world was finite. Travel far enough in any direction and you came home again, and except for the polar regions the world was being visited by people like yourself. It contained strange folk and an occasional griffin; there was a fountain of youth somewhere and the Garden of Eden might still be found, but all these were just dots on a map that was starting to take a definite form. The Age of Exploration was nearly over.

• • •

There was, however, unfinished business. Even after the Portuguese, Spanish, French, and English had sailed and traded in the South Pacific, nobody knew what lay still farther south. Two centuries went by, and always there were rumors of a Great Southern Continent waiting to be claimed by a European power willing to make the effort. This continent was not all snow and ice, for the map of figure I.10 shows that it extended almost to the equator.

In England, the gentlemen's club known as the Royal Society recommended an expedition to find and claim this last great landmass, if it existed, and in 1768 there came a wisp of evidence that it did. An expedition commanded by Samuel Wallis, just back from a voyage of discovery in the South Seas, reported some new islands, which were named the Society Islands in honor of the club. Tahiti was the largest of them and there, looking south from a mountain peak, Wallis thought he had seen cloud-capped mountains in the distance.

Captain James Cook knew the central Pacific. He had already charted the Australian coast, and in 1772 he was sent off to look for the unknown land. He provisioned in New Zealand and then for two years he combed the South Pacific, making several forays into Antarctic waters. In January 1774 he reached latitude 71°10'. By his own account,

> The Clowds near the horizon were of a perfect Snow whiteness and were difficult to be distinguished from the Ice hills whose lofty summits reached the Clowds. The outer or northern edge of this immence Ice was composed of

loose or broken ice so close packed together that nothing could enter it; about a mile in began the firm ice, in one compact solid boddy and seemed to increase in height as you traced it to the South; In this field we counted Ninety-Seven Ice Hills or Mountains, many of them vastly large.

With "ropes like wires, sails like boards or metal plates, sheaves frozen fast in the blocks," Cook turned his ship northward. After further explorations of the western Pacific, he headed west, past Cape Horn and across the South Atlantic as far down as practical. He found a few cold and stormy islands, nothing more, but from the enormous size of some of the icebergs he saw, and the fact that all of them gave fresh water when melted, he reasoned that somewhere to the south was land. Thus perished hopes of a great southern continent. Antarctica was there, of course, behind the ice fields, but no one saw it until 1820. And that, until mankind made its next giant leap, ended the Age of Exploration.

SEVEN

Toward a New Astronomy

> Philosophy is written in that great book, the universe, that
> forever stands open before our eyes, but one cannot read it until
> one has first learned to understand the language and recognize
> the symbols in which it is written. It is written in the language
> of mathematics, and its symbols are triangles, circles, and other
> geometrical figures without which one cannot understand a
> word, without which one wanders through a dark labyrinth
> in vain.[1]
>
> —Galileo

AT THE END OF the fourteenth century astrology was still the queen of the
sciences, but the great project of completing Aristotle's work was beginning to reveal weaknesses in its foundations. In universities, conservative
professors still took Scripture and Aristotle as absolute. Their teaching consisted mostly in training students to read, understand, and defend Aristotle
against new ideas that were being discussed outside the classroom. Students
were reading the Spanish-Arabian commentator Averroës who claimed that
philosophic truth, based on strict rules of reasoning, was the only reliable
truth, and that religion, based on revelation and authority, provides a
subordinate version suitable for simple people. But what philosophic
truth? New trends in medicine, chemistry, and mathematics arrived from
Arab lands and swept across Europe, knowledge that was delivered
without ancient authority and therefore had to be accepted or rejected on
its own merits. This brought an increasing consciousness of experience
and simple common sense as sources of knowledge, and it also opened
the way for explanatory schemes based on the concept of law.

Look back three centuries to when William's judges had to remind him
that his theories were a waste of time because whatever happens in nature is what God wills should happen, that is, what *ought* to happen. But
there is another possibility: that the world was created so that, other
things being the same, certain events must *always* occur. The pores of Lucretius's horn lantern are large enough so that some atoms of light will always go through; a stone dropped from the hand always falls toward the

---

[1] The omission of numbers from this list reminds us that whereas numbers only measure,
geometrical figures bring philosophical enlightenment.

center of the Earth; and if a competent astronomer calculates where Mars will be a year from tonight, you can bet it will be there. These *musts* were established at the beginning of time,[2] but was it by a million separate decrees or does the veil of nature perhaps cover one law from which the others follow? In the fourteenth century no one thought beyond the million decrees; by 1700 Isaac Newton had lifted a corner of the veil.

7.1 THE SUN STANDS STILL

Niklas Koppernigk (1473–1543) was born in a bit of Polish territory called Ermland, on the Baltic Sea. His uncle, who was bishop and administrative head of the district, intended that Niklas should succeed him and gave him a good education. He studied first at the University of Cracow, where he learned his Latin and became Nicolaus Copernicus, and then in Italy, where he attended classes but took no degree. In 1500 he found himself in Rome and spent the Holy Year giving informal lectures on mathematics and astronomy. Then he went home and was elected a canon in his uncle's cathedral. Like most canons he was never ordained as a priest, but he was expected to help manage diocesan affairs, maintain church property, sing the responses in divine service, and live in some style, maintained by at least two servants and three horses. Assured of a life of peace and comfort, he decided not to succeed his uncle but to spend his free time studying stars and planets. He moved to a house attached to the cathedral of a coastal town called Frauenberg (Frombork in Polish), and there, except for a few business trips, he spent the rest of his life.

There is little record of Copernicus's studies, but an early biographer says, "He had as friends . . . Cracow astronomers, formerly his fellow-students, with whom he corresponded about eclipses and observations of eclipses." For some time he had been ripening an idea in his mind, and in about 1512, not long after he settled down in Frauenberg, he circulated a few handwritten copies of a six-page manuscript, now known as the *Commentariolus* or Little Treatise, in which, modestly but with conviction, he sketched the Copernican System.

Everyone interested in astronomy was familiar with Aristarchus's idea—that it is actually the Sun, not the Earth that stands still, that Earth is one of six planets that circle around it, and that the daily motion of all the heavenly bodies is only an illusion because the Earth spins like a top. Everyone could also quote Psalm 93, "The world also is established, that it cannot be moved." In conservative Poland, this sounded like a fact, but

---

[2] See the quotation from Augustine's *On the Trinity* in section 5.2.

in Italy by this time the sacred and secular realms were drawing apart. The remarkable thing is not that a Polish churchman decided to argue for a Sun-centered world, but that nobody in Italy had done it first.

There were philosophical difficulties also. Copernicus was taking away the natural motions of the five elements that compose the universe and replacing them with violent ones. If the stars do not revolve around the Earth there is no reason to imagine them mounted on a sphere, and the fact that some stars are much brighter than others suggests they are scattered in space. Though Copernicus did not say so, he was reducing natural philosophy to a pile of smoking ruins. Why believe him at all? The Sun-centered system had a certain simplicity, but astronomers were little concerned with that. Most of them made their living calculating where a given planet had been or would be on a given night. The Ptolemaic system, though not perfect, allowed them to do this with reasonable accuracy. Could a Sun-centered system do it as well? Another technical difficulty was this: Suppose you look out the window at a nearby tree due north of your house, then move to another window. The tree will not be exactly to the north any more, an effect called *parallax*. The tree's apparent direction depends on where you stand. An eye looking at the stars sees nothing of this kind as the Earth swings in its great loop around the Sun. If the Earth really moves and we see no parallax, the stars must be unimaginably far away, and the cosmos must be constructed on a scale vastly larger than anyone had thought. If skepticism is a virtue, and in scientific matters it is, the Sun-centered system is not as plausible as it might seem at first sight. At any rate, a manuscript of six leaves would not change the opinion of anyone who happened to see it.

Though the *Commentariolus* does not show Copernicus's calculations, it quotes enough results to show that he has convinced himself that his system allows calculations as accurate as Ptolemy's and at some points better. I have mentioned that Ptolemy's Moon moves so that its size varies by a factor of almost 2. Copernicus mounts the Moon so that its apparent size changes by only about one-fourth. This is better, though still twice what it is observed to be.

The *Commentariolus* promised a larger work that would show the calculations, but it was thirty years before that took its final form. In the meantime, word got around. In 1532 a Swiss publisher put out a world map that had an angel at the North Pole busily cranking the Earth around (fig. 7.1), and another at the south. If angels moved the planets in their orbits, they could spin them as well. A year later, after a papal secretary had explained the Copernican idea to Pope Clement VII in the Vatican garden, he was thanked and rewarded, but Protestants were less tolerant. A volume of Luther's table talk has him saying, "The people gave ear to an upstart astrologer who tries to show that the Earth re-

Figure 7.1   An angel turns the Earth. From a map in Grynaeus, *Novus orbis,* Basel, 1532.

volves, . . . the fool wants to turn the whole art of astronomy upside down."

Why did Copernicus wait so long to publish the big book? Historians choose from a menu of reasons: press of business, fear of offending the Church, war against the Teutonic Knights whose territory bounded Ermland's, fear for his own reputation. At last a young enthusiast named Rheticus attached himself to Copernicus and got him to complete a manuscript. Figure 7.2, in the author's handwriting, shows the crucial page describing his system. Rheticus arrived just in time, for while the book, *On the Revolutions of the Celestial Orbs,* was with the publisher in Nürnberg, Copernicus suffered a stroke that left him physically and mentally weakened. In May 1543, as he lay dying, the first copy came from the printer and was put into his hands.

Copernicus put the Sun at the center of his system, but his work was otherwise Ptolemaic. He assumed with Aristotle that the components of the cosmic machine are perfect spheres, though since the only purpose of the planetary spheres was to carry a planet, he didn't need the whole sphere, and in fact he uses the terms sphere, circle, and orb almost interchangeably. He never talks about the spheres as things; their only property is to carry something in a circle at a constant rate. From Greek astronomy, particularly Ptolemy, he borrows some of the tricks that make the calculations work better. There are two kinds of assumptions here. I think that spheres, for Copernicus as for Ptolemy, were computational tools, but the assumption that they move uniformly even though they don't exist, which causes so much trouble because none of the observed motions is uniform, is philosophy; it pertains to the essential nature of God's perfect

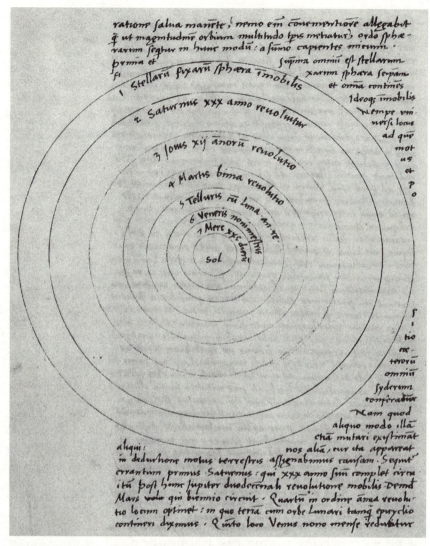

Figure 7.2 The heavenly spheres with the Sun at the center as drawn in Copernicus's original manuscript. Compare fig. 6.2. (Jagiellonian Library, Kraków, Poland.)

universe. As to the rotating Earth, he assumes it, but his argument leans on Aristotle:

> We hold it as a fact that the Earth, enclosed between poles, is bounded by a spherical surface. Why then do we still hesitate to grant it the motion appropriate by nature to its form rather than attribute a movement to the entire uni-

verse, whose limit is unknown and unknowable? Why should we not admit, with regard to the daily rotation, that the appearance is in the heavens and the reality is in the Earth? The situation closely resembles what Vergil's Aeneas says: "Forth from the harbor we sail, and the land and the cities slip backward." [*Aeneid* III.72]

For when a ship is floating calmly along, the sailors see its motion mirrored in everything outside, while on the other hand they suppose that they are stationary, together with everything on board. In the same way, the motion of the Earth can unquestionably produce the impression that the entire universe is rotating.

According to Aristotle the rotation of the spheres has a formal cause in their shape as well as a material cause in the element of which they are made, for Ether naturally moves in circles. Copernicus is claiming that for the Earth to rotate, the formal cause must override the material one, and if pressed he could have pointed out that while we really know very little about Ether, anyone can set a ball spinning. Answering Ptolemy's claim that a rotating Earth would produce high winds and other kinds of damage, he argues that this motion is natural, that natural motion is never destructive, and that in this case the Water and Earth in the atmosphere are enough to pull it around at the same speed as Earth itself.

If Copernicus was able to look into the book that was held in front of his dying eyes, he would have been surprised, for the first page was not written by him. His assistant Rheticus had had to leave Nürnberg and had confided the details of publication to Andreas Osiander, a friendly and well-intentioned Lutheran clergyman of the city. We know nothing of Osiander's real thoughts, but apparently he was afraid for the book's reception and he inserted an unsigned one-page preface that denied its central thesis. He says he has no doubt that some scholars "are deeply offended and believe that the liberal arts, which were established long ago on a sound basis, should not be thrown into confusion." He points out that no one is so naïve any more as to believe that the spheres of Ptolemy or Copernicus really exist, and it is just so with the moving Earth: think of it as a convention that makes computation easier, and don't worry. As for the real truth, we shall not know it unless it is divinely revealed.

Until Copernicus's friends revealed the authorship of this note, careful readers had the impression of a man who gives with one hand and takes with the other. But to working astronomers, who were almost the only readers of this highly mathematical book, the main point was that it contained a new way of calculating planetary positions that was both simpler and more accurate than the old. From the beginning, therefore, professionals were its most enthusiastic supporters. Copernicus had never tried or intended to create a new astronomy, but only to present the Ptolemaic system in a new and better way. By putting the Sun in the center and re-calculating some of the numbers that other astronomers were using, he produced a computational system that gave better results than the old one, but for most astronomers there was really nothing new. The Copernican system uses thirty-four spheres, and Otto Neugebauer has shown that if one omits a few nonfunctional spheres from the Ptolemaic system and counts only those that are actually used, the number in Ptolemy's system is the same.[3] This is not very surprising—after all, both systems describe the same motions, but whereas Copernicus' spheres provide small adjustments to circular motions, Ptolemy's cause the planets to trace great loops in the sky. It must have been this philosophical simplification that persuaded Copernicus that he had found the way the solar system actually is, even though no one could prove it.

Astronomers were soon convinced. Erasmus Reinhold, professor at the University of Wittenberg, praised Copernicus as the new Ptolemy and calculated a set of planetary positions, the *Prutenic* (i.e., Prussian) *Tables,* which appeared in 1551 and set the standard for the next eighty years. The public, having an older view of things, followed more slowly. In 1572 appeared a supernova that measurement showed was located somewhere outside the lunar sphere in the region where nothing was ever supposed to change. In 1577 came a comet that passed through several planetary orbits as it came and went. Astronomers were surprised, but the educated public, brought up inside a universe of crystalline spheres, was shocked and dismayed. The Catholic Church, on the whole, viewing the *Revolutions* in the light of Osiander's preface, took the affair lightly. From 1616 to 1835 the book was in the index of forbidden books, though with the note *donec corrigatur,* until corrected. This meant that, after a few strokes of a pen through paragraphs that argue Earth's motion, it would be safe to read. Very few copies, almost none outside Italy, were ever corrected.

The Protestant world, brought up on the Bible, was on the whole more offended and more skeptical. The new theory was not easy to swallow, and many sniffed at it and left it alone. It used the implausible Ptolemaic machinery, and philosophically it was absurd. Sir John Davies, English lawyer, poet, and member of Parliament, defended tradition. In a work called *Orchestra,* or A Poem of Dancing (1596), he writes

> Only the Earth doth stand forever still;
> Her rocks remove not, nor her mountains meet—

---

[3] One occasionally reads that by Copernicus's time astronomers trying to make more accurate tables were using up to eighty spheres. Owen Gingerich (1992, 61; 1993, 125) has made a computer study of the Alfonsine Tables, which were standard at that time, and found no sign of any such scheme.

Although some wits enricht with learning's skill,
Say heaven stands firm, and that the Earth doth fleet,
And swiftly turneth underneath their feet.

A few years later, John Donne found his London friends taking Copernicus more seriously. In *An Anatomie of the World: The First Anniversary* (1611) he complains,

And new Philosophie calls all in doubt;
The Element of fire is quite put out;
The Sun is lost, and th'Earth, and no mans wit
Can well direct him where to looke for it.

The trouble was not that the new philosophy contradicts Scripture. The Bible assumes the Earth stands still but doesn't make a big point of it, and it nowhere mentions the four elements. Donne's objection is that the new philosophy doesn't make any sense. The old one placed the Earth at the center of the celestial sphere, but where is it if there is no sphere? If the Earth isn't in the middle but moves around some randomly selected spot, what does the new cosmos imply about the relation between man and God? What is its plan? What does it mean? Unknown to Donne, the beginnings of an answer had already been published in Austria.

7.2 THE MATHEMATICAL PLAN

Contemplating the Ptolemaic sky, people must have often wondered why God ever installed such a huge and complex machine just so that humans could watch Sun, Moon, and five bright dots dance along the zodiacal pathway. Copernicus had made the machine easier to imagine, but questions remained. Why, for example, are the orbits of the planets laid out as they are, and why is it that the further a planet's orbit is from the Sun, the more slowly it moves? These are reasonable questions, and it was not long before a natural philosopher answered them with a system designed according to mathematical principles.

Johannes Kepler (1571–1630) was born into a poor Protestant family in Weil der Stadt, a small free city of the Holy Roman Empire; his father seems to have been an itinerant mercenary soldier. Through the generosity of the local duke, he was sent to study for the ministry at the University of Tübingen. He also heard lectures on mathematics and astronomy, and when he graduated and was offered a job teaching mathematics at a Protestant boys' school in Graz, he decided to take it. Graz was ruled by a Catholic prince but the administration and much of the population were Protestant, and there seemed to be a good chance that he could live there unmolested. Thus, at age twenty-three he found himself teaching mathematics and astronomy and, ex officio, appointed *mathematicus* of the province. *Mathematicus* translates as astrologer, and each year it was his duty to produce an almanac predicting weather and significant events. Kepler never thought that earthly matters were unconnected with the sky, and heaven showered him with blessings. His first almanac, in 1594, predicted an invasion of Austria by the Turks and an unusually cold winter. Both happened, and in Graz his reputation was as solid as the ice on the river Mur that flows past it.

Kepler impressed his colleagues but baffled his students. The first year there were few of them; the second, none, but the school paid his salary and he had time to think. His thoughts started this way: Pythagoras has shown us, once and for all, that the fundamental principles of harmony rest on small whole numbers. The architecture of the universe must be based on some similar harmony. Kepler had Copernicus's numbers for the mean radii of the planetary orbits. If we take Saturn as 1, the rest can be expressed as fractions:

Saturn	Jupiter	Mars	Earth	Venus	Mercury
1	0.569	0.166	0.109	0.078	0.041

If Kepler could fit these numbers with tidy fractions, he would know something about the Almighty's plan for the universe. The Mars-Saturn ratio is 1:6 but the others are hard to fit. "I wasted a great deal of time on that toil, as if at a game, since no agreement appeared either in the proportions themselves or in their differences." So he dug into his bag of mathematics and found geometry, and with that he had better luck. Two years after he arrived in Graz he had finished a book called *The Secret of the Universe,* whisked it past the ecclesiastical censors by presenting its Copernican framework as a picture and not a fact, and, with the help of the Tübingen faculty, got it published. The reader learns that God designed the Solar System around the five Platonic solids (fig. 4.2). This is how it was done: God assigned to each planet a thick spherical shell whose inner radius is the radius of its orbit and whose thickness is measured by how much the orbit departs from a circle. To determine the spaces separating these six shells, God inserted Plato's solids tightly between them. Figure 7.3 shows the system with Saturn marked on the outermost shell. So that we can see the construction, Kepler has cut away half the shell, and the edge of the remaining half is what he calls the circle. He writes,

> The Earth is the circle which is the measure of all. Construct a dodecahedron round it. The circle surrounding that will be Mars. Round Mars construct a

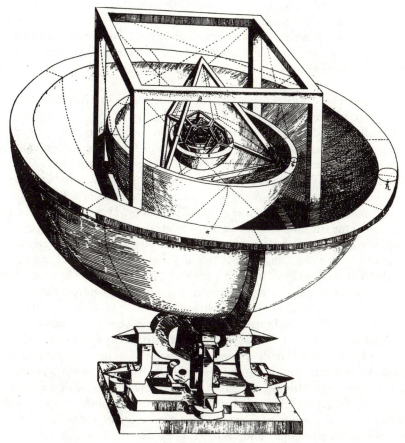

Figure 7.3 Kepler nests Copernicus's planetary spheres inside the Platonic solids. (From Kepler's *Secret of the Universe.*)

tetrahedron. The circle surrounding that will be Jupiter. Round Jupiter construct a cube. The circle surrounding that will be Saturn. Now describe an icosahedron inside the Earth. The circle inscribed within that will be Venus. Inside Venus describe an octohedron. The circle described within that will be Mercury.

Each Platonic solid is used once. Why did God install them within their spheres in this particular order? He had the best of reasons. The cube, which comes first, is the solid that looks out at the rest of the universe symmetrically in all three dimensions of space, and the other four choices are similarly justified. Now he calculates the empty intervals between

successive spheres and compares their sizes with the radii Copernicus calculated. Mars and Venus come out perfectly; the others have errors of less
then 10 percent. Next, since God makes no mistakes, come arguments
that adjust the results so that the errors disappear, but we needn't follow
them. And now, if you have absorbed that immense hypothesis, you
know why there are six planets as well as the reason for the size of their
orbits.

The Counter-Reformation gathered momentum. In Graz the Protestant school where Kepler taught was closed down. Kepler and his family
went sadly off to Linz, where he had thoughts of studying medicine and
opening a practice. But during his six years in Graz, Kepler had corresponded with Tycho Brahe (1546–1601), Emperor Rudolph II's Imperial
Mathematician. The emperor was a strange, unworldly man, immersed
in his collections of coins, sculpture, paintings, and clocks. He admired
astronomers but was totally uninterested in money. Tycho had pushed astronomical measurement to a new kind of accuracy and was interested in
using his accumulated data to establish his own version of the Solar
System in which the Sun and the outer planets travel around the Earth
and the two inner planets travel around the Sun. Tycho was impressed by
Kepler's energy and mathematical knowledge and offered the young
schoolteacher a job in a new observatory outside Prague. In the spring of
1601 the Keplers moved into a little house in the street now named after
him, and in October Tycho died. The observatory was without a director,
and Kepler, who had made a good impression at court, was named Imperial Mathematician, with the expectation that he would continue
Tycho's work. The honor was great but the salary was small and rarely
paid. His first act was to publish a short book titled *On the More Certain
Foundations of Astrology.* Three years later he wrote a book on optics
that explained telescopes and showed, once for all, how light forms an
image inside the eye. During this time he also worked on a problem that
Tycho had left him: to determine, using Tycho's matchless observations,
the path that Mars follows as it moves around the Sun.

• • •

Consider what the problem was. Tycho and his assistants had observed
Mars every clear night, measuring the position of the reddish dot against
the background of fixed stars. From night to night the dot traces a path
across the sky. They stood on Earth, whose motion they knew roughly,
observing Mars, whose motion they knew no better. How much of the
apparent motion of the dot was due to Mars's motion and how much to
the Earth's? Kepler knew a few things exactly: he knew that an Earth year
is 365¼ days long. That is, if two observations of Mars are taken 365¼

days apart, the Earth is in exactly the same place each time and the observations show only the motion of Mars. On the other hand, Mars's year is 687 days long. If two observations of Mars are taken 687 days apart, we are looking from two different places in Earth's orbit at something that has not moved. If Mars is east of us, we are west of Mars. To find out where Earth is as seen from Mars, imagine the measuring instrument pointed the opposite way. This is how Kepler was able to plot the apparent motion of Mars as seen from Earth and of Earth as seen from Mars. The real problem was now to find out from the positions of the red dot in the sky how Mars actually moves through space: what the shape of its orbit is and how fast it travels.

Most scientists finish a piece of research and then write it up. Kepler began writing his book *Astronomia nova* almost as soon as he started work, and it turned into a treatise shaped like a novel, a record of stops and starts, of good and bad ideas, during the years he toiled on the problem.[4] He starts with epicycles and thinks he has solved it. To make sure, he tests his solution using some observations he has not used before, but they don't fit. He concludes that no epicycles will fit the orbit and starts to look at the shape of the orbit itself. Helped by a lucky coincidence,[5] he guesses that the shape may be an ellipse. Finally he stumbles out into daylight clutching two simple propositions he has guessed but not proved:

I. The orbit of Mars is an ellipse with the Sun at one focus.
II. Mars travels around its orbit with a speed that varies so that a line drawn from the Sun to Mars sweeps out equal areas in equal times.

Figure 7.4 shows an ellipse and how it can be drawn with a pencil, a loop of string, and two pins. The points of the pins mark the foci of the ellipse, and one of these is where the Sun is. The drawing greatly exaggerates Mars's ellipticity; actually the orbit's length is greater than its width by less than one part in 200. For Earth, the figure is about one part in 10,000, so Mars was easier to study.

The ellipse is a simple geometrical form; its properties had been studied by the ancients, but no one before Kepler had thought to find it in astronomy. The orbit so simply described by the first law could have been replicated by a carefully chosen epicycle, but Ptolemaic and Copernican dogma required that circular motions be at constant speed, and Kepler's second principle makes that impossible; therefore, there can be no celestial orbs of any kind, real or mathematical. Kepler found the principles

[4] A modern study (Stephenson 1987, 90) argues that the story told in *Astronomia nova* is not a step-by-step account but a narrative constructed so as to show how Tycho's observations forced Kepler to abandon the Ptolemaic theory.
[5] Briefly explained in Park 1988, 151ff.

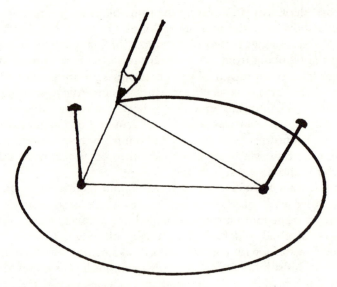

Figure 7.4   Drawing an ellipse.

that fix Mars's orbit and finished his book, but there was a dispute with Tycho's heirs over his use of Tycho's data. The emperor, who had promised to meet the costs of publication, was in the meantime having a war and couldn't find the money to pay Kepler's salary, but *Astronomia nova* finally appeared in 1609. Kepler never doubted, of course, that the two principles established for Mars govern the motions of all the other planets, but it was several years before he published an *Epitome of Copernican Astronomy,* which treated the whole planetary system.

● ● ●

Kepler's announcement made little noise. The only people who could read this dense and tortuous book, based on a hypothesis that in many parts of Europe could not safely be discussed, were also those who had a large investment in the classical theory. They were already confronted with a new and baffling question: What supports the planets and what drives them? It had been easy enough when planets were lights attached to crystalline spheres, but the comet of 1577 and a more recent one in 1604 had come and gone through a series of spheres without breaking anything. And if the Earth is a planet, maybe the other planets also are huge masses of rock, and what could make such things move in an absolutely unvarying ellipse at a varying rate, again and again, year after year? Even Kepler was puzzled by these questions. He believed in the

Aristotelian principle that, to keep an object moving, a continual force must be applied, but to drive something the size of the Earth the force must be gigantic. Nothing except the Sun is big enough to exert it, but if the Sun is at one focus of the ellipse, then it must drive the planet toward and away from itself as well as around the orbit. Chapter 34 of *Astronomia nova* proposes that the Sun drives the Earth just as, on a smaller scale, the Earth drives the Moon. The Earth rotates; the Sun must rotate too, and each must produce some kind of emanation, like the beam from the rotating lamp in a lighthouse, that pushes its satellites along. There is no longer any question of angels. "The bare and solitary power residing in the body of a planet itself is not sufficient for transporting its body from place to place, since it lacks feet, wings, and feathers by which it might press upon the aethereal air."

According to ancient optical theory, when you look at a tree a sort of image of the tree, known as a *species,* peels off the tree and comes to your eye, which somehow absorbs it and communicates the tree's appearance to your soul. Kepler's emanation is another kind of *species,* perhaps more like the mysterious force by which a magnet moves a nail. In 1600 William Gilbert, Queen Elizabeth's physician, had published a brilliant book, *De magnete,* in which he showed that the whole Earth acts as a magnet. Kepler ends chapter 34 with the words, "It is therefore plausible, since the Earth moves the Moon through its *species* and is a magnetic body, while the Sun moves the planets similarly through an emitted *species,* that the Sun is likewise a magnetic body." Two years after *Astronomia nova* appeared, sunspots were discovered telescopically, and a few days' observation showed that as Kepler had predicted, the Sun actually rotates.

So much for the driving force, but what guides a planet? Perhaps, as the ancients supposed, it has intelligence and guides itself, but if so it must have senses that tell it where it is. Kepler studies the possibilities but in chapter 57, after carefully weighing the possible roles of Nature and Mind, he decides that in a sky empty of Ptolemaic spheres an angel has no way of taking its bearings. A natural mechanism is more likely, and he proposes one involving a magnet in the Sun and another in the planet.

Kepler's contemporaries were uncomfortable with his attempts to find physical explanations for the motions in the heavens. Astronomy is mathematics, they said; physics is for philosophers. But Kepler persisted. He dreamed of a mathematical physics that would explain his philosophy and did some calculations but, impeded by Aristotle's theory of motion, he could not get anything to come out right.

Nine years after his first two laws of planetary motion, Kepler discovered one more mathematical principle that applies to the entire Solar System. I mention it here to keep it close to the other two and will illustrate it by example. For convenience, measure time in Earth years and distance in what are called astronomical units (A.U.), defined as the average distance between Earth and Sun. According to Kepler's calculations, the radius of Mars's orbit is 1.524 A.U. and the length of its year is 1.881 years. Now watch:

$$1.524^3 = 3.5396, \qquad 1.881^2 = 3.5382.$$

They are the same to about one part in 2500. Is this a coincidence? Try it on the Earth: $1^3 = 1^2$, obviously, and it works the same for all the other planets. This is Kepler's third general principle, announced in anecdotal fashion in a book, *The Harmony of the World*,[6] which extends and elaborates the geometrical model introduced in *The Secret of the Universe*. He tells us he found it on May 15, 1618, but he gives no numbers to support it. In modern terms, it says:

III. If you divide the cube of the average radius of a planet's orbit by the square of the length of its year you get a number that is the same for every planet.

Kepler's *Secret* was no bit of youthful enthusiasm. A glance at the *Harmony* book shows that its central theme is that the universe was laid out by the Great Geometer on mathematical principles. The harmonics are Pythagorean relations he has found between the greatest and smallest angular velocities of each planet as it sweeps out its elliptical orbit. To the end of his life, he believed that *The Secret of the Universe* was his greatest discovery. But these remarks have jumped ahead of the story. We now return to the year that *Astronomia nova* came out.

7.3 THE WORLD OBSERVED

Galileo Galilei (1564–1642), professor of mathematics at the University of Padua, could have helped Kepler with his theory of planetary motion, for careful experiments had convinced him that, in some cases, Aristotle's principle—that nothing moves unless something is moving it—is false. (Jean Buridan, sec. 6.5, had reached the same conclusion long before.) Galileo had rolled metal balls along a level surface and found that the smoother the surface, the further they went. Obviously, neither air nor anything else was keeping them going. The two mathematicians were not in touch. Kepler had sent Galileo *The Secret of the Universe* when it was published and was politely thanked, but the older man, busy experi-

---

[6] This is the title of the modern translation of *Harmonices mundi*. It would render both Kepler's title and the book's contents more accurately if it were *The Harmonics of the World*.

menting with pendulums and little balls to find out how the world actually is, must have been appalled. He was a closet Copernican at the time, but even when he came out it was to acknowledge the Copernican version, orbs and all. This was not conservatism. If he saw *Astronomia nova* he understood at once that the two laws it announced were grand generalizations based on a small number of calculated points. It just wasn't his kind of science.

In July 1609, Galileo heard a piece of news that changed his life: a lens grinder in the Netherlands had invented a spyglass that made distant objects appear close. Since one lens will not do that, more than one must have been involved. Galileo experimented with spectacle lenses and found that if he mounted a convex (farsighted) lens at one end of a tube and a concave (nearsighted) one a suitable distance away at the other end, he could point the tube at a distant house and, looking through the concave lens, see an enlarged though fuzzy image of the house.

At that time spectacle lenses were ground by apprentices from glass of uneven texture with clouds and spots in it. Twenty miles away, in the workshops of Venice and nowhere else, Galileo could get glass that was clear and uniform. He used it to make a series of telescopes that gave sharper and larger images, and soon he had an instrument that magnified nine times.

Galileo had never been much interested in astronomy, but one night he pointed his telescope at the Moon. He saw what looked like jagged mountains and volcanic craters and began to make drawings of what he saw. Figure 7.5 shows engravings made from two of them. Because the Moon always keeps its same face toward the Earth, it revolves only once a month and its dawn comes thirty times more slowly than ours. But watching for several hours along the shadow line, Galileo saw the process that takes a few minutes in terrestrial mountains: first the light touches only a few peaks; then it moves down the slopes and finally it spreads into the valleys below. This was bad news to people of a philosophical frame of mind for whom the Moon was a smooth and uniform ethereal sphere and its surface markings were some kind of optical illusion.

Everywhere Galileo pointed his telescope, he saw things that no one else had seen. A familiar constellation like the Pleiades, in which most people see six stars, revealed itself to have thirty-six of them, and the Milky Way turned out to be not milky at all but composed of countless little stars. In January 1610, using a new instrument with a magnifying power of about 30, Galileo discovered three tiny stars close to Jupiter. Their positions changed from night to night. Figure 7.6 shows a few lines from his notebook. He concluded that Jupiter has four moons circling around it at different rates. In March he described these discoveries in a sober little pamphlet titled *The Siderial Messenger.* "Variously moving about most

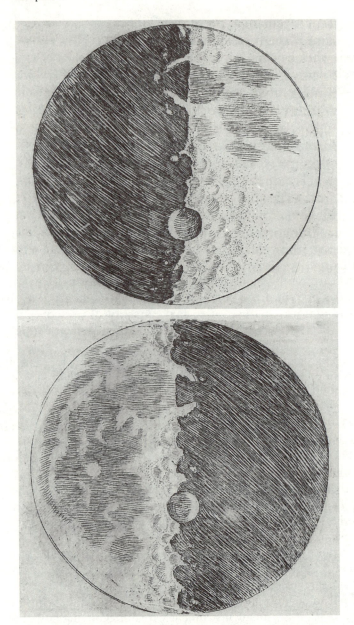

Figure 7.5   Galileo's drawings of the moon. (From *Sidereus nuncius*, ff. 30r–37v, 1610. Firenze, BNC. By concession of the Ministero per i Beni e le Attività Culturali della Repubblica Italiana. Reproduction of image prohibited.)

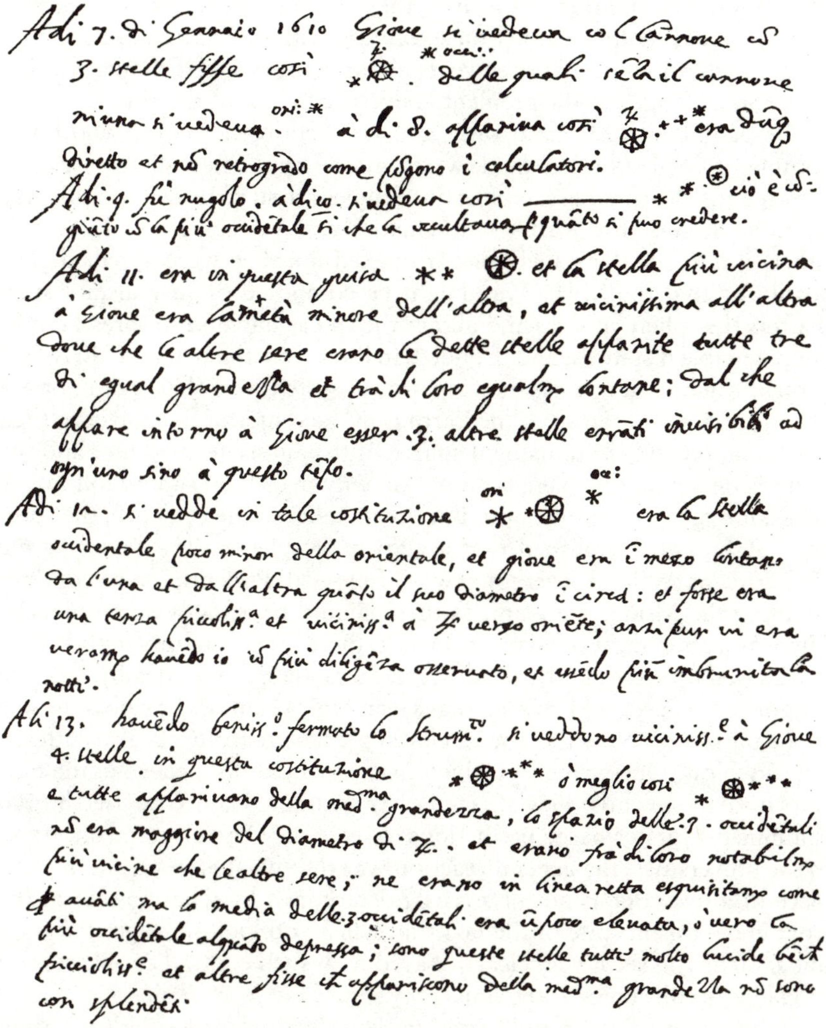

[Galileo's handwritten observational notes in Italian, reproduced in facsimile]

Figure 7.6   Galileo's notes on the appearance of three little stars next to Jupiter on successive nights. On Jan. 7, 1610 they lined up * * o * . On the eighth the alignment was o * * * . The ninth was rainy but on the tenth Galileo saw only two little stars * * o and guessed that Jupiter was masking one. On the eleventh and twelfth he saw * * o and * * o * ; on the thirteenth he realized that there were actually four little stars, not three. (From Ms. Galileiano 48, *Osservazioni siu pianeti medicei*, f. 30r. Firenze, BNC. By concession of the Ministero per i Beni e le Attività Culturali della Repubblica Italiana. Reproduction of image prohibited.)

noble Jupiter as children of his own, they complete their orbits with marvelous velocity—at the same time executing with one harmonious accord mighty revolutions every dozen years about the center of the universe; that is, the Sun." This was the moment Galileo came out of the closet.

The pamphlet sold out at once and was reprinted. People who knew nothing of Galileo's other work were amazed by the new discoveries and admired his clear and graceful Latin. Scholars didn't know what to make of it, for it certainly wasn't mathematics and yet it wasn't philosophy either, and as for the possibilities it opened up, they were appalled. The possibility that the Milky Way might be composed of very distant stars shatters the sphere of stars and allows the stellar universe to spread out to unimaginable distances, and the new satellites were the worst of all. An argument often used against the Copernican system was that it did not explain the Moon's motion. If the Earth is at the center of the universe, then the Moon revolves with its own sphere. If the Sun is at the center and the Earth is only a planet, there is no reason why the Moon should follow the Earth; it ought to orbit the Sun. But if you believed the evidence, the spectacle of Jupiter demolished that argument. Kepler, in *The Secret of the Universe,* as well as everyone who followed Ptolemy, believed that, counting the Moon, there are, as there should be, exactly seven planets. There are, after all, seven days in a week, there are seven stars in the Great Bear, seven openings in the head, seven sacraments, seven deadly sins, the list goes on. Cicero had written that seven "might almost be called the key to the universe." Now there were eleven planets. Astrologers also suffered a heavy blow. One wrote to Galileo that he would not believe in the new stars until he saw how they affected earthly affairs; Galileo advised him to examine his Jupiter-based predictions to see if they had all come out right. In the universities, the general reaction was that what issues from the eyepiece of a telescope is not necessarily determined by what goes in at the front end—this in spite of the fact that when you look through it at a familiar sight you see it very clearly. Faced with such contradictions, several of Galileo's colleagues at the university refused to look at all.

In the emperor's court in Prague, everyone waited for Kepler's reaction, and soon he produced a pamphlet of his own. Kepler, seven years younger than Galileo, had risen from nothing. On the title page of his *Conversation with Galileo's Sidereal Messenger,* he is the Imperial Mathematician and Galileo is a mathematician from Padua. In a lordly and patronizing tone he praises the discoveries and their author. Most of what he says tries to relate Galileo's work to his own, and the implication throughout is that, now that the telescope has been invented and someone has looked through it, thinkers like himself will start the serious work of saying what it all means. Galileo thanked him for his interest and their paths once more diverged.

Kepler, in his *Conversation,* gives an example of a philosophical conclusion that follows from Galileo's discovery: "Our Moon exists for us on the Earth, not for the other globes. Those four little moons exist for Jupiter, not for us. Each planet in turn, together with its occupants, is served by its own satellites. From this line of reasoning we deduce with the highest degree of probability that Jupiter is inhabited." Then the argument gets sticky: "If [other] globes are nobler, we are not the noblest of rational creatures. Then how can all things be for man's sake? How can we be the masters of God's handiwork?" He saves mankind's face by several arguments, of which the most plausible is that the Earth is in the midst of the Solar System while Jupiter is out in the suburbs. Galileo, on the other hand, believed that the whole universe was created for mankind and so he had no use for plurality. He regarded "as false and damnable the view of those who would put inhabitants on Jupiter, Venus, Saturn and the Moon, meaning by 'inhabitants' animals like ours, and men in particular." This philosophical disagreement could not easily be settled, and so we can leave it.

• • •

Galileo and Kepler went on with their lives, each suffering the torments that religion can inflict. Galileo became a partisan in the battle between the few Copernicans and the many upholders of the status quo. The story of his prosecution "on vehement suspicion of heresy" has been told often and well, and there is no space for it here. He spent the last nine years of his life under house arrest in Arcetri, outside Florence, writing and making inventions. He died, blind, in 1642.

Emperor Rudolph, having allowed most of his power to slip away while he fussed with his collections, was pushed aside in 1611 and died the following year. He had never concerned himself with other people's religion, but now Kepler was no longer safe. He moved to Linz as district mathematician; then the Thirty Years' War broke out, and in 1625 Protestants were expelled from the city. He moved his family to Ulm and finally to a place called Sagan, now Zagan in western Poland but at the time it was on the forgotten northern fringe of the empire, where he hoped to be safe for a few more years. He died at age fifty-nine in Regensburg, where he had traveled, alone on horseback through freezing weather, to try to collect the salary the emperor owed him. A few years later, armies tore up the churchyard where he was buried; what remains is the epitaph that he wrote for himself,

Once I measured Heaven, now the shadow of Earth.
The mind belonged to Heaven, the body's shadow lies here.

7.4 A World Invented

In the early 1600s Ptolemy's invisible spheres were pretty much all the astronomy there was. God had put them there, but why? Neither Scripture nor Aristotle said why. The spheres were always hard to imagine, and after they had been pierced by comets they began to seem less real. In France, members of the thinking world who looked for fresh ideas were ready for new thoughts but Kepler's books, full of mathematics and written by an obscure Protestant in a land to the east, were not what they were looking for and seem to have passed unnoticed.

René Descartes (1596–1650; fig. 7.7), celebrated as a mathematician and philosopher, was born into a landed family in Brittany. He took a law degree at Poitiers but drifted to Paris, where he studied geometry in seclusion. The Thirty Years' War was in full swing. With the idea of seeing the world, but reading and writing all the time, he joined the Catholic side and served in the rear echelons of armies in France, Holland, and Bohemia, returning to Paris in 1628. Parisian conversation would have been interesting and by this time he had much to say, but the Parliament of Paris, responding to an appeal from the Faculty of Theology, had recently decreed that under penalty of death no person should either hold or express any doctrine opposed to Aristotle. In search of freedom he moved to Holland, where he spent most of the rest of his life.

Descartes is best known for a heroic attempt to replace the whole Aristotelian system of explanation (soul + substance + qualities + urge toward completion) with a version more solidly based on experience. How to do this is set out in his *Discourse on Method*, written when he was forty-one years old. He starts with a question that takes up Aristotle's search for certainty, "What do I know that must necessarily be true?" He answers with a famous example, "I think, therefore I am."[7] This passes the test, but you can't build a whole philosophy—religion, ethics, knowledge of nature, and principles of civic life—on one proposition. Therefore, on the next page he expands his list:

> Having remarked that there is nothing at all in the statement *I think, therefore I am* which assures me of having thereby made a true assertion, excepting that I see very clearly that to think it is necessary to be, I came to the conclusion that I might assume, as a general rule, that the things which we conceive very clearly and distinctly are all true—remembering, however, that there is some difficulty in ascertaining those which we very distinctly conceive.

---

[7] The same formula, but in the context of faith, not doubt, is found in Augustine's *City of God* XI.26.

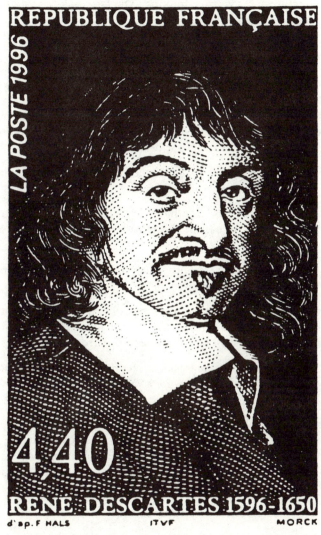

Figure 7.7   René Descartes, 1596–1650, from the portrait by Frans Hals.

Because different people very distinctly conceive different things this opens the gate to a very personal worldview, but it also encourages readers to seek truth in their own minds rather than in tradition and authority as Descartes had been taught to do. There is more to his method than this: divide a complicated problem into simpler parts and investigate them separately, always proceeding from the simple to the complex; but it is enough to introduce his way or thinking. The *Discourse on Method* is the beginning of modern philosophy.

• • •

Descartes' picture of the cosmos is surveyed in a book, *Principles of Philosophy*. It was published in Latin in 1644 and in French three years later, reworked and expanded. Because his ideas dominated European ideas of the physical world during much of the seventeenth century and even afterwards, I will select a few relevant points.

Part I 28. That we must not examine the final causes of created things, but rather their efficient causes.

This contradicts Aristotle's claim that everything in nature has a purpose. In simple terms, says Descartes, the world creates things as a machine does.

43. That we never err when we assent only to things that are clearly and distinctly perceived.

This repeats what he said in the *Discourse*. Here perception is either sensory or an act of the mind, and "since God is not a deceiver, the faculty of perceiving which he gave us cannot lead toward what is false."

Part II 11. That space does not in fact differ from material substance.
16. That it is contradictory for a vacuum, or a space in which there is absolutely nothing, to exist.

Imagine a stone; now take away from it all its accidental properties—size, shape, color, weight, temperature, etc., all those that are not absolutely essential to its being a stone. It can now neither be seen nor touched, but it still exists. What remains is the space it occupies, and therefore 11 is true and 16 follows.

21. That the world is infinitely extended.

Because one cannot imagine a wall, this follows from I.43.

22. That the matter of Heaven and Earth is the same, and that there cannot be a plurality of worlds.

Since space and matter are the same (11), there is only one kind of matter, and if that is so and there is no empty space, there would be no way of separating one world from another.

36. That God is the primary cause of motion, and that he always maintains an equal quantity of it in the universe.

A thing's motion is here defined as its size multiplied by the speed with which it moves, and the motion of each of the world's innumerable particles is in God's continual care.

> 37. The first law of nature is that each thing, as far as it is in its power, always remains in the same state; and that consequently, once it is moved, it always continues to move.

Here Descartes dismisses Aristotle's theory of motion, but since Galileo's discussion of the point had been published six years earlier, this announcement was no longer news.

Now for the cosmos. Descartes seems to have intended a Copernican universe, but a letter to a friend in Paris says he has heard that teaching it had got Galileo into trouble. He thought (even in the safety of Holland) of burning all his papers, but in the *Principles of Philosophy* he found a clever way out.

Descartes' cosmos, explained at great length in Part III, can be simply described. The region between the stars is full of matter (II.16)—"the Heavens are fluid," he says—that moves ceaselessly (II.37) like a vortex around each star, carrying with it any planets that the star may have. Now he tosses a bone to the Aristotelians: the Earth does not move with respect to its vorticial Heaven, and must therefore be said to be at rest:

Part III   28. That the Earth, properly speaking, is not moved, nor are any of the planets, although they are carried along by the Heaven.

   29. And that no motion is to be attributed to the Earth, even if we use "motion" improperly, according to the common usage; but that it would then correctly be said that the other planets are moved.

Figure 7.8 shows the Sun S and some of its neighbors, surrounded by vortices tilted at different angles. The figure does not show any planets, perhaps because to do so would have provoked the question: If S has planets, why not F, f, Y, and the rest? Nevertheless, a few years later, answering questions about his theory, Descartes wrote, "But what do we know about anything that God may have produced somewhere away from this Earth in stars, etc.? About whether he put there other creatures that look different from us and lead different lives—I could call them other men, or at least beings analogous to men?"

Thus Descartes has explained it all, insisting

> 43. That it can scarcely be possible that the causes from which all phenomena are clearly deduced are false.

"And certainly, if the principles which I use are very obvious, if I deduce nothing from them except by means of a mathematical sequence, and if

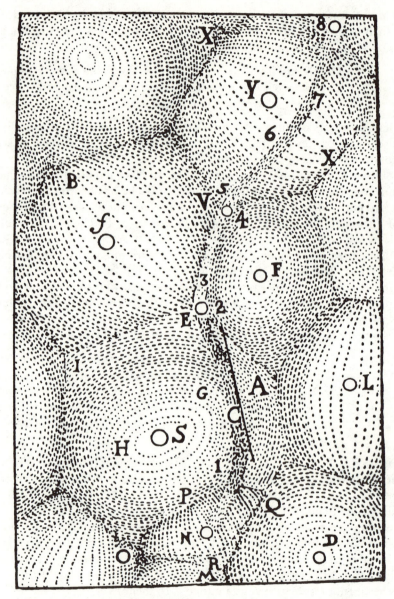

Figure 7.8   Stars surrounded by Cartesian vortices. S is the Sun. (From Descartes, *Principles of Philosophy*, plate VI.)

what I thus deduce is in exact agreement with all natural phenomena; it seems to me that it would be an injustice to God to believe that the causes of the effects which are in nature and which we have thus discovered are false." Thus God himself reinforces the principle that the things we conceive very clearly and distinctly are all true, and it is not necessary to ask, for example, whether planets so driven would actually obey Kepler's three laws of planetary motion.

All this is worked out at great length, and I would not have spent so much time on it here except that this model of the universe seized the imagination of European thinkers, and for most of a century, it was propped up and defended against the penetration of other ideas. Some of the reasons for its success are obvious. The language of both Latin and French versions is entirely free from scholastic jargon. A society that had been taught that Aristotle's is the only philosophy that explains everything was delighted to learn that there is another one, more modern, more complete, and more clearly described. French readers liked to read about French ideas. But best of all, perhaps, was the pleasure of being able to lean back in one's chair with eyes shut and imagine the whole universe, understanding not only its layout but able to throw away Ptolemy's crazy machinery and perceive without effort why the system moves as it does.

The devil was in the details. Descartes knew nothing of Kepler's ellipses, and his successors found it hard to explain them with a vortex. Nowhere in Descartes' book can one find a calculated number that can be compared with observation, but he speaks of "deductions in a mathematical sequence." Mathematics was the only part of his education that he could not doubt. Euclid starts from a few indisputable principles, for example that "between any two points one and only one straight line can be drawn." He then progresses in an orderly sequence of deductions, each one depending only on propositions already assumed or proved. Medieval theologians argued from assumption to conclusions; Descartes recommends the procedure in the *Discourse on Method* and follows it in *Principles of Philosophy*. It is structured like mathematics except that it isn't mathematics, for that draws necessary conclusions while in Descartes' writing, as in theology, few of the conclusions are indisputable if anyone chooses to dispute them.

And the star S, our Sun, where is it? If the stars, each with its vortex, are scattered through space at great distances from one another, there is nothing special about the Sun's place in the universe; it could have been any one of those stars. This is what Cusanus had written and Donne had lamented, but few in Europe had read.

• • •

In 1649 Queen Christina of Sweden invited Descartes to Stockholm for conversations on philosophy, but the cold was too much for him and after a few months he died. His remains were taken to France where they were blown here and there by the winds of religion and policy until in 1819 they came to rest, marked by a little bronze plaque, in the old Parisian church of St.-Germain-des-Prés.

Descartes' works, published in France and Holland, covered a range of subjects that includes mathematics, physiology, psychology, meteorology, optics, and law, and though not popular with the establishment, they were widely read. He challenged not Catholic doctrine itself but the Church's claim to be the only source and interpreter of truth, and by mechanizing the universe he seemed to shut the door on Providence. By 1663 his works were on the Index, and in 1671 Louis XIV signed an order that forbade teaching from any of them in Paris.

Nevertheless, Descartes' great achievement in the work I have been describing is his mechanized universe. The educated public was sick of living in a world governed, or at least tinkered with, by unseen powers. Like Epicurus's followers, they longed for assurance that each person's effort, large or small, would succeed or fail on its merits. Success is its own reward, and even from an honest failure we can learn something. This is how it would be in Descartes' model, but otherwise the model has little to do with the real world, for Descartes had overlooked the importance of exactness. In presenting his speculations in the pattern of theorem, proof, theorem, proof, . . . , he thought (or claimed) that he was building a mathematical theory, whereas he was just sketching a fantasy. Only Kepler, Galileo, and a few others insisted on a natural philosophy validated by number and measurement. When Descartes died Isaac Newton was seven years old, a schoolboy in Lincolnshire. The next few pages will show what happened when Newton built a world out of real mathematics.

It is hard for a scientist to be fair to Descartes, so I will let Voltaire say good-bye:

> He was wrong on the nature of the soul, on the proofs of the existence of God, on matter, on the laws of motion, on the nature of light, . . . but it is not too much to say that he was admirable even in his fantasies. He was wrong, but his method was logical. He slaughtered the absurd chimeras that had entertained children for two thousand years; he taught men of his time to reason, giving them the weapons with which they attacked his own work. If he didn't pay in good money, he deserves credit for denouncing what was counterfeit.

7.5 Isaac Newton

Isaac Newton (1642–1727) was born a tiny baby, probably premature, on Christmas Day in the manor house of Woolsthorpe in Lincolnshire. The family was still mourning the death of his father, also Isaac, a prosperous yeoman farmer. At some point in his early life, perhaps as early as grammar school, the boy received what he seems to have thought was a personal message. Take his name as he had been taught to write it in Latin class, Isaacus Newtonus, and write it in Roman style, Jsaacvs Nevvtonvs, then rearrange the letters into Jeova Sanctus Unus, the Holy God is One. Of course messages of this kind can often be found, and it takes a good deal more to raise a mortal above his fellows, but there are signs in his strong antitrinitarian views, in the enormous amount of work he did, and in the wide range of his interests that he felt, perhaps since early in his life, that on that Christmas day God had set him down with a special mission.

Isaac Newton Senior could not write his name, but his wife Hannah had relatives a little farther up the ladder who had studied at Cambridge. When Newton was eighteen he was sent there and enrolled at Trinity College, living the life of a poor student who supported himself by running errands for young gentlemen who passed a few carefree years at the university before they embarked on life's serious amusements. Their future was bright, for this was the year of the Restoration; the Commonwealth was gone and a Merry Monarch lolled on the throne. Cambridge in those days was not like universities we know today. Students were supposed to read Aristotle and the Church Fathers in the original tongues; mathematics and science could be studied but were not formally taught; students graduated as bachelors of arts if they stayed around long enough (most didn't) and passed some fairly simple examinations, and there were no other requirements. Newton started the assigned readings but with no reason to keep on with them he was soon following his own inclinations. He discovered Descartes' geometry book, but knowing only beginners' mathematics he bogged down and had to study Euclid for the theorems that Descartes assumed. In 1664 he was granted a scholarship, and no longer needing to work as a servant, he moved quickly through the mathematics then known. In two years he had mastered all of it and was making new discoveries. The evidence for this is in a thousand pages that he left when he died, but no one, except possibly for a mathematically inclined professor named Isaac Barrow, had known much about what was in them.

In the summer of 1665, plague struck Cambridge and the university closed for two years. Newton spent most of the time in Woolsthorpe working out his own ideas. He kept on with mathematics, putting together a version of what is now called differential calculus. Some of the ideas were already floating among European mathematicians, and the notation Newton used was not very useful, but he was now doing calculations that had been impossible for anyone a year or so earlier. Back in Cambridge, he performed experiments which convinced him that white is not a separate color but a mixture of all the colors of the rainbow. People had long known that if a prism is held in sunlight the rainbow colors come out, but they thought they had somehow been produced in the glass. Newton showed that they are in the light all along. He thought that light probably consists of particles corresponding to different colors and that a prism only sorts them out. He also thought about planetary motion and had some ideas about gravity, but whereas he communicated his work on light to the Royal Society, he kept those thoughts, whatever they were, to himself. He may have been thinking about gravity as early as the plague years, but he seems to have favored the Cartesian theory for a long time. A paper written in 1675 mentions "the vortices of the Sun and planets" and a private note dated 1681 says, "The matter of the heavens is fluid, and revolves around the center of the cosmic system in the direction of the courses of the planets," but the following year astronomers watched in astonishment as Halley's comet traveled around the Sun in the wrong direction. This was impossible in the Cartesian theory. A vortex turns one way or the other but not both, and Descartes' vortices were therefore fictional. The first mention of an attractive gravitational force in Newton's surviving papers occurs two years later.

• • •

When Newton was in his forties, his attention leapt from one subject to another. Chemistry, including the alchemical kind, fascinated him and he spent days and weeks in his laboratory; more about this in the next chapter. His studies of Scripture had reinforced his inner conviction that the Holy God is One, sole and supreme in the world, and that the doctrine of the Trinity dims his glory. He wrote a book to show that the Trinity has no scriptural basis, but he wisely kept it in a box.

In London the talk among gentlemen of the Royal Society was of planetary theory, stimulated by the ideas of Halley and of Robert Hooke, a professor of mathematics who also worked for the Society as curator of experiments. Hooke had proposed that the Sun holds each planet in its orbit with a force whose strength varies inversely as the square of the planet's distance from the Sun.[8] He could not calculate the shape of the

---

[8] This is easy to prove (Park 1988, 182) if one assumes that planets move in circles, and Newton had assumed it, but they don't.

resulting orbit because the necessary mathematics did not exist. Newton's friends feared he would be scooped if he did not get to work on the problem himself. For two years he put his other interests aside in order to clarify his thoughts, make the necessary calculations, and write up the results in a book.

In 1687, when the gentlemen opened their new copies of *Philosophiae naturalis principia mathematica,* Mathematical Principles of Natural Philosophy, they expected to be told how the machine of the universe actually works. Those who hoped for something in the style of Descartes' inventions, easy to read and imagine, found that the *Principia* is a densely mathematical work, written in Latin, that tends to avoid physical questions. It consists of three books. Book I, after some preliminary definitions, sets down three laws of motion that determine how a compact object such as a ball or a planet will move and where it will go if it is started in a given way and guided by some specified force. Even for simple examples the mathematics is not easy—how do you use geometry to describe something that is moving?—but special cases are worked out, and there is one that is important in the sequel. It turns the problem of motion upside-down: If a body moves in an ellipse around a fixed point that is one of the ellipse's foci, and if the only force that is acting is directed toward that focus, how does the strength of the force depend on the distance from the focus? Everyone knew that Kepler had shown that planets move this way around the Sun; now Newton proved that as Hooke had guessed, the force must vary inversely as the square of the distance. The problem that had stumped Hooke, given the force to find the orbit, stumped Newton also, and this was the best he could do.[9]

Book II is mostly about the motion of fluids, and at the end of it comes Newton's demolition of the Cartesian vortices. Kepler's third principle says that the square of the length of a planet's year varies as the cube of its distance to the Sun; Newton calculates that in Descartes' theory it would vary as the fourth power. Furthermore, vortices can in no way explain elliptical orbits, so that "the hypothesis of vortices can in no way be reconciled with astronomical phenomena, and serves less to clarify the celestial motions than to obscure them."

• • •

Book III, "The System of the World," is the one people were most interested in, for here he states the law of gravity, "Gravity exists in all bodies universally, and is proportional to the quantity of matter in each." That

---

[9] It was solved in 1710 by the Swiss mathematician Johann Bernoulli using mathematical methods that Newton had invented but not used in the *Principia* (see Park 1990, 74).

is, the Moon exerts a tiny force of attraction on my right hand, and my hand attracts the Moon with a force that's exactly equal. Then Newton sets out the theory of the Solar System. It explains Kepler's three principles of planetary motion and then goes farther, giving, for example, a conceptually sound though mathematically shaky calculation of the precession of the equinoxes which shows it is an effect of the Moon's gravity that exists because the Earth is not a perfect sphere. What he does not explain is *why* every two bodies in the universe attract each other with a gravitational force. Descartes had done this with a rather implausible argument, so why not Newton? The Continental philosophers were by this time heavily committed to Descartes, and they attacked. They admitted that Newton had shown himself to be a very clever mathematician, but they claimed that the famous law of gravity is nothing but marks on a piece of paper if gravity itself is not explained. Leibniz, one of the principal critics, wrote, "'Tis also a supernatural thing, that bodies should attract one another at a distance, without any intermediate means," implying that Newton has given up explaining the world in terms of matter and motion and is resorting to occult influences. A simple mechanism like Descartes' vortex was what they wanted, and when Newton left his universal gravitation unexplained, it seemed that he had given a clever but superficial description of the facts. They scoffed at the word "philosophy" in his book's title, for that was exactly what it lacked. This is why, for at least four decades after Halley's comet went the wrong way and the *Principia* proved that vortices cannot explain planetary motion, Cartesian astronomy remained, on the Continent, the astronomy of choice. It may sound as if people were being stupid, but there is a better explanation. A large part of everyone's education was religious; students were expected and required to accept the truth of what they were taught without asking for factual evidence to support it. When confronted with a scientific claim, that is exactly what one must ask for, but in 1700 few were used to asking it. (Three hundred years later, Americans have managed a partial separation of church and state, but a comparable separation of church and school has not been achieved.)

Bruised by disdain of Continental intellectuals, Newton prepared a second edition of the *Principia* that appeared in 1713. At the end of it comes his answer to Leibniz: "I have not as yet been able to deduce the reason for these properties of gravity from the phenomena, and I do not 'feign' hypotheses.[10] For whatever is not deduced from the phenomena must be called a hypothesis; and hypotheses, whether metaphysical or physical, or based on occult qualities, or mechanical, have no place in ex-

---

[10] The translators' quotes around "feign" are supposed to indicate that the word is used contemptuously.

perimental philosophy." But science is not all deduction from observations of nature, for nature shows us only what happens, and sometimes how it happens; if anyone is going to tell us why it happens that way, hypotheses are needed. Descartes' astronomy is all hypothesis: space full of matter, planetary vortices, and the rest. Newton's writings show a mind overflowing with hypotheses. He tries to distinguish them from what he considers demonstrable facts, but great questions demand great answers.

What causes gravity? The question shows up in records of Newton's conversations, and in a General Scholium a page or so before the words just quoted, the *Principia* nibbles at it. It is a passage on the nature of God: "He is omnipresent not *virtually* only [Latin *virtus* means force] but also *substantially*; for action requires substance. In him all things are contained and move, but he does not act on them nor they on him. God experiences nothing from the motion of bodies; the bodies feel no resistance from the omnipresence of God. It is agreed that the supreme God necessarily exists; and by the same necessity he is *always* and *everywhere.*" That is, God actually (substantially) exists and without actually exerting any force he is able to cause matter to move, a distant echo of Aristotle's Prime Mover. In his *Opticks*, Newton says more specifically that God can move bodies "by his Will." Does that mean that at every moment God is guiding the planets through space? It seems unlikely if in the beginning he established Newton's laws that make guidance unnecessary, but this does not mean that the whole universe is self-sufficient and can continue to run without his care. For example, if every star in the sky attracts every other, Newton asks, "What hinders the fix'd Stars from falling upon one another?" He also suspects that as time goes on, the gravitational interactions within the Solar System may slowly begin to perturb its beautifully arranged orbits "until this system wants a reformation." Leibniz, speaking for the opposition, writes, "Nay, the machine of God's making is so imperfect, according to these gentlemen [Newton and his followers], that he is obliged to clean it now and then by an extraordinary concourse, and even to mend it, as a clockmaker mends his work." Newton, who by this time had entered public life as Master of the Mint, replied through an intermediary: "The case is quite different, because he not only composes or puts things together, but is himself the author and continual preserver of their original forces or moving powers; and consequently 'tis not a diminution, but the true glory of his workmanship, that nothing is done without his continual government and inspection." Newton believed to the end of his life that all his inventions were only rediscoveries of knowledge that God had disclosed to Adam and his sons in the beginning, and to Moses and his sister Miriam, who is perhaps Mary the Prophetissa who shows up in old alchemy books. So when the mathematician David Gregory asked him what the ancients thought was the cause of gravity and he answered that he thought it was God, we can take that as his best guess. During the seventeenth century, religious people had been appalled by the new philosophy that sought to replace the old world and its spiritual forces with a vast machine. Now at the end of it, Isaac Newton, alone, declares that God acts throughout the universe to maintain it by his continual care.

Most thought otherwise. There were some, known as deists, who claimed that having made and started the world, God left it to function by itself (except possibly for an occasional miraculous intervention), with every person free to plan his life in a rational way just as Epicurus had taught. Life, they said, should be lived face-to-face with Nature, in obedience to the laws by which it governs the natural world. Then what is God's role in the universe? In Europe, especially in France, well-intentioned people looked out on the societies on which they floated and found them anchored in regulations that were cruel and unjust toward those below. Sovereigns ruled by force, supported by churches that ruled by threats, but there is another law, based in reason and quietly in effect wherever one looks, provided that one looks: the law of Nature. This was an idea that steered the intellectual currents of the eighteenth century, that the whole universe, from top to bottom, ought to be governed, as plants and pendulums and animals are, by "the laws of Nature and of Nature's God." To say that the world is a great machine containing many small ones is to say that it is just, that it favors everyone equally, and that it is up to the world's nations to copy and extend that justice in the institutions they set up. There is no grandeur in a machine? Let the machine function in a way that liberates and encourages the grandeur in every human spirit.

In Newton's hands the solar-planetary system evolved pretty much to its present form almost overnight. He had also shown for the first time how to calculate (provided the mathematics does not get too difficult) the motion of any system of moving objects. The planets, for example, move in ellipses. What ellipses? What shape? How big? How oriented in space? All that depends, Newton said, on how they were started off, and a mathematical theory doesn't talk about that. But why do the planets and their satellites move around the Sun in almost perfect circles in almost the same plane? The answer came easily: "This most elegant system of the Sun, planets, and comets could not have arisen without the design and dominion of an intelligent and powerful being." It is not for us, Newton said, to guess God's motives. God had a reason, and his reason is the system's final cause. Descartes (*Principles*, I.28) banished final causes from his mechanized universe. Newton, writing almost seventy years later, enshrines them: "A god without dominion, providence, and final causes is nothing other than fate and nature."

The distinction between laws of motion and initial conditions is of course obvious when you think about it: where a ball goes depends on the laws of gravity and motion but also on how it is thrown; yet somebody had to be the first to see the relevance of this fact in explaining the universe. The importance of Kepler's three laws is that they are true no matter what the initial conditions of a planetary orbit are. Before Newton published the *Principia* he made sure of his own laws by applying them to the motion of small things such as pendulums. The laws predict that if you make a pendulum by tying something to the end of a piece of thread and swinging it through a small amplitude, the pendulum's period will be the same whatever (within reason) you choose to tie onto it provided the length of the thread is the same. Having verified this and other predictions, Newton then applied the same principles to the Solar System, billions of times larger. His universe contained nothing larger than that, but today Newtonian principles are used to investigate the structure of galaxies and clusters of galaxies, a billion times the size of the Solar System. (Is this reasonable? Some think not.) One of the stars in one of the galaxies is our Sun, and one of its planets is the Earth. For a long time many people have assumed that there is a God who created and now rules the universe and that he has a special concern for Earth's inhabitants. As scientific discovery enlarges God's domain it becomes harder to imagine that our speck of dust has any cosmic significance, but if it has none, why the special concern? We may say with Tolstoy that the kingdom of God is within us, but this excludes him from the larger domain. Is there any other way out? In the seventeenth century the universe had not grown to anything like its present size, but already, with the Earth decentered, the tension was being felt.

• • •

Joseph Addison (1672–1719), best known as a British essayist, wrote a poem, "The Spacious Firmament on High," that affirms God's sovereignty over the world he created. At the end he refers to the planets:

> What though in solemn silence all
> Move round the dark terrestrial ball;
> What though nor real voice nor sound
> Amidst their radiant orbs be found?
> In Reason's ear they all rejoice,
> And utter forth a glorious voice;
> Forever singing as they shine,
> "The hand that made us is divine."

Addison, born 201 years after Copernicus, lived and worked in the heart of London's intellectual culture. There must have been talk of the vastness of the universe and the possibility of other inhabited worlds. He knew that the Earth is one planet among several, and that, once started, the Solar System runs like a machine. Then what is he saying? I think he has recognized that the old cosmos, centered on Earth and the descendants of Adam and Eve and governed by a supernatural power, belongs to the world of the sacred. It exists for a purpose, and humanity's place in it is defined by philosophy, that is to say by reason, and by Scripture. The new mechanical cosmos must be separated off by an epistemological firewall if the sacred is not to be overwhelmed by facts. The so-called war between science and religion is not really a war. Many scientists are sincerely religious and many people who follow a religion respect what scientists think. For them, what is known as a war is at worst a boundary dispute.

So much for the cosmos, but what about structure at the other end of the scale? After the 1670s there were excellent microscopes that disclosed a vast new world of life. That was being actively explored, but the world at the atomic level, if in fact it was there at all, was wrapped in darkness. The next chapter describes attempts to map that invisible world.

EIGHT

What Is the World Made Of?

> We are to admit no more causes of natural things than such
> as are both true and sufficient to explain their appearances.
> To this purpose the philosophers say that Nature does nothing
> in vain, and more is vain when less will serve; for Nature
> is pleased with simplicity, and affects not the pomp of
> superfluous causes.
>
> —Isaac Newton

THOUGH IT MAY SEEM like only a few minutes, it has been five centuries since we have heard anything about atoms. The ancients knew them in several varieties. Anaxagoras had "seeds," while Leucippus and Democritus had little things that are uncuttable by definition. Wandering through space in gigantic numbers, they occasionally combine to form worlds that last for a while and then disperse again. Those old atoms were of many kinds, perhaps an infinite variety, and Epicurus adopted them, adding that they are in ceaseless motion. Either they rush through space at a great speed that is the same for all, or, if they are caught up together as in a solid body, they rattle around. This is not an arbitrary assumption. Plants grow, leaves change color: something in them must be moving.

These theories, as well as the newer versions produced by William of Conches and his colleagues in Chartres, were condemned by the Church for several reasons. The imagined atoms were supposed to be eternal and uncreated, their random motions left no room for Providence, and the suggestion that there are many worlds contradicted Christian assumptions. Epicurus taught that the soul is made of atoms that disperse when the body dies, and that whether or not the gods now exist or ever existed, they are no longer in the picture. And finally, in addition to these heresies, there lingered around Epicurus the smell of a man too fond of pleasure. Nevertheless, it was that unlikely pagan who reintroduced atoms, first to France and then to the wider European community. This chapter will start with the rediscovery of Epicurus and stop at a happy moment when scientists began to think they knew what the universe is made of.

8.1 ATOMS REBORN

By the beginning of the seventeenth century, intellectuals were giving up the attempt to salvage Aristotelian metaphysics by amending it. The doctrine of substance and qualities survived among chemists because the experience they had accumulated did not suggest anything better. It also survived in provincial universities, where teachings were handed down from generation to generation of scholars like the crown jewels of a crumbling empire. Elsewhere people were trying to think in new ways. Some, like Galileo and Kepler, conducted research, as Aristotle had done in his biological studies. In France and England, among churchmen and nobility and the upper bourgeoisie, was a population of educated men and women who welcomed new ideas into their salons and discussed them eagerly. Among them were publicists like Marin Mersenne (1588–1648), a Minimite friar and minimal (according to Voltaire) mathematician, who corresponded with this new generation, circulated copies of their letters, and hosted discussion meetings in his cell. He wrote a number of books himself, including one called *Truth in the Sciences* and another consisting of *Unheard-of Questions* that encouraged readers to think for themselves. He lists thirty-seven of them, that give a good indication of what people talked about in those meetings. Two of them, together with his opinions, are:

17. Are there necessarily four elements? (Too early to say.)
18. Is there any certain truth in physics or mathematics? (Physics, no. Mathematics, only in the imagination, since there are no perfect circles, etc.)

He says he has a thousand more.

One of the people Mersenne introduced to the Parisian world was a Provençal priest named Pierre Gassendi (1592–1655) who had circulated letters disputing Aristotle. Mersenne urged him to publish his thoughts, and in 1624 appeared the first part of a long critique, but only the first part because that was the year when the Parliament of Paris issued the decree that upheld Aristotle and drove Descartes out of France.

The decree did not cause Gassendi or anyone else to change his mind, and if Aristotle was declining a successor must be found. Gassendi looked for a philosophy anchored in the authority of Greek tradition but not Aristotelian. He found it in Book 10 of Diogenes Laertius's *Lives of Eminent Philosophers,* which contains most of what is known about Epicurus, and for the rest of his life he took Epicurus as his guide. He announced that Epicurus's doctrinal errors arose from his ignorance of Scripture and could be corrected just as Aristotle's had been, and that anyhow, none of those errors affect the important things that Epicurus tells us. Of course, God created the Heaven and the Earth, but after that he let history take its own course. As to the objection that in a world whose atoms rush at random in every direction there is no room for Providence, that is not how it is at all. Gassendi rescues Providence just as Augustine had done almost 1,200 years earlier (sec. 5.2): God's foreknowledge allowed him to put every atom in motion at Creation so that afterward, at the proper times, it will be where it is needed in order to help the great plan for the salvation of the world: "It may be supposed," writes Gassendi in 1658, "that God gave each atom the ability to move, to impart motion to others, to roll about, so that they can disentangle themselves, free themselves, leap away, knock against other atoms, and likewise they can take hold of each other and join together, that he did this in such a way that each atom would serve every purpose for which he had destined it."

For Gassendi, there can be no doubt about God's plan. It controls everything we see, but since we cannot possibly perceive it at the atomic level, everything seems to happen as if atoms moved at random. And as to God's exclusion from the world's government, we know from Scripture that he can and does intervene when he pleases.

In 1645 Gassendi was named professor of mathematics at the Collège Royale and spent most of the rest of his life in Paris. There, under the protection of powerful friends, he published a life of Epicurus and also a summary of his own Christianized version of Epicurean philosophy, intended to guide readers who seek a peaceful and ethical existence. In its details, his atomism is not much different from the theory that Leucippus and Democritus had put forth and Aristotle had contemptuously dismissed. Gassendi claims that each sensory message—color, odor, sound, heat, cold and the rest—is carried by its own atoms and is received in an appropriate sense organ. The universe is finite and contains only one world; therefore everything is finite: the total number of atoms and the variety of their shapes and sizes. To explain the structure of matter he assumes that there is at least one level of size between the atomic and the visible, for the observable properties of a substance do not resemble the properties assigned to atoms. As William of Conches said, the properties of an element appear only when there is enough of it to form a little lump. To denote this level of size Gassendi needed a new word. The Latin word for a lump of something is *molēs,* and he called a little lump a *molecula.*

Gassendi only sketches the atomic idea, citing little evidence to support it, but as a visualizable model it had the same charms as a Cartesian vortex and it quickly became popular. Still, he had no clear idea of the relation between the simple properties of an atom and the properties of the substances we see around us, not to mention new ones that chemists were continually producing for medicine and industry.

Atoms were not useful in the mental pictures that guided the work of a chemist. All he had to work with were the properties of different materials and the visible changes that can take place when they are combined. Take mercury and sulfur for example, the liquid metal and the yellow crystals. Heat them and they slowly combine into a bright red substance called cinnabar. What happened to the mercury's metallic quality? Where did the sulfur go? Nobody knew, not the traditional natural philosophers, not the little band of atomists, not the artisan who used cinnabar to mix the paint called vermilion. Yet transformations of matter are everywhere, all the time. A sheep changes grass into mutton, an acorn becomes an oak, flies and worms appear in dead meat; even in the nineteenth century many people thought that mice are generated by decaying refuse. Note that each of these transmutations is an improvement. A sheep is better than grass; it is more complex and has more possibilities. An oak is better than an acorn, a mouse trumps a rotten apple. Aristotle's principle that all nature strikes after the better is certainly true, but it doesn't tell us how to predict what will happen when materials are combined or how to form new materials with properties that might make them better. As Epicureans ventilated their tentative thoughts in Father Mersenne's cell, they must often have thought what a long way they had to go, for their world of cakes and wine had nothing to do with the work of those who actually sought to study and control the transformations of matter.

8.2 Transformations

Most medieval chemists were pharmacists, distillers, dyers, potters—people whose arts involved inventing and producing special materials. A few driven by greed turned silver into what seemed to be gold; others, driven by curiosity, turned gold into silver. To see what was involved in processes like these, we have to go back to the ancient doctrine that everyone knew more or less by heart and say in plain words what had to be done.

A sample of matter, according to Aristotle, consists of substance and its qualities. If we assume there is only one basic material substance, the various forms of matter are distinguished only by the qualities added to it. Viewed in this way, transmutation of one material into another is straightforward in principle: remove undesired qualities from the first material and substitute those of the second. Gold is heavy, yellow, hard but malleable, incorruptible, and impermeable to most acids. To produce a lump of metal with some of those qualities was easy, but kings and bishops wanted all of them at once, and that turned out to be much harder.

The art of chemistry is very old. Egyptian papyri give recipes for dyes and drugs. Much later there was a great Arab chemist named Jabir ibn-Hayyan, known in Latin as Geber, who may have lived in Baghdad in the eighth century, though perhaps he did not live at all and was only the name used by a group of searchers. Whoever he was, he identified two fundamental principles, a dry one, called sulfur, and a moist one, mercury. Each of these is substance combined with certain qualities. All metals, he said, are made of them in varying proportions; gold, for example, is half and half. The obvious way to transmute metals is to change the amount of mercury and sulfur in them. But there is a catch: if actual mercury is combined with actual sulfur the result is cinnabar, not gold, and therefore Jabir's mercury and sulfur are not what most people know by those names. His followers called them philosophic mercury and sulfur, and if anyone hoped to make gold, they first had to make these two substances. They could then be combined into a single object called the Philosopher's Stone (fig. 8.1), which helps base metals ascend the ladder of being to the better state toward which they strive.

In addition to forms of matter, all space was imagined full of *quintessences,* forms of a fifth element that connects Earth, stars, and planets and participates unseen in every transformation. In this way the practice of chemistry became controlled by the clock and the calendar, and alchemists, those who reached after the great mysteries, regulated their thoughts and their habits of life so as to keep them in harmony with a vast cosmic scheme.

In the sixteenth century arose Theophrastus Bombastus von Hohenheim, who called himself Paracelsus. He lurched across Europe, arrogant, abusive, often drunk, teaching a new approach to medicine. Rejecting the doctrine of the four humors, ridiculing the traditional curative powers of cow dung and precious stones, he told his students to look carefully at the symptoms of each disease, try to see what the suffering body needs, and use wisdom and common sense to treat it accordingly. He moved among the common people, wrote in German not Latin, and introduced a new chemical principle, salt. Any substance, he claimed, could be analyzed into mercury, sulfur, and salt, and from them any substance could be made. As an example he mentioned wood. Burn it: the flames indicate sulfur; the moisture that comes off is a form of mercury, and the ashes that remain are salt. He turned the practice of medicine in a new direction, but I mention him because (according to himself and his followers) he completed chemical theory and vastly extended its range.

There is no need to describe here the procedures used to make chemical transformations, but every one of them was based on a living prototype.

Figure 8.1 Symbolic representation of the Philosopher's Stone as a combination of Sulfur and Mercury. (From Stolcius's *Viridarium chymicum,* 1608.)

The governing text was from the Bible. Jesus speaks: "Verily, verily I say unto you, except a corn of wheat fall into the ground and die, it abideth alone: but if it die it bringeth forth much fruit" (John 12:24). Jesus refers to his own approaching end, but the metaphor assumes, and medieval philosophers believed, that every appearance of a new thing is a sort of improved resurrection of something that has died. In animal reproduction, for example, conception happens only after semen has been dissolved and destroyed by a female's menstrual fluid. Thus an acid or any other solvent was known as a *menstruum,* and every transformation of a piece of matter started with a process called putrefaction, the killing of undesired qualities (often by prolonged heating) before the implantation of new ones. Thus it was that alchemical symbolism was full of images of death, copulation, and birth, and that procedures were timed so that each new substance was born when the stars were just right.

Chemists had recipes, some still used today, for preparing useful and health-giving substances, but they had no clear picture of what was happening. Descartes, Newton, and their followers had explained the world mechanically. Is there a mechanism that underlies chemical change?

8.3 A THEORY OF MATTER

In May 1669 one of Isaac Newton's Cambridge friends talked of going to Europe. Young Newton (who never in his life left England) wrote him a letter advising him how to behave himself and stay out of trouble, but he also told him to keep his eyes open. "Observe the products of nature in several places . . . and, if you meet with any transmutations out of one species into another (as out of iron into copper, out of any metal into quicksilver, out of one salt into another, . . .), those above all others will be worth your noting." Learned men thought changes like these were possible. Science was leaving Aristotle behind, but the vision of ancient wisdom remained like the glow in the evening sky. There were searchers in England and Europe who studied battered books of alchemy and were always on the lookout for one of the solitary adepts said to exist here and there, communicating with one another through secret channels. Every year there were fewer of them, but perhaps the tattered pages might still yield some hints of knowledge which, now that God no longer speaks directly to man, human wit will never find again.

In Cambridge, Newton had a laboratory where he worked in alchemy to rediscover what the ancients had known, and when he died he left a large and expensive alchemical library. Like no other student before or since, he sought for whatever truth remained in these books in order to strip it of mystical, allegorical, and pictorial trappings and write it out in clear language. Here a few words from his only purely chemical published writing; they appear in Harris's *Lexicon Technicum,* an encyclopedia published in 1710: "If Gold could once be brought to ferment and putrifie, it might be turn'd into any other Body whatsoever. And so of Tin, or any other Bodies, as common nourishment is turn'd into the bodies of Animals and Vegetables." He does not say that one could make gold. Alchemy's reputation with the reading public was already bad enough without bringing that up (see Ben Jonson's play *The Alchemist,* 1610), but it is clearly implied. And his comparison shows that at this level of inquiry he sees no fence between the organic and inorganic worlds. As he wrote in a commentary on an alchemical work called the *Emerald Tablet,* "All matter duly formed is attended with signs of life."

• • •

Others spent less time in the library. Newton's older colleague Robert Boyle (1627–1691) was a chemist whose fortune (he was a son of the Earl of Cork) allowed him to set up a laboratory in London where he spent his time experimenting and writing. He was a founder of the Royal Society, and his repetitive and disorganized works fill six heavy volumes. Inspired by the Cartesian goal of a mechanical explanation for everything, he found what he sought in the works of Gassendi and his colleagues, the Epicurean atomism which he modified as he went along.

Boyle's theory of matter assumes a hierarchy of corpuscles. First come the *minima naturalia,* the smallest natural particles. They seem to be all alike. Each of them, "though it be mentally, and by divine Omnipotence divisible, yet by reason of its smallness and solidity, nature doth scarce ever actually divide it." Next come clumps of these *minima,* which he calls primary concretions. Though they can be broken up, they are normally stable. They correspond, at least roughly, to Gassendi's *moleculae,* and if one clump sticks to another they keep their individuality and can be separated again. "Quicksilver, for instance, may be turned into a red powder, or a fusible and malleable body, or a fugitive smoke, and disguised I know not how many other ways, and yet remain true and recoverable mercury." This particular clump corresponds to what we would now call an atom of mercury.

But what, then, is the red powder that forms when mercury is heated with sulfur? Like Epicurus, Boyle allows his corpuscles to differ only in size, shape, mass, and motion. They have no color; what has happened? He invents hypotheses, but he doesn't know; nevertheless it is not hard to imagine—and Boyle and his contemporaries did imagine—that by a suitable rearrangement at the corpuscular level it might be possible to change one metal into another and change it back again.

As the Aristotelian categories sank, transmutation swam on, and for many years Newton swam with it. By the time he was fifty he had created in his Cambridge laboratory some of the chemical substances—the Net, the Oak, the Doves of Diana, the Star Regulus of Mars—which the old masters claimed led the way to transmutation. But after all that work nothing happened, and abruptly he gave it up and turned to Boyle's mechanical model of particles, clusters of particles, and clusters of clusters. In 1704 the first edition of his *Opticks* appeared. At the end of the book he prints a few hypotheses concerning matter and light that he has not been able to confirm or disprove by experiment. To avoid being accused of using hypotheses to prop up sagging theories, he phrases them as queries; in subsequent editions he adds many more, and their subjects range far beyond optics. They are his bequest to searchers who would follow him. The last of these queries, no. 31 in the third edition, is almost as long as all the others, and it contains some of the author's ideas on the structure of matter. Assuming the reader knows about atoms, he asks, "Have not the small particles of Bodies certain powers, Virtues, or Forces, by which they act at a distance . . . ?" and he continues over many pages to speculate on theses forces:

> Now the smallest Particles of Matter may cohere by the strongest Attractions, and compose bigger particles of weaker Virtue; and many of these may cohere and compose bigger Particles whose Virtue is still weaker, and so on for divers Successions, until the Progression end in the biggest Particles on which the Operations in Chymistry, and the Colours of natural Bodies depend, and which by cohering compose Bodies of a sensible Magnitude.

This hypothesis is written with Newton's usual clarity, but what is the evidence that leads to it? There really isn't any; that is why the words are part of a query. His great successes depended on measurements that he and others had made, of the swing of pendulums, the speed of sound, the motions of stars and planets. I think that even when he had given up alchemy, he imagined chemistry in terms of strange colors and smells and crystalline forms and did not think how a program of measurements might have aided his search.

8.4 ATOMS AND NUMBERS

By the mid-eighteenth century, the atomic theory, such as it was, piled hypothesis on hypothesis with no hope of verifying any of them. Skeptics abounded; among them was Pierre-Louis Moreau de Maupertuis (1698–1759), a French mathematician, president of the Berlin Academy of Sciences, and one of the founders of the science of genetics. In his *Système de la nature* (1751) he has no problem with an atomic theory as long as it is not based on the inert mechanical corpuscles of Newton and Boyle, for consider the development of an embryo: "A uniform and blind attraction, spread in all parts of matter, could in no way explain how these parts could arrange themselves to form a body with even the simplest organization. If they were all subject to the same tendency, why would some form an eye and others form an ear? Why this marvelous arrangement? And why do they not unite haphazardly?" He wants each of the smallest parts of matter "to have recourse to some principle of

intelligence, to something akin to what we call desire, aversion, memory."
As long as one thinks of biological matter in terms of little lumps, the argument is unanswerable; the theory Maupertuis developed in response to this objection will be described in section 9.3.

According to Newtonian philosophy the world consists of atoms that move, if they move, in accordance with mathematical laws. There is nothing else (except, of course, for God's occasional interventions). But is this mechanical model really adequate to explain the harmony we sense when we contemplate a flowered meadow or a kitten trying to wash itself? Is nature only a machine? The atomic model is absolute. Either it is right or it is wrong, but at best it explains a very small part of our experience. What about the rest? Maupertuis dreamed of particles with something like life and desire, and as the century went on there was a reaction against the idea of a mathematically ordered world. Opposing the Classicists, who saw the highest beauty in the measurements of a Greek temple, arose Romantics who saw it in a mountain thunderstorm. For them the universe is a living and sentient being in which every created thing, however small, contributes consciousness and activity to the whole. The unity of nature was that of a vast concourse of sentient beings and not that of a pile of sand. Romantics saw no value in a science of external qualities expressed as numbers, which, as Wordsworth wrote in *The Prelude* (1805),

> . . . enslave the mind—I mean
> Oppress it by the laws of vulgar sense
> And substitute a Universe of death,
> The falsest of all worlds, in place of that
> Which is divine and true.

The world is one, Romantics claim, and they look to imagination, not experiment, for insights that make nature clear. I use the present tense because many people still think this way.

• • •

In the seventeenth century, particles of matter had nothing to do with the four classical elements, but both were essentially philosophical notions. The four elements began to seem less elemental when Carl Scheele in Sweden and Joseph Priestly in England showed that air is a mixture of oxygen and nitrogen (the names came later), and Henry Cavendish discovered that water can be decomposed into hydrogen and oxygen. At that point, attention swung to the question, If not the traditional four, then what actually are the elements? Robert Boyle did not talk much about them because he had not definitely identified any such "primary

TABLE OF SIMPLE SUBSTANCES.

Simple substances belonging to all the kingdoms of nature, which may be considered as the chemical elements of bodies.

New Names.		Correspondent old Names.
English.	*Latin.*	
Light	- -	Light.
Caloric	Caloricum	{ Heat, Principle or element of heat, Fire, Igneous fluid, Matter of fire and of heat.
Oxygen	Oxygenum	{ Dephlogisticated air, Empyreal air, Vital air, or Base of vital air.
Azot	Azotum	{ Phlogisticated air or gas, Mephitis, or its base.
Hydrogen	Hydrogenum	{ Inflammable air or gas, or the base of inflammable air.

Oxydable and Acidifiable simple Substances not Metallic.

New Names.		Correspondent old Names.
Sulphur	Sulphurum	} The same names.
Phosphorus	Phosphorum	
Carbon	Carbonum	{ The simple element of charcoal.
Muriatic radical	Murium	} Still unknown.
Fluoric radical	Fluorum	
Boracic radical	Boracum	

Oxydable and Acidifiable simple Metallic Bodies.

New Names.			Correspondent old Names.
Antimony	Antimonium		Antimony.
Arsenic	Arsenicum		Arsenic.
Bismuth	Bismuthum		Bismuth.
Cobalt	Cobaltum		Cobalt.
Copper	Cuprum		Copper.
Gold	Aurum		Gold.
Iron	Ferrum		Iron.
Lead	Plumbum	Regulus of	Lead.
Manganese	Manganum		Manganese.
Mercury	Mercurium		Mercury.
Molybdena	Molybdenum		Molybdena.
Nickel	Nickolum		Nickel.
Platina	Platinum		Platina.
Silver	Argentum		Silver.
Tin	Stannum		Tin
Tungstein	Tungstenum		Tungstein.
Zinc	Zincum		Zinc.

Q 3   Salifiable

Figure 8.2   Lavoisier's table of the elements. (From Lavoisier 1796.)

concretions" but he was sure that they exist and give each kind of matter its special characteristics.

The first chemist to publish an empirically based list of elements was Antoine Lavoisier (1743–1794). His cautious definition, in the preface to his *Elementary Treatise on Chemistry,* is: "Those substances which we have not by any means been able to decompose are, for us, elements." Figure 8.2, from the English edition, shows his list. It begins with light, which, as Newton had persuaded almost everyone, consists of particles. They are emitted by hot objects, plus a few cold sources such as glow worms and phosphorescent seawater, and they are absorbed by almost everything except air or glass or a mirror. Are light emission and absorption chemical processes? Lavoisier seems to have thought so. The next element is heat. Galileo, Newton, and several others had suggested that heat is not a substance but the rapid vibration of the particles composing a piece of matter, but Lavoisier wanted to keep clear of the invisible world. He treated heat as a substance that can be passed from one thing to another and manifests itself in fire, though as you see at once if you hold a spoon in a candle flame, the flame is not pure heat. Lavoisier's list then continues with twenty-three substances all of which are regarded as elements today, plus three more that he was unable to decompose, containing respectively chlorine, fluorine, and boron.

For Lavoisier atoms were philosophy and not part of his science, but after his death one of the assumptions mentioned in his book led straight to an atomic theory of matter. The assumption is that when chemically active substances are mixed together and allowed to react, the one observable quality that never changes is their total weight. Implication: if for example a piece of tin is dissolved in acid, its constituents and those of the acid have changed their arrangement, but they are all still there. This notion led the chemist John Dalton (1766–1844) to pay attention to the weights of the substances that went into and came out of the reactions he studied. As one example among many, he found in 1803 that when oxygen combines with nitrogen, the ratio of their weights was sometimes 1.7 to 1 and sometimes 3.4 to 1, depending on how they were combined, but nothing in between. Suppose now that there are atoms of oxygen and nitrogen. A simple notion that explains these ratios is that sometimes one, sometimes two oxygens connect with one nitrogen: (NO and NO_2 in modern notation), and further, that an oxygen atom is 1.7 times as heavy as a nitrogen atom. (The number is inaccurate and Dalton improved it later.) The explanation of course was only a guess, but he strengthened his guess. Nitrogen combines with hydrogen to make ammonia (NH_3); oxygen combines with hydrogen to make water, which he thought was HO: are the weights of hydrogen calculated from these two reactions the same? Actually they weren't, since hydrogen is very light and hard to weigh, and it was a long time before anyone realized that water has two hydrogens (H_2O). Finally, a web of reactions was found that connected the known elements, and it was possible to say that, to reasonable accuracy, a single set of atomic weights describes them all.

Of course those are only relative weights; Dalton and his colleagues had no idea what an individual atom weighs or how big it is, but in 1816 Dr. Thomas Young (1773–1829), a London physician, produced a number. The doctor was called, a bit derisively, Phenomenon Young because as a boy he had seemed to know everything, including a dozen or so languages, most of them ancient. Now he studied the capillary attraction that draws water up a narrow tube and found that he could reason his way to the size of a water molecule: between 0.3 and 1 billionth of a centimeter. The modern value is closer to 15 billionths, but for a first try that was not too bad. The conclusion was that the clumps of matter that make the world are smaller than anyone had ever tried to imagine—far smaller than even the little animals a microscope reveals.

By about 1825 one could talk sensibly about how the world is composed of atoms without ever having seen one. A sample of one of the elements consists of atoms that are all alike. Compounds such as water consist of molecules, each made of atoms stuck together according to a definite recipe, and when a chemical reaction occurs, such as the fizz when baking powder is dropped into vinegar, it is because the atoms of the ingredients have rearranged themselves. Why do they do that? And what about Maupertuis's question: How do the particles that assemble to form a fetus know where to go?

When we last looked at the atomic theory in the eighteenth century, it was promising its believers a better life, but the properties and changes of matter were as mysterious as before. There were a few facile explanations—water boils because some of the fastest molecules break away from the surface and escape into the environment—but Lucretius would have said the same thing. And even the atomists had to admit a serious weakness in the idea: it appears that those water molecules move at random, so that no exact theory of processes like boiling can ever be made. Whatever happens does not have to happen and might have happened differently. To put it bluntly, where randomly moving atoms are involved there can be no absolutely certain physical laws, only tendencies.

During the nineteenth century, the opposition to atoms became stormy and contemptuous. There seem to have been three reasons for it. The first is a simple calculation which showed that the molecules of air move a little faster than the speed of sound, hundreds of meters per second. This seemed preposterous—what gives them such speed, what keeps them going, why don't they just slow down and stop? The second was, What has happened to the principles that God laid down in the beginning to govern the physical world? Surely they weren't just statistical tendencies. Third was the Romantic view of nature. Its followers, especially in Germany, disparaged the kind of knowledge that is teased bit by bit out of experiments; instead, they looked for broad principles, drawn by imagination from experience that nature presents to our senses. Such principles would describe nature as a whole, explaining its visible changes in terms of general statements that do not depend on atoms or any other hypothetical machinery. One example of such a principle is that it is impossible, by any means whatever, to make a perpetual-motion machine. This principle still holds, and from it flow two more general principles known as the first and second laws of thermodynamics.

Among the Romantics, the role of Sir Galahad was played by the physical chemist Wilhelm Ostwald (1853–1932). He admitted that imagining atoms and molecules make chemistry easy to think about and remember, and his first-year text, *Outlines of General Chemistry* (1890), was all written in terms of atoms. Nevertheless, his atoms are only a pedagogical device. They have never been shown necessary to explain anything, he says, and to deduce that they actually exist is to make a mistake in logic. Ostwald's reasoning was perfectly correct; he maintained it with passion tempered by courtesy, and for a while he attracted a considerable following. But those who hoped for fundamental insights based on seeing and feeling waited in vain to be told anything they didn't already know, and as the century turned so did opinion, propelled by a crucial experiment.

• • •

Albert Einstein (1879–1955) was born in Ulm, southern Germany, where his father was an electrical contractor. He was a thoughtful child, unusually slow in learning to talk. Sent to high school in Munich, he and the school disliked each other equally. He quit and went to Italy where his family had moved, and finally he was accepted at the Federal Institute of Technology in Zurich. There too he did not impress his teachers. When he graduated he was not offered any kind of academic appointment and ended up, through the kindness of a family friend, with a job in the Swiss Patent Office. In intervals of work and in the evening, he had time to think about some of the conceptual problems of physics.

There is no doubt that Einstein believed in atoms from the beginning; the point was to find evidence that would convince a doubter. In 1905 he published a paper which pointed out that what was needed was a process at an intermediate level of smallness, something visible through a microscope but directly related to actions on the molecular scale. A good example is Brownian motion, the endless random twitching of microscopic particles of pollen or dust suspended in water. One of these particles, if it is big enough to be seen with a microscope, is much too big to be moved by the impact of a single atom; it would be like throwing a grain of sand at a boat. But if the particle gets billions of such impacts from every direction each second, there are statistical fluctuations in those billions that push it one way or another and give it a perpetual jitter. Here was the missing link between the world of experience and the world of atoms, and Einstein showed that if he assumed water molecules in random motion he could calculate how far, on the average, the zigzags of one particle would transport it in a given time.

In Paris, Jean Perrin and his students sat down at their microscopes and measured the wanderings of those barely visible specks. The agreement with Einstein's theory was perfect and in 1909 Ostwald, faced with this and other developments, lowered his flag.

The year 1905 is a long time ago and much has happened since to clarify what atoms are and what they do, but this knowledge would contribute little to our story, and we can let it go.

8.5 ETHER AND THE NATURE OF LIGHT

For Aristotle the space beyond the Earth and its atmosphere could not be empty, for nothingness can have no properties at all, not even *when* or *where,* and therefore nothing can be said about empty space, not even that there is such a thing. The fifth element, Ether, replaces that nothingness. As a sort of vaporous fluid, it fills the region out to the sphere of stars, and compacted, it forms that sphere as well as the planets and the smaller spheres that carry them. Its natural motion is in circles. Aristotle consecrated it: "The superior glory of its nature is proportionate to its distance from this world of ours." Newton, seeming to connect ether with the spirit that binds atoms together and animates living bodies, saw it as an instrument of God's power. He had another purpose for it: it is by ether that "light is emitted, reflected, refracted, inflected." And what exactly is light? If we can't answer that question our theory is incomplete.

It was clear to Newton that sound is a wave of pressure in the air, and almost as clear that light consists of rapidly moving particles. I can hear you if you are around the corner of a building from me, but I can't see you. Anyone who spends time near the water has seen how waves curve around a breakwater or an anchored boat, but light travels in straight lines. In Newton's picture, different colors of light corresponded to different kinds of particles and white light was a mixture of all, but there were problems. If you use light from a small source to cast the shadow of a knife edge on a piece of paper, the shadow isn't sharp. Instead, there is a pattern of light and dark fringes. Newton thought the knife edge might be pulling the particles of light one way or another, and his successors, for want of any better ideas, followed along. But in the dawn of the nineteenth century, almost a hundred years after Newton's *Opticks,* Thomas Young gave a lecture to the Royal Society, "On the Theory of Light and Colors," which argued that waves explain more about light than particles do. He associated color with wavelength, and a few years later he was able to explain how a wave bends as it passes an obstacle such as a knife edge or the corner of a building. Roughly, he showed that the intensity of a wave such as light or sound depends on how far one is into the shadow zone, measured in terms of the wavelength of the wave. The wavelength of the sound of a human voice is five or ten feet, whereas that of light is about 1/50,000 inch, so that though sound seems to bend much more than light, when reduced to the same scale it doesn't.

Young's wave, of course, was a wave in the ether—what else was there to wave?—but as soon as anyone started to talk about ether a puzzling question arose: Why don't we feel anything as our Earth races around the Sun, and how can planets move through such a fluid without being slowed down and ultimately stopped? Careful experiments failed to detect the slightest sign of an ether wind.

There was more for ether to do than occupy space and transmit light. When a magnet approaching a nail on a table makes it jump up, how is that force transmitted through empty space? There are also electric forces that make two pieces of dry paper stick together; how does that happen and how is the force of gravity exerted? There must be something in what is apparently empty space that transmits force from one thing to another. A magnet sets up some kind of tension in the space around it. Is it a tension in the ether, or is it something entirely independent?

In the mid-nineteenth century, this tension, whatever its nature, was named a field, and many experiments involving electricity and magnetism were conducted to learn about it. In 1864 the results of these experiments were summarized in mathematical form in a very long paper by a Scottish professor, James Clerk Maxwell (1831–1879). He arrived at two main results. The first is a set of eight equations that tell how these fields are produced and how they are related to each other but not how they are related to ether. The second is a particular solution of those equations in which the two fields unite to form a composite that Maxwell called electromagnetic. It would be produced by a rapidly alternating electric current, and once produced it detaches itself from its electrical source and flies off into space in the form of a wave. Maxwell's equations tell how fast the wave travels: using the numbers from those experiments he calculated 310,740 kilometers per second. In 1849 Armand Fizeau had measured the speed of light as 314,858 kilometers per second.[1] Maxwell was not bothered by the discrepancy; he wrote, "The agreement of the results seem to shew [*sic*] that light and magnetism are affections of the same substance, and that light is an electromagnetic disturbance propagated through the field according to electromagnetic laws." The "same substance" is of course ether, but Maxwell is careful not to claim that light is a wave in it. It is a wave in the field. As to the connection between field and ether, those who fought their way through Maxwell's paper (it is the foundation of all electrotechnolgy from that day to this) were left to deduce his opinion from his silence: he didn't know; there may not even be a connection. The question remained. Scientists who had spent their careers constructing mathematical theories of ether did not want to be told they had wasted their time. For them, Maxwell's equations and the physical processes they explain dealt only with the surface of things, but during the rest of the century no amount of thinking and experimenting yielded any insight into the nature of ether.

As ether's role in the working of nature became doubtful, some of its partisans found work for it on higher planes of being. I digress for a moment to describe one way this was done.

• • •

In about 1802, lecturing at the Royal Institution on the essential properties of matter, Thomas Young strayed from his subject long enough to consider the possibility of "absolutely immaterial and spiritual" beings that transcend ordinary existence:

> We know not but that thousands of spiritual worlds may exist unseen forever by human eyes; nor have we any reason to suppose that even the presence of matter, in a given spot, necessarily excludes these existences from it. Those who maintain that nature always teems with life, wherever living beings can be placed, may therefore speculate with freedom on the possibility of independent worlds; some existing in different parts of space, others pervading each other, unseen and unknown, and others again to which space may not be a necessary mode of existence.

There wasn't much to reply except "Well, maybe," but in 1875 appeared an anonymously published book called *The Unseen Universe* that quoted Young's words. The authors were said to be Balfour Stewart and Peter Guthrie Tait, two British physicists. They replace Young's "immaterial" with "not grossly material," for they are preparing their readers for a

---

[1] It is now clocked at about 299,800 kilometers per second.

new cosmology in which ether plays a leading part. Their ether is infinite in space, eternal, and uncreated in time, and in it our material world occupies a short interval beginning with the Creation and ending with the gradual extinction of life as one by one the stars grow cold. God, they claim, did not create the atoms of the material world out of nothing, for he had ether, and to make them he had only to stir it so as to establish an infinity of little vortices which are atoms. Perhaps at the world's end he will erase them, but the universe will not be the same as if they had never existed, for it will contain the visual images of every thing and every event that has ever been: "A large portion of the energy of the universe may thus be said to be invested in such figures. . . . And we may even imagine that the same thing will hold true of those molecular motions which accompany thought." In this way the unseen universe contains now, and will always contain, a record of our lives, thoughts, and memories which, if they gather together, will form a "spiritual body," an immaterial replica of each of us. To sustain these bodies, the universe is storing up the immense quantity of energy that stars have been radiating into space since their formation—this, the authors imply, was the reason they were created in the first place—and in this sense we shall inherit eternal life. I mention the episode and its mechanical explanation of soul and immortality not because it started a movement but because it shows how important it was to accommodate science to other beliefs.

• • •

Quickly on the Continent, slowly in England and the United States, the ether blew away. What remained was the electromagnetic field; perhaps it didn't even have to be explained. When Heinrich Hertz, a professor in Karlsruhe, wrote "Maxwell's theory is Maxwell's system of equations," no dogs barked. Maxwell had introduced a new physics of fields alongside Newton's physics of matter and apparently independent of it. Are fields as real as matter? I try to avoid metaphysical questions, but step outside for a moment. Hold your hand in the sunlight and feel the warmth.

In 1905, while scientists debated the properties of ether and whether in fact it exists at all, Einstein wrote another paper that was ultimately the one that won him the Nobel Prize. Wilhelm Wien, a physicist in Berlin, had studied the changes in color of a piece of metal when it is heated. The color is a mixture of colors of the spectrum and in 1896 Wien found an approximate formula that described the mixture and how it depends on temperature. Einstein studied this formula and found that surprisingly, he could translate it into a well-known formula that describes the thermal properties of a gas of fast-moving molecules. He concluded that there must be situations in which a beam of light produces effects indistinguishable from those of a stream of gas molecules. "When a light ray is spreading from a point, the energy is not distributed continuously over ever-increasing spaces, but consists of a finite number of energy quanta[2] that are localized in points in space, move without dividing, and can be absorbed or generated only as a whole." To those who thought of light as an electromagnetic wave this was harmless babble; only a handful took it seriously, and it was several years before experiments showed a beam of light at short wavelengths (where Wien's law works best) producing the same effects as a jet of gas. Had Newton been right all along? The shorter the wavelength of a wave the higher its frequency, and according to Einstein the higher the frequency of a light wave, the larger the energy of its associated particles, now called photons. X-rays, for example, are high-frequency light, and their particle aspect is shown in the effects, harmful or therapeutic, they produce as photons plow their way through tissue.

At this point, a question occurred to everyone who thought about these developments. The experiments with which Young and his successors proved that light is a wave are still perfectly valid, so how can Einstein or anyone else maintain that something can be a wave and a particle at the same time? The question puzzled everyone then and still puzzles careful thinkers today. Theories can be wrong; the experimental situation is what counts. It seems that if you make an experiment to find out whether light is a wave, nature answers "Yes." If you make a different experiment designed to detect the impact of particles, nature again says Yes. And the world seems to be constructed so that nobody can invent an experiment that forces nature to answer both questions at once; at any rate, nobody has. Mathematics can explain what happens—that a bridge collapsed because it was only designed to support six tons and the truck weighed eight—but it doesn't say what a bridge is. That is a different kind of knowledge.

Twenty years after Einstein's proposal, a French graduate student named Louis de Broglie turned the picture upside down: Einstein had said that the electromagnetic field of light has a particle aspect. De Broglie now proposed that electrons, which everybody knew as particles of matter, have wave aspect and are describable in terms of a field. Out of this and other inventions was born quantum mechanics, the modern theory of atoms and much else; we need not follow it further but it will turn up when we discuss the origin of the universe.

Our imagined world started as a great rock slab inhabited by gods, devils, and people and surrounded by stars. Thinkers invented ether to

---

[2] In Germany the word "quantum" was commonly used in those days to denote a small portion of something.

explain the stars and fill otherwise empty space. The idea was followed up until it turned out that ether explains nothing, but it led to the idea of a field. Planetary motions are controlled not by ethereal mechanisms but by gravitational fields. What is a field? Mathematics doesn't say. People try to explain the idea in terms of something else but the road is not clear. For the moment, a field is just what it is, a convenient way to describe matter and its interactions, but I talk about it here because although there may be other physical things in the universe than matter and fields, we know nothing about them. Fields are part of every modern theory of the world. The electromagnetic field lights our days and nights, it holds solid matter together so that we don't all fall through the floor, and Scripture says that light heralded the Creation. The next chapter will tell of light, still detectable, that is nearly that old.

NINE

The Universe Measured

> Thou hast created all things in measure and number
> and weight.
>
> —King Solomon

COMPARED WITH SUN AND MOON and the other planets, is the Earth large or small? How does the Sun stack up against thousands of stars that we can see? The Solar System—is it a large or a small part of the universe? And what is the universe anyway? Is it infinite? Does it really go on, one star after another, forever, or is it somehow boxed in? Questions rise in an inquiring mind: On a large scale or scales, what's out there beyond the Earth? Similar questions arise if we wonder about how old everything is: Did it all start at once? If so, when? And if not, what happened? This chapter surveys the Earth, its environs, and its history, and at the end are some remarks on how human beings came to live on it. Then, in chapter 10, comes the whole shebang.

9.1 SURVEYORS AT WORK

Having already seen how measurements helped chemists to puzzle out what they could not see, let's now look at some measurements at the other end of the scale. In the third century BCE, Eratosthenes of Alexandria (sec I.1) measured the Earth's circumference as roughly 23,000 miles (part of the roughness is because we don't know what unit he used). A century later, Hipparchus of Nicaea calculated the distance to the Moon. Probably in the year 189 a solar eclipse occurred that was total near the Hellespont, while in Alexandria the Sun was only about four-fifths covered. From this and some clever geometry Hipparchus calculated that the distance to the Moon is about sixty-eight times the radius of the Earth. Eratosthenes' value for the radius, which I shall call r_e, was about 3,600 miles, so that the Moon's distance comes out as 245,000 miles. As time went on and techniques improved, the value of r_e rose to about 4,000 miles, which for simplicity I will use from now on.

But there is a better way to make the measurement. In figure 9.1 an observer named Alice stands on the Earth's equator and sights across the

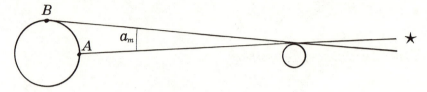

Figure 9.1   Measurement of the Moon's distance by parallax.

disk of the Moon at a distant star that appears just at its edge. At the same moment a colleague named Bob stands on the North Pole and makes the same observation. Evidently the two lines of sight are not the same, and Bob sees the star floating above the Moon's disc. The angle a_m, known as the *lunar parallax,* is about one degree, and a simple calculation shows that the distance to the Moon is about $57r_e$, or 230,000 miles. Nobody needs to travel to the North Pole and one person can make the whole measurement. Make the first observation when the Moon rises in the evening and the second just before it sets at dawn and let the Earth's rotation do your traveling. The nearer to the equator you are, the better.

The obvious next step is to try this method on the Sun, but the problems are overwhelming. The solar parallax a_s is very small and also it is impossible to see a star right next to the Sun. The project of surveying the cosmos stood still until Copernicus and Kepler worked out the *relative* dimensions of the Solar System. Now it was not necessary to look at the Sun, for if just one planetary distance was measured all could be calculated. Mars was the best candidate. It is closest to the Earth when the Earth is between it and the Sun, and in 1672 Jean Richer led a French expedition to Guyana, where (among other measurements) the astronomers tried to measure Mars's parallax. The result was nearly swamped by uncertainties of measurement, but after a decade of delay the number was announced, very tentatively, as 25 arc seconds. Since the distance from Earth to Mars at that moment was known to be 38 percent of the distance from Earth to Sun, the solar parallax was .38 × 25" or 9.5", which is the angular size of a dinner plate seen from 3.5 miles away. The Earth-to-Sun distance came out about 87 million miles. This distance, which has grown a few percent since 1672, is a convenient unit for measuring distances within the Solar System and as already mentioned, it is known as the Astronomical Unit, or A.U. Saturn was at that time the outermost known planet, and the diameter of the Solar System, which was that of Saturn's orbit, came out as 19 A.U., more than 1.5 billion miles. The logical next step in establishing the scale of the heavens was to measure the distance to another star, but as we shall see, that took a long time. In the

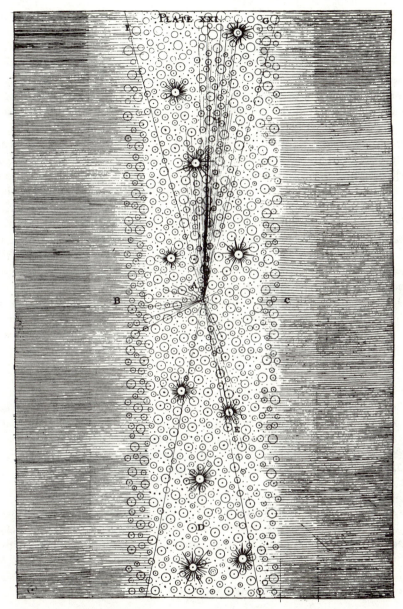

Figure 9.2   Thomas Wright's model of the Galaxy as seen from a point inside it. (From Wright 1750, 1971.)

meantime, there was another interesting question to answer: How are the stars arranged in the sky?

Galileo's telescope showed that the Galaxy, that is, the Milky Way, consists of thousands of dim stars. The news can be read two ways: either there is a band of little stars that stretches across the firmament of larger ones, or else the dim stars are ordinary stars very far away. Neither explanation seemed likely, but what else was there? It was time for someone to make a guess, and one of the first to do so was a self-taught English religious enthusiast named Thomas Wright (1711–1786). His father, a carpenter, owned a little land near Durham. His schooling ended at thirteen when he had to go to work, but he studied when he had time and made himself a local reputation as a teacher of physics and astronomy. For years he pondered how the order of the universe reflects the order of God, angels, and man, and in 1750 appeared *An Original Theory or New Hypothesis of the Universe,* which proposes a model that explains the Milky Way. Figure 9.2 shows that if many stars are clustered in a flat slab shaped like a grindstone with the Earth somewhere near the middle, they will appear from the Earth to lie in a band across the sky as the Milky Way does, and we will see many more distant stars if we look along that band than if we look across it. Wright proposes that the divine power dwells at the center and the slab turns at a speed such that centrifugal force exactly cancels gravity that would otherwise cause the stars to drift inward.

> It is here, and here only, as in the Center of his infinite Creation, where he resides in sensible Magnificence, and in the midst of these Splendours, which can Effect the Imagination of his Creatures; and though the most sacred and supreme Divinity be allowed as essentially present in all other places as well as this, . . . yet it is here only . . . where he manifests his corporeal Agency, as in the Foci of his infinite Empire over all created beings.

Newton had been careful not to say plainly that gravity is the pull of God, but Wright, whether or not the force is Newton's gravity, knows who exerts it. Each of his ideas was prophetic of future developments, except that at the center of the disk, instead of "an intelligent principle, from whence proceeds that mystic and eternal power productive of all life, light, and the infinity of things," it has turned out that there is nothing but a gigantically massive black hole.

• • •

Galileo's telescope had also revealed that "whitish clouds are seen; several patches of similar aspect shine with faint light here and there throughout the aether, and if the telescope is turned on any of these, it confronts us with a tight mass of stars." Wright and also Immanuel Kant, who in his youth wrote about science, guessed that they are distant galaxies like our own (Kant called them island universes), but in general the objects attracted little interest.

As to our own Galaxy, Wright's rotating grindstone was nothing but the fantasy of an obscure person until William Herschel (1738–1822), director of George II's Royal Observatory, built himself telescopes and studied the distribution of the stars. A native of Hannover, Germany, Herschel landed his excellent job because he played the oboe. Let me explain. As a boy, during the Seven Years War, he joined the band of a German regiment that was later brought to England by its Hannoverian king to strengthen the British army against a possible French invasion. He learned English and started reading, not novels but authors such as Locke and Newton. The regiment returned to Germany where it was cut to pieces by the French. When the band disbanded, Herschel went back to England and set himself up as a violinist, conductor, and composer. (Several of his symphonies have survived.) While doing all this he got interested in astronomy, and because he could not afford a telescope he made several for himself with the help of his sister Caroline, who later became a well-known astronomer. In 1781 he discovered a planet beyond the orbit of Saturn, and with a bow to the king he named it the Georgian Star. This got him his job, which he kept even though the planet ended up as Uranus.

During the 1780s Herschel began a long study of the arrangement of stars in the Galaxy.[1] He started where Wright and Kant had started, assuming that it is large and pretty flat, and that at least roughly, its stars are evenly distributed throughout. This being so, if you see more stars in one direction, it is because a line drawn in that direction from your telescope to the edge of the Galaxy is longer. In this way, taking the Earth as somewhere near the center and counting stars night after night, Herschel arrived at the picture shown in figure 9.3. The double tail corresponds to the two branches of the Galaxy that are seen on a clear night, and the dot in the middle is the Sun. Herschel even provides a scale for his picture: assuming that Sirius looks bright only because it is very close, he estimates that in this picture its distance from the Sun would be $1/150$ inch. In the same paper, he gives another estimate that is perhaps more impressive than a big number: Imagine, on the same diagram, a sphere with a radius of a twentieth of an inch centered on the Sun.[2] "Probably all the stars, which in the finest night we are able to distinguish with the naked eye, may be comprehended within [this] sphere." For the rest you need a

---

[1] Astronomers use the capital letter to distinguish our own galaxy from others.
[2] The number is corrected for the smaller size of this reproduction of Herschel's illustration.

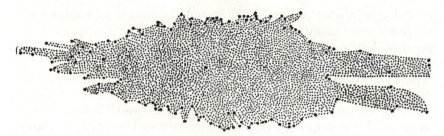

Figure 9.3   William Herschel's map of the Galaxy. The dot near the middle represents the Earth. (From Herschel 1785.)

telescope. Since no one had any idea how far it is from one star to another, these measurements were all relative. The next step in the survey was to measure the distance to some other star, and that step was taken fifty-five years later.

The first stellar distance was measured in 1840 by Friedrich Wilhelm Bessel (1784–1846). He came from a poor Westphalian family. At fifteen, working as a merchant's clerk in Bremen, he started teaching himself languages, astronomy, and mathematics. Five years later he attracted attention by calculating the orbit of Halley's comet from observations made a century earlier, and a wise director invited him to join the staff of a provincial observatory to learn the arts of astronomy. His work was so skillful that when the Prussian city of Königsberg decided to build a major observatory, this twenty-five-year-old with little formal education was selected to design the instrumentation. Given the chance to devise new instruments and procedures and to hold technicians to unheard-of standards of precision, he was able to measure the distance to a barely visible but fast-moving little star called 61 Cygni (the sixty-first-brightest star in the constellation Cygnus). He used the parallax method, but with an important change: while the baseline of the earlier measurements was the Earth's radius, for this one a much longer baseline was available. If one looks at a nearby star in April and then again six months later (when the temperature and other atmospheric conditions would be similar), the telescope will have moved to the other side of the Earth's orbit, two A.U. away. The parallax was still very small, but Bessel measured it.

The parallax of 61 Cygni came out to be 0.3483 arc seconds (though it is hard to take the last couple of figures very seriously). It follows that the star's distance is 592,000 A.U. Since one A.U. is 93 million miles, those who enjoy writing zeros may figure out the distance in miles, but there is a more convenient way to express it. It is about 10.4 light-years, meaning that a flash of light, which travels from Sun to Earth in about 8 minutes, would reach Earth from 61 Cygni after 10 years and 5 months.

Now we can finish Herschel's measurement of the Galaxy. Sirius is actually twenty-three times as bright as the Sun, and its distance is 8.6 light-years; with these numbers, the galactic diameter comes out as 36,000 light-years, a little less than half the modern figure. The number is interesting, but the method depends on assuming that the density of stars is constant throughout the Galaxy and that Herschel's telescope allowed him to count stars all the way to the edge; later he came to doubt his own result.

Even without the numbers, for the first time it was possible for curious people to make a mental image of the universe with a sense of where we are in it: here, at this spot in this galaxy, which is isolated in space but surrounded by others like lily pads in a pond. So many stars in our own Galaxy, thousands of galaxies, our Earth reduced to a dot. It was becoming hard to read the beginning of Genesis as anything more than a fable. If Genesis were true, and if the universe was created as a setting for the salvation of mankind, why was it made so large? And in those years another question concerning the Creation was beginning to disturb ordinary people: When did it happen? Experts had been debating the question for some time, but now, with our position in space so much changed, it was natural to ask about our position in time.

9.2 The Age of the Earth

As to the age of the world, the Bible has a ready answer. Adding its numbers for the generations of mankind gives about four millennia from the Creation to the year 1. Jews count 3761 years and other reckonings roughly agree. In 1650 James Ussher, primate of all Ireland and archbishop of Armagh, made a careful calculation and wrote in his *Annals of the Old Testament*: "In the beginning God created Heaven and Earth which happened at the beginning of time (according to our chronology) in the first part of the night which preceded the 23rd of October in the year of the Julian period 710." For reasons too complicated to explain here, the year zero of the Julian calendar is 4713 BCE. Subtracting 710 gives 4003 BCE, but because the Christian calendar has no year zero this becomes 4004. The day was supposed to be the Sunday before the autumnal equinox. The importance of this chronology is that someone in charge of an edition of the King James Bible used it to date every event, so that for the next two centuries these dates were regarded in England and North America as the result of some sacred calculation that must not be challenged.

Verifying or refuting the biblical reckonings turned out to be difficult, but the Earth tells its own story, and as people began to travel and observe landscapes it was natural to look for signs of its history, especially the Flood. There are, for example, what appear to be fish bones and seashells in many inland regions, some lying on mountain tops ready to be picked up and others buried within layers of solid rock and brought to light by miners. If they were the remains of living creatures they seemed to record some gigantic catastrophe, perhaps a flood combined with volcanic eruptions, though what had actually happened was not clear. Some thinkers proposed that these were never marine creatures at all. Looking at the remains they found not only shells and bones but crosses and miniature papal tiaras, all made of stone, and concluded that they were produced by planetary influences at the beginning of history. Others tried to reason from the evidence. The notebooks of Leonardo da Vinci (1452–1519) show that he thought carefully about this. He worries about seashells found on mountains far from the sea. Rainwater flows down a mountain slope much more strongly than surf washes up on it. "I do not see therefore in what way the said shells could have come to be so far inland unless they had been born there. . . . If the Deluge had carried the shells for distances of three and four hundred miles it would have brought them with the various different species all mixed up, all heaped up together; but even at such distances from the sea we see the oysters all together and also the shell-fish and the cuttle-fish." He recognizes that the mud around the shells petrified at the same time as the shells and that the layering of these deposits shows that they were formed not in the few weeks of the Flood but over many years. Even in the privacy of his notebooks Leonardo does not question the sacred chronology, but he shows that the more closely you examine the facts the harder it is to make them fit.

• • •

More than a century later, Nicholaus Steno (1638–1686), a Danish prelate, published an introduction to a projected book on fossils. Its title is *The Prodromus of Nicolaus Steno's Dissertation Concerning a Solid Body Enclosed by Process of Nature in a Solid,* and in it he sets out two principles that have guided geological research ever since. The first is that fossils are the remains of formerly living creatures that were covered by some fluid that solidified around them, so that older strata lie below newer ones. The second is that since fluid surfaces are necessarily level, the tilted strata that are often seen must be the result of geological upheavals. There is more in the *Prodromus,* much of it relating to the forms and origins of crystals, but Steno never had time to write the promised *Dissertation.*

These two men—and there are several others—wondered about the birth of geological formations, but no one questioned the Bible. Still, it must have seemed strange that, given the Earth's relatively uneventful history during the last two millennia, such immense changes could have occurred during the previous four. Was it really only four?

• • •

Georges-Louis Leclerc, later Comte de Buffon (1707–1788), was an ambitious young man from the French provinces who, by energy and good luck, got himself appointed director of the Royal Garden of Medicinal Herbs, and in this economically and politically secure position he lived a long life of hard work. He is best known for his studies of natural history, but he interests us here for experiments that led to a theory of the Earth's thermal, and hence biological, history.

Buffon's *Natural History* was published in forty-four volumes from 1749 to 1767. The first three appeared in 1749 and contain the earliest version of his *Theory of the Earth*, which he claims is based on the actual behavior of actual substances. There is no place in it for the Flood— where would all that water have come from, where would it go? "Above all," he writes, "bad physics [must] not be mixed with the purity of the Holy Book." How, then, does good physics explain the Creation? He starts with the Sun as a white-hot liquid mass of iron and various other substances that compose the planets. It is struck by a wandering planet that knocks off a spray of liquid matter that gravity slowly gathers into orbiting planetary spheres. The spheres shrink slightly as they cool. The outer layers solidify first; then, as the inside continues to cool, they wrinkle and mountains are formed. The Earth's later history is determined by its continued cooling,[3] and after several years' experiments on the cooling of small spheres of rock and metal and wild extrapolation of the results, he calculates from the Earth's present temperature that its age is 74,832 years. By the same reasoning he concludes that by the year 168,123 it will be too cold to support life. Life began when Earth was about 36,000 years old, but how, in the 40,000 years since then, could huge masses of calcareous rocks have been formed by accumulating seashells? His notes show him beginning to realize that his estimates are impossibly short, that his thousands should perhaps be millions. Thousands or millions, whatever they were, contradicted Moses, and the doctors of the Sorbonne, by now used to the voices of Enlightenment, objected pro forma. Buffon issued a pro forma retraction of everything he had written that conflicted with Holy Writ, and kept on working.

---

[3] The idea that the planets began as balls of molten matter was already in circulation; see, for example, a passage from Leibniz's *Protogaea*, c. 1680, translated in Mather and Mason 1939, 45.

Though *On the Epochs of Nature* was clearly written and widely read, his explanations showed how shaky were the foundations on which it rested, and few students of geology took them seriously. Buffon is not even mentioned by James Hutton (1726–1797), a Scot who spent many years mapping European geological formations and speculating about their age and history. In 1785 Hutton presented a paper, "Theory of the Earth" to the Royal Society of Edinburgh, which tells how he tried to date the Earth under the assumption that the rocks of its crust are accumulations of materials washed into the sea by erosion of the land. At the bottom of the sea, he imagined, they are solidified in strata by the Earth's heat. From time to time, the bottom rises catastrophically and the layers are broken and tilted, some of them raised high above sea level. Once raised, weather begins again to wear them down, and the cycle continues. Hutton's general principle was not to assume in the past any geological process that we do not see in action today. (In a sense this was Buffon's principle also, but Hutton made better use of it.) To estimate the rate of erosion, he tried to find ancient records that could be compared with modern measurements, but none were precise enough. The process he describes must be very slow, but he can draw no other conclusion. "If the succession of worlds is established in the system of nature, it is in vain to look for anything higher [i.e., older] in the origin of the Earth. The result, therefore, of our present enquiry is, that we find no vestige of a beginning,—no prospect of an end." Hutton was perhaps the first person who pushed Creation into an unfathomable past. Some of his critics interpreted these words as claiming that the Earth is eternal, but it is not what they say, and "the succession of worlds" implies the opposite.

Like Buffon, Hutton was able to publish his conclusions without serious opposition, but they raised deep feelings all the same. In 1809 Jean André Deluc, in his *Elementary Treatise of Geology*, wrote: "Certainly no conclusion from the natural sciences can be more important to men than that which concerns Genesis: for to place this book in the class of fables would be to throw into deepest ignorance that which is most important for them to know: their origin, their duty, and their destination." For many people it was impossible to surrender the biblical chronology. Truth, for them, was One.

Hutton's conclusion could not possibly be the end of the story, for the fact remained that as Buffon had argued, the Earth, immersed in the cold of space, must lose heat. Earlier writers speculated about fires in its interior that keep it warm, but fire requires air; where would that come from, and what was there to burn? Buffon's numbers could be questioned, but what exactly is the matter with his argument? In 1864, armed with a rigorous mathematical theory of cooling that replaced Buffon's extrapolations, the Scottish physicist William Thomson, afterwards Lord Kelvin

(1824–1907), returned to the question. Assuming, like Buffon, an original molten state and allowing for large uncertainties in the Earth's composition and structure, he calculated that the Earth solidified between 20 and 400 million years ago, so that some number in that wide range represents its present age.

As to the future, he concluded that unless the laws of nature change, "Within a finite period of time past, the Earth must have been, and within a finite period of time to come the Earth must again be, unfit for the habitation of man as at present constituted." The Sun also cools as it expends its energy, and after another analysis he writes, "It would, I think, be exceedingly rash to assume as probable anything more than twenty million years of the Sun's light in the past history of the Earth, or to reckon on more than five or six million years of sunlight for time to come." The finality of these words of a leading physicist was shocking to a public which for most of a century had regarded science as a force that would cure human societies of their irrational notions and guide them into a safe and happy future. Now the same science told them that the world was scheduled to end in twilight, and that there would be a time when, in the frozen silence of what were once cheerful homes and famous libraries and galleries, there would be no more human presence, ever.

But the devil, as always, was in the assumptions. Standing on the Earth's surface, what do we really know about what is going on inside it? And is the Sun's heat only what is left of the blazing heat with which it started, or is something keeping it hot? If we know nothing about this, how can anyone estimate how much time we have? Some missing information on the Earth shows up in the next section, and one source of the Sun's heat will be mentioned in section 11.2, but there is much more to say about the nineteenth century. Next come a few remarks concerning the development of life on this ball of rock.

9.3 THE LONG DESCENT OF MAN

As amateur geologists looked more and more closely at the countryside, following strata they could identify by the kinds of fossils they contained, everything they saw pointed to a timescale much longer than the biblical. An explanation was called for, and in 1819 William Buckland, a clergyman and reader in geology at Oxford, offered a suggestion. The Bible, he wrote, is a work of moral teaching, and its story really begins on the sixth day with the creation of Man. Genesis condenses everything before that into five days and describes it in words that Moses' simple audience would understand. Buckland then shows by careful study of bones and rocks that the time before the sixth day must have been very long, perhaps millions of years, and that during it the Earth suffered great changes. Thinking in terms of these immense spans of time, he is not surprised that except for effects of the Flood, the six thousand years since the time of Adam and Eve have changed the Earth very little.

But this reasoning raised many questions. Caves and rock strata contain fossils of creatures that no longer exist, but there are no remains of the animals the Bible mentions as passengers in the Ark. If Noah's animals were not created in the beginning, where did they come from? Did God go on and on creating new species, and is it really certain that the creation of humankind was an entirely separate act? The story of Creation in the first two chapters of Genesis required readers to gnaw the hard biscuit of belief, and an increasing number of educated people wanted a softer diet. The man who opened the possibility of new species to a wide world of discussion and debate was of course Charles Darwin, but he was not the first to propose a theory of evolution. To show the nature of his achievement, I will select three of his many predecessors. Let's start in the eighteenth century.

There were, then and for the next two hundred years, many sketchy theories of animal reproduction that agreed on only one point: that human souls are immortal. Reincarnation is not mentioned in Scripture and is hard to square with Judeo-Christian belief, but few people believed that a new soul was created for each child. More likely, God in the beginning had created the "germ" of the soul and body of everyone who would ever be born. These germs resided somewhere in male bodies to be dispensed as needed. Since both parents seem to contribute to a child's visible qualities, a theory arose that in sexual union female as well as male releases a semen; they mix in an undifferentiated gelatinous mass together with the germ that organizes it and gives it a soul. In this mass the separate structures of the fetus gradually form, but how they form was not clear, and even Anton Leeuwenhoek's historic discovery of microscopic "little animals" swimming in the male semen of many species did not really explain anything.

The first and most talented of the three forerunners of Darwin I will mention is Pierre-Louis Moreau de Maupertuis (1698–1759), already cited in section 8.1. He was born in Normandy, son of a merchant and shipowner whom Louis XIV had ennobled for his piratical disruption of English commerce. He gravitated to Paris where he studied mathematics and, buoyed by the family fortune, joined the city's lively and intellectual café society. He was the first French mathematician (forty-five years after the *Principia*) to take seriously Newton's theory of a Solar System governed by gravity, and he slowly persuaded its critics to take a second look. Later, after Frederick the Great had invited him to join the Academy of Sciences in Berlin, he started to think about the mechanisms of reproduction and, particularly, heredity. In the spirit of his age, he set out to explain it by a mechanical model.

Maupertuis assumed that male and female contribute equally and in the same way to their offspring's physical qualities. Looking for an unmistakable example that tested this claim, he studied the record of four generations of a Berlin family many of whose members had six fingers on each hand. It showed that as he expected, fathers and mothers who carried this trait were equally likely to hand it on. He reasoned also that if obvious features like the number of fingers, not to mention racial traits, are inherited, it is probable that none of the hundreds or thousands of distinguishable features of every human body is laid out in advance on a divine blueprint but that every one is supplied by one parent or the other, in a sort of competition. And since all this information is carried in such a small amount of fluid, it must be expressed in the form of matter on the smallest level of size. He imagined both semens as containing a huge variety of submicroscopic corpuscles in many versions inherited from parents and grandparents and even farther back. Carrying different plans of the same organ, they combine somewhere in the female's genital tract; the problem was to explain how they line up so that each of the offspring's parts is properly located in the microscopic fetus that is being formed. In the 1740s, when Maupertuis had these thoughts, chemists were discovering a phenomenon they called *affinity:* some chemical compounds are bound more tightly than others so that, as the chemists imagined, a corpuscle confronted with two others will bond with the one with which it has the greater affinity. Maupertuis, in his *Vénus physique* (1745), proposes that affinity between generating particles is so delicately adjusted that each one fits into the developing fetus exactly where it belongs; an occasional misalignment might produce either a monstrosity or a new species. But there could be gradual changes too. A hot climate, for example, might encourage the selection of corpuscles that produce dark skin, so that in time a whole population would be dark.

Maupertuis loved animals of many kinds and his household was full of them; visitors reported embarrassing experiences. He bred members of the same and different species and studied the heredity of varietal differences. He applied mathematics to genetics a century before Mendel and explained how a species might gradually improve by adapting better to its environment, but he didn't mention natural selection. With many corpuscles competing for the same place in the fetal structure, he assumed that chance, weighted by environmental factors, picks the winner. Maupertuis could reasonably be called the greatest biologist of his time, and his model, which encodes genetic information in particles of molecular size, is a startling anticipation of DNA.[4] But he had few followers and there is no sign that Charles Darwin, a century later, ever heard of him.

• • •

Erasmus Darwin (1731–1802), well known in England as poet, medical man, and bon vivant, was an English gentlemen who pushed the outer envelope of Christian piety. Like Maupertuis, he proclaimed no doctrine but only proposed ideas that he thought were worth studying. In 1794 he published a medical treatise called *Zoonomia; or, The Laws of Organic Life* which contained some speculations about how life arose on Earth. It started "millions of ages ago," with the formation of perhaps "one living filament," a single cell that was able to reproduce itself, changing a little in each generation. First plants developed, then animals and all the other forms of life. Concerning animals, he writes that their development was guided by "lust, hunger, and security." Competition for females determined that "the strongest and most active animal should propagate the species, which should thence become improved," and the search for food and safety would have the same effect. None of this was supported by any evidence that such changes had actually occurred. Erasmus Darwin had studied fossils, but if he drew any conclusion from them he does not mention it. His speculations occupy seven pages of a two-volume work. Churchmen objected, but a little further on they found the author humbling himself before God, and it would have been hard to speak of blasphemy. For a few years *Zoonomia* was considered a leading medical text and went through several editions, but there is no sign that those few pages changed anyone's opinions about the origin of life.

In 1844 appeared a book called *Vestiges of the Natural History of Creation*. It was small, cheaply printed, and sold for 7s 6d, a price that aimed at an audience in the middle and even the artisan class. By the end of ten years, 20,000 copies had been sold, a very good number in those days. It was anonymous and one of the parlor games of the time was guessing the author. A few years later he revealed himself as Robert Chambers, a partner with his brother in publishing books and newspapers who was also a student of the sciences. I won't write much about his book because it isn't very good and hasn't lasted, but it set the stage for Erasmus Darwin's grandson.

---

[4] In Berlin, Maupertuis turned from biology to physics and gave a new form to Newton's laws known as the principle of least action. Generalized by Joseph Louis Lagrange in his *Mécanique analitique* [sic] (1788), it is the foundation of almost all the later development of Newtonian dynamics and, in the hands of Paul Dirac in the 1930s and Richard Feynman in the 1950s, it opened new paths in quantum mechanics.

Chambers begins with the origin of the Solar System and suggests ways it could have begun naturally. Once Earth was formed, living creatures appeared. Were they divinely created? Perhaps, in a sense, but when God says, "Let the Earth bring forth living creatures," it does not sound as if he intended to create them himself, one by one.

> We have seen powerful evidence, that the construction of this globe and of its associates, and inferentially that of all the other globes of space, was the result, not of any immediate or personal exertion on the part of the Deity, but of natural laws which are expressions of his will. What is to hinder our supposing that organic creation is also a result of natural laws, which are in like manner an expression of his will? . . . How can we suppose that the august Being who brought all those countless worlds into form by the simple establishment of a natural principle flowing from his mind, was to interfere personally and specially on every occasion when a new shell-fish or reptile was to be ushered into existence on *one* of those worlds? Surely this idea is too ridiculous to be for a moment entertained.

Among the new species that organic law produced is, of course, Man. Embryologists had discovered that every fetus goes through a series of changes that seem to recapitulate its evolution from more primitive forms, and that this is equally true of Man. "His organization gradually passes through conditions generally resembling a fish, a reptile, a bird, and the lower mammals," all of these "before we see the adult Caucasian, the highest point yet attained in the animal scale." ("Caucasian" was an anthropological term referring to that branch of the human race to which, by the greatest possible luck, Robert Chambers himself belonged.)

All this was directed toward, and consumed by, readers of a social class very different from those who read *Zoonomia*. The upper ranks, still shuddering at the memory of the French Revolution, didn't think that radical ideas should be peddled at seven-and-sixpence, and Chambers's propositions were opposed by churchmen as well as scientists of every kind, who considered that he was long on invention and short on evidence. In everyone's mind was the question, How could gradual evolution governed by the operation of natural law produce immortal souls at some particular point in the line of human descent when there were none before it? Humanity's identity and its place in the universe, as well as the truth of religion, were at stake. Book after book ruminated these questions; friendships were made and broken. It is not surprising that when Charles Darwin's much better book, known as *The Origin of Species*, appeared in 1859, the printing sold out in a day. That story, though, should be told more slowly.

• • •

If you go to a mountainous region where geological strata can be seen, they are tilted and crumpled and bent in ways that suggest that a huge catastrophe obliterated an older landscape, and this seems to have happened several times. In the geologic strata laid down after such a catastrophe, the fossils are usually quite different from those before, suggesting a wholesale kill-off of living species and a fresh creation of new ones. As the geologist Hugh Miller wrote in 1841, "Whole races became extinct, through what process of destruction who can tell? Other races sprang into existence through that adorable power which One only can conceive, and One only can exert." People who interpreted the facts this way were known as catastrophists, and they had an argument in their favor: it was possible to imagine, and some did, that the few thousand years of sacred history refer only to the time since just before the latest disaster, the Flood. Those who took the contrary view were called uniformitarians. Following Buckland, they claimed that great changes are always in progress, even now, but that except for an occasional earthquake or volcanic eruption and, of course, the Flood, we don't notice them because they are so slow.

In 1830 appeared the first volume of a book, *The Principles of Geology,* by Charles Lyell (1797–1875). It was so complete, so consistent and well argued that after a few years, except in Gothic chambers, it had settled the debate. It describes catastrophism in scathing terms:

> Never was there a dogma more calculated to foster indolence, and to blunt the keen edge of curiosity, than this assumption of the discordance between the ancient and existing causes of change. It produced a state of mind unfavorable in the highest degree to the candid reception of the evidence of those minute but incessant alterations which every part of the Earth's surface is undergoing, and by which the condition of its living inhabitants is continually made to vary.

"Evidence" is the key. Put away argumentative texts and go out into the field. Dig and pry and get your clothes dirty; then you will see from a hundred little signs that the Earth has changed slowly and is changing as you dig.

In 1831 the Royal Navy decided to send out a scientific expedition on the two-masted ship *Beagle* to survey the southern coasts of South America, then to go into the South Seas and on around the world. The expedition needed a naturalist, and the unlikely choice was a twenty-three-year-old named Charles Darwin (1809–1882), a recent Cambridge graduate who had prepared for the ministry in a desultory way but had also attracted the attention of several scientific professors. With their recommendations plus strong family connections Darwin got the job, and in December he set out on what turned out to be a five-year voyage of discovery. He had studied catastrophic geology, but during the voyage

Lyell's first volume was sent to him and it made him begin to think historically. Everywhere the ship anchored, he looked for signs of slow process even where the panorama seemed to show violent upheavals. He found the signs and became convinced that Earth's history is very long. He collected specimens of birds and animals, compared fossil bones with those of creatures still alive, and in the long days at sea, when not helplessly seasick, he wrote to his friends in England. These letters were read at scientific meetings and made a name for him even before he reached home. They show him mostly concerned with geology, but during a month spent in the Galápagos Islands, about 600 miles west of Ecuador, he made some observations which, as he absorbed what they meant, contributed to the theory that later made him famous.

He noticed that birds, notably finches, looked different on different islands of the group, and that the differences seemed larger when the islands were farther apart. He noticed particularly that the birds on different islands had beaks of different shapes and sizes (fig. 9.4). Between the largest and the smallest beaks, "there are no less than six species with insensibly graduated beaks. . . . One might really fancy that from an original paucity of birds in this archipelago, one species had been taken and modified for different ends." These words, from his account *The Voyage of the Beagle*, were written after experts at home had looked at his specimens and assured him that these really are different species; at first he had thought they might be startlingly different varieties of the same bird. The birds' characteristics had drifted in the course of time, but when he writes "has been taken and modified" the reader wonders, by whom, or by what?

Back in England, Darwin read Thomas Malthus's *Essay on Population*, a book that warned that if humans continue to multiply so that each pair produces more than two children who survive and reproduce, the Earth must inevitably contain more people than it can feed. Darwin, thinking about birds, did his own calculation:

> Suppose that in a certain spot there are eight pairs of birds, and that *only* four pairs of them annually (including double hatches) rear only four young, and that these go on rearing their young at the same rate, then at the end of seven years (a short life, excluding violent death, for any bird), there will be 2048 birds, instead of the original sixteen. As this increase is quite impossible, we must conclude either that birds do not rear nearly half their young, or that the average life of a bird is, from accident, not nearly seven years.

Consider a tree that for fifty years produces bushels of fruit, each one containing several seeds. In a region where tree populations are stable, how many of those seeds, on the average, germinate and produce a tree that survives? One. Birds do better, but a recent study of Galápagos

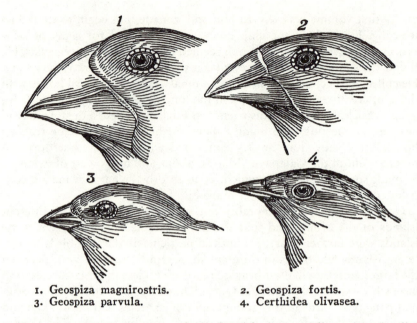

1. Geospiza magnirostris.
2. Geospiza fortis.
3. Geospiza parvula.
4. Certhidea olivasea.

Figure 9.4   Darwin compares the beaks of finches on different islands of the Galápagos group. (From Darwin 1859.)

finches found that on the average, only one out of a hundred eggs produces a breeding adult that produces a breeding adult. Nature catches up with the rest; they don't make it into adulthood. I have remarked several times that until recently, and in some regions it is still true, most people lived on the edge of destruction by violence, disease, or starvation. This is why, in old times, the Malthusian population explosion never occurred, and the same is true of nearly every species in stable surroundings: life is a constant struggle for survival and reproduction. Mostly, survival is a matter of luck, but you might expect also a slight tilt in favor of those best able to cope with the challenges of their environment. They will reproduce, and just as breeders produce new varieties of dogs and horses by selective breeding, selective breeding on each of the Galápagos Islands has produced drifts in the characteristics of its inhabitants.

The islands are not all alike. Their soils are different and support different kinds of fruits, seeds, and insects. On each island, every creature is nourished by its food sources and threatened by its predators, all evolving together, linked in complex interdependence. The last sentence of the passage from Lyell printed above shows that Lyell had grasped this much of the theory of evolution; I quoted it because Darwin had the book with him.

"Survival of the fittest" is a popular tag for the law of evolution. Survival is easy to define, but which are the fittest? There is only one reasonable answer: those that tend to survive, and therefore Darwin preferred "natural selection." Both phrases suggest that a species must change or risk elimination, but neither suggests what determines the kind of change that actually takes place. That, Darwin says, is determined by the small variations that actually occur as creatures reproduce. In a litter of kittens there are differences of color and size, but all are obviously cats. It is from among these variations, he says, that nature makes its slow selection. Conditions on the Galápagos were ideal for the process. The populations were small, so that a successful variant had a good chance to establish itself in the genetic pool, and geographical separation discouraged the populations from mixing. The physical types on the various islands seem to have drifted apart until, in the judgment of Darwin's experienced colleagues, they belonged to different species.[5]

Darwin identified another mechanism that drives evolution, which he called "sexual selection"; his grandfather had called it lust. If a female is able to choose strong and enterprising males over weak ones, the breed will obviously be strengthened. And how does she choose? If two males fight in front of her she can crown the winner; this behavior is seen in animals, birds, fish, and insects. She may choose the male she thinks is the best dancer or, among bower birds, who builds the prettiest alcove. And there seem to be other criteria at work: trimmings. The male elk grows inconveniently large antlers; male birds grow conspicuous plumage while females wear modest colors that make them safer from attack. Excesses such as these are hard to explain except as part of the reproductive process.

Darwin had read Malthus and arrived at the principle of natural selection in 1838; it was another twenty years before he unwillingly made it public. First he put out several volumes of zoological and geological studies stemming from the expedition. Soon afterward, Chambers's *Vestiges*, contradicting revealed religion and only weakly supported by evidence, appeared and was struck by lightning from every side. Not only would Darwin attract the same lightning, but he could be indicted under the blasphemy laws. He circulated his ideas in letters to friends and went on accumulating evidence.

It is hard to say when the world would have learned about natural selection if it had not been invented independently by a younger man. Alfred Russel Wallace (1823–1913), an explorer and collector of specimens in the South Seas, had become convinced that species come somehow

---

[5] In the *Origin of Species*, Darwin several times says what he means by species, but experience and judgment are what it boils down to.

into existence and after a while die out. In 1858, as he rested on the island of Halmahera in the Moluccas, his mind made the same connection with Malthus that Darwin's had. He wrote out his version of evolution and sent it to Darwin, asking him to communicate it to Lyell at the Linnaean Society if he thought it was worth publishing. Darwin did so, but Lyell, knowing Darwin had been mulling over the same idea for twenty years, urged him to submit his own version at the same time. Wallace, a perfect gentleman, did not object and went on to make great discoveries in the zoology of the South Pacific. Darwin, meanwhile, stimulated by this narrow escape, sent a short paper to Lyell and produced *On the Origin of Species by Means of Natural Selection or The Preservation of Favored Races in the Struggle for Life* (I will call it the *Origin*) in the following year. The reception was as might have been expected. Darwin had done well to wait and sift his words and his evidence until it was hard for scientists to find specific points to object to, for as he expected, they were among his severest critics.

Viewed superficially, Charles Darwin's theory seems to echo his grandfather's ideas, but there is a great difference. Some of the earlier evolutionists (but not Maupertuis) had seen the replacement of one species by another as the execution of a master plan in the mind of God. For Darwin there is no plan. Once the variations have been born, he claims that death and survival happen at random, pushed one way or another by changing factors in the environment. From the cumulation of small differences slowly, inevitably, statistically, the organism changes. One thing is clear: the mating of two animals of a given species always produces offspring of the same species, and therefore what changes in evolution is the species itself. When several populations of the same species develop along different lines for a long time, they may become so different that it is silly to call them all by the same name, and if certain criteria are met someone thinks up a new one. It is at this moment, by the edict of members of a profession, that a new species originates. It sounds simple but it isn't, for a naturalist sees only an animal's present state and to make a rigorous judgment would need to know its past equally well. For some creatures, fossils give blurred snapshots of past states from which continuous change can reasonably be inferred, and other evidence makes evolution plausible, many would say obvious, but none of it actually proves it; those who favor purpose produce no proof either, and there it sits.

Lyell "pleaded pathetically with Darwin to introduce a little divine direction or 'prophetic germ,'" but Darwin never wavered. In his autobiography he admits that his religious faith was that of a deist at the time he wrote the *Origin*, "and it is since that time that it has very gradually and with many fluctuations become weaker." This is the voice of someone for whom the topic is not important, but perhaps it is. Deists are people who believe that God may have created the world but that since then he has not intervened in any way to affect what happens there. If final causes operate, one can never build an argument on randomness.

In the *Origin* Darwin had not said much about his own species. The world waited for the other boot, and in 1871 came *The Descent of Man and Selection in Relation to Sex.* It is a long book of which only the first quarter treats the descent of Man. As one might expect, natural selection explains the attributes that distinguish humankind: erect posture, dexterous hands and skill in using them, and, of course, brains. In the *Origin,* Darwin could base his conclusions on the study of many species; here he has only one, and he can't study it the same way. Since there was (and is) no evidence that man is descended from any known kind of ape he made no such claim, though his opponents often said he did. Nevertheless, he is specific: "There can [] hardly be a doubt that man is an offshoot from the Old World Simian stem." He quotes Thomas Huxley's verdict, based on careful dissections, that the brain of an orangutan or a gorilla resembles a human brain more closely than it resembles that of any of the lesser apes, but when he gets to moral and social qualities, concerning which evidence is only now beginning to emerge, he is guessing. The book was met with yells of rage and derision from pedigree-conscious Britons who resented being told that a monkey was hiding in their family tree. Among the scientists, some, including Wallace, doubted that Spirit, or Soul—call it what you will—could arise naturally by little steps, but they couldn't argue their position any better than Darwin, and so they held their peace. Most of the *Descent* concerns sexual selection, and one purpose of it is to explain the appearance of different races as the effect of different conceptions of beauty. It does not seem to have made many converts. Darwin died in 1882, worn down by endless symptoms of illness that never led to a diagnosis. He is buried in Westminster Abbey, where Isaac Newton is buried as well.

Darwin's bequest to us is not a proven theory but a viewpoint that invites us to make sense of nature. All of nature? There are fish that catch their dinner by producing a 200-volt pulse of electricity that stuns smaller fish in the neighborhood. Did this talent develop in little steps: ¼ volt, ½ volt, . . . ? But what would its survival value be? (I suggest: Could the volts have started as a sexual attractor?) But how about the coconut crab, which I have observed on the South Sea island of Éfaté in Vanuatu. Although it is only a crab, it can recognize a coconut palm, climb it, walk out along a branch to where there is a coconut, saw off the coconut, climb back down, walk to where the coconut fell, and have lunch. How did it learn to do that? Another objector might mention the coelacanth, a species of fish that was thought to have died out in the Jurassic era 130 million years ago until it showed up in a fisherman's net off South Africa in 1939. Doesn't it need to evolve, or is it already the best of all possible fish? Among people open to evidence there is little doubt that evolution has occurred, but it remains a program. There is, as yet, no Theory of Everything.

• • •

As to the history of the Earth, after Darwin the fossil record could be read more clearly than before, as a sequence of species. They endure long enough to leave traces set in stone; then slowly, or sometimes quickly, they depart. Clearly the time must be very long, but how long, and even how to guess how long, was not clear for another half-century after the *Origin*.

In Paris, in 1896, Henri Becquerel, professor of physics at the École Polytechnique, left a crystal of potassium-uranium sulfate standing for a few days on a photographic plate securely wrapped in black paper. When he developed the plate it showed an image of the crystal. Further experiments showed that any substance containing uranium would produce this effect, and soon other heavy elements were found that had the same property: they radiate something that passes through black paper. Thus radioactivity was discovered, and in a few years this led to the discovery that the atoms of these substances have a nucleus that shoots off particles. It turned out that when an atom of uranium decays a long process begins. The emitted particle, called an alpha particle, is the same as the nucleus of a helium atom, and it leaves behind an atom of an element called thorium. This in turn soon decays into protoactinium which is itself radioactive, and after fourteen successive decays the result is finally a stable atom of lead.

Laboratory measurements revealed that each of these decays has its own speed: a sample of uranium is half gone after 4.5 billion years, thorium lasts 24 days, protoactinium lasts 1.18 minutes, and so on down the chain—some quick, some slow, but none as slow as uranium. Each of these atomic species occurs in several forms called isotopes, and by paying attention to them and measuring their relative amounts in a given bit of rock, it is possible to deduce the age of the rock. Different rocks turn out to have very different ages, and determining the age of the Earth amounts to finding the oldest rock you can, measuring its age, and reasoning backward from its position and probable history. The presently accepted figure, 4.6 billion years, was established in 1953 by an American geologist named Clair Cameron Patterson. Then what was wrong with the much smaller numbers that Lord Kelvin derived from an exact theory of the Earth's cooling? Only this: the nuclear reactions just mentioned, and others as well, slowly generate heat inside the Earth and keep it from cooling. After the Earth's size had been measured, it was another two thousand years before anybody measured its age.

Now that a piece of rock can be dated, it is possible to date specific geological events. This has led in recent years to a new kind of catastrophism, but this time the catastrophes are mostly biological. The end of the Cretaceous era 66 million years ago is marked by the rapid decline of dinosaurs and the sudden disappearance of many plants and sea creatures, totaling at least half of all species on Earth. In the 1970s the geologist Walter Alvarez discovered at Gubbio, Italy, a thin layer of clay with an unusual amount of the rare element iridium and dated it close to that time. Below this layer the rock is rich in fossils; above it there are almost none. Later research has turned up traces of the iridium layer on every continent and in cores drilled from the ocean floor. It is taken as the boundary between the Cretaceous Age and the Tertiary. Since iridium is more common in meteors than on Earth, Alvarez proposed that the Earth was hit by a comet or an asteroid that produced clouds of dust, ash, and smoke that obscured the Sun long enough to kill plants and the creatures that depended on them for food. Calculation showed that the projectile must have been several miles in diameter and traveling at one or two hundred miles per second. If it had landed on dry land, the crater would still be visible, but the preferred spot now is in an area of the Gulf of Mexico off Yucatan, where there is fractured rock and other evidence of a major impact. The cause of the Permian-Triassic extinction is less clear. Some say it was also a projectile, others say there was a period of great volcanic activity. The Ordovician extinction may have been caused by radiation from the explosion of a nearby supernova.

• • •

Creatures die all the time, but only a few get preserved as fossils, and in spite of the number we see in museums, the record is full of gaps filled by guesswork and interpolation. Nevertheless, a story of life has been worked out, doubtless to be amended in the future. The Earth is about 4.5 billion years old, but numbers like that are not easy to imagine, so imagine instead a 900-page Book of the World that tells the story at a steady pace, 5 million years per page. Here are entries from the Table of Contents that date some beginnings and a few endings in the march of life:

Page(s)

200	Bacteria? Algae?
450	Algae
780	Little marine animals

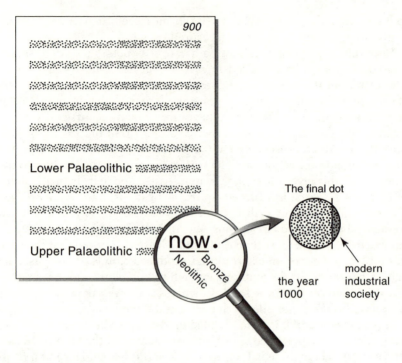

Figure 9.5   Last page of the Book of the World.

790   Trilobites
810   Land plants
812   Ordovician extinction, most species disappear
830   Land animals
832   Reptiles
850   Permian-Triassic extinction, almost all species disappear
865–87 Dinosaurs
870   Birds
887   Cretaceous-Tertiary extinction
890   Mammals dominant
899   First hominids
900   See figure 9.5

TEN

The Exploding Universe

Soleil, soleil! . . . Faute éclatante!
. . . .
Tu gardes les coeurs de connaître
Que l'univers n'est qu'un défaut
Dans la pureté du Non-être!

—Paul Valéry

AFTER FRIEDRICH BESSEL'S MEASUREMENT of a stellar distance, astronomers continued to survey the Galaxy while questions piled up concerning the universe beyond it. The most urgent of them was whether the hundreds of nebulae faintly seen through a telescope are large and far away, as Thomas Wright and Immanuel Kant had supposed, or comparatively small and nearby. Measurement by parallax was hopeless and an entirely new technique was needed. It came from the New World, and I will explain it in a minute. In a few years most astronomers were convinced that the universe is very much bigger than our Galaxy and is getting bigger all the time. Since then, and especially in the last fifty years, there have been many advances in the design of telescopes, and every one of them has revealed something surprising. This chapter will survey some of the new knowledge of how the universe is and how it got that way. Though it will be mostly about theories and observational results, everything I describe depends on the work of brilliant men and women who have invented the new instruments and those who have used them to expand our knowledge—and, I should add, on the forbearance of taxpayers who have financed most of them.

10.1 THE COSMOS IN MOTION

Henrietta Leavitt graduated in 1902 from the "Harvard Annex," soon renamed Radcliffe College, and went to work at the Harvard Observatory. Once established there, she launched a study of some intriguing stars known as Cepheid variables. These are stars whose brightness fluctuates regularly with periods from one to about eighty days; they are not the only variable stars, but they are distinguished by changes in their color as they go through their cycles. They were first noticed in our own Galaxy, but there is a nearby galaxy called the Small Magellanic Cloud that contains some. Leavitt studied them because the cloud is small enough that all its stars are at about the same distance from us, and therefore if one seems brighter than another it really is brighter. By 1908 she had measured the period and luminosity of sixteen Magellanic Cepheids and discovered that the brighter the star, the longer the period. Because the parallaxes of a few nearby Cepheids, and therefore their distances, were known, it was now possible for Harlow Shapley, also at Harvard, to compute the distance to the Cloud. His result was 75,000 light-years, a first modest glimpse of the scale of the cosmos.

Shapley's work was extended by Edwin Hubble (1889–1953), the man who can fairly be credited for discovering the universe. He was born in Missouri, studied mathematics, astronomy, and boxing at the University of Chicago (he was one of the very few astronomers who ever climbed into the ring with a world heavyweight champion), and at Oxford, as a Rhodes scholar, he took a degree in law. A short spell of legal practice in Kentucky sent him back to astronomy, and after World War I he found a job at the Mount Wilson Observatory near Pasadena, which had a new telescope with a mirror more than eight feet in diameter. He found Cepheid variables in the Andromeda nebula, the nearest large galaxy, and in 1924 he announced that its distance is 930,000 light-years. Later, Walter Baade showed that there are two kinds of Cepheids that differ slightly in brightness, and when this was cleared up the distance became, and remains, about two million light-years. The measurement opened the possibility that the hundreds, even thousands (and now billions) of more distant nebulae that telescopes reveal are galaxies more or less like our own, scattered all around us in space, each one containing millions or billions of stars. As the vast size of the universe began to sink in, urgent questions arose: Is the universe finite, or does it go on forever? Has it always existed, or did it start at some moment? And if so, how did it start? The investigation proceeded along the usual two lines of observation and theory but with this difference: usually an observation comes first and then someone makes a theory to explain it. In this case it was the other way around.

• • •

In 1905 Einstein published a paper with the unpromising title "On the Electrodynamics of Moving Bodies," inaugurating what became known as the theory of relativity. It proposed a radical revision of traditional notions of space and time; experience presents them to us as entirely separate categories, but within a few years people were learning to think of them as the four dimensions of a geometrical entity called *spacetime*. It allowed a tidy formulation of Einstein's and Maxwell's theories, but our distinct experiences of space and time did not suggest that it was otherwise a very useful notion. Einstein labored on, and after ten years he produced his masterpiece, the general theory of relativity, which explains the force of gravity as an effect of curvature of spacetime in the neighborhood of a gravitating object like a star or a planet. The time part of the curvature amounts to this: how fast a clock keeps time depends on its closeness to the object. This is easy to visualize. What is less easy, because we have no experience of it, is that it is mathematically possible for three-dimensional space to be curved. We are all used to curvature in two dimensions, so start by imagining a mattress with a flat surface. Put a heavy stone on it and the surface will curve, strongly near the stone and less so further away. In an analogous way, a heavy object curves space.

General relativity has been verified in several ways, but the cosmic part of the story begins with a paper "Cosmological Considerations on the General Theory of Relativity" that appeared in 1917. To create a simple theory of the universe, Einstein started with the astronomers' observation that, although the visible matter in the universe is lumpy—there are stars and galaxies and clusters of galaxies—the lumps seem, at the very largest scale, to be spread pretty evenly. He asked himself: Does the universe actually go on this way, galaxy after galaxy, to infinity? And if not, is it possible that the spatial dimensions of spacetime are finite but unbounded? That *all* of space is curved? Einstein made the simplest assumption, that the lumps are smoothed out and that our three-dimensional space can be mathematically described in terms of a motionless four-dimensional sphere in the same way that we can understand an ant's experience on an ordinary ball's finite but unbounded two-dimensional surface in terms of a three-dimensional sphere. His theory told him that the four-dimensional sphere couldn't exist. He changed the theory's basic equation by adding an extra term known as the cosmological constant and it all came out neatly. Given the average density of matter that astronomers had measured, he could even calculate the universe's size. The radius R of its three-dimensional curvature came out as 10 million light-years—a very small figure nowadays, though at the time some astronomers thought it was too big. Hubble's measurement of the distance to the nearby Andromeda nebula was still seven years in the future.

After Einstein's paper, things moved slowly. In 1921 Alexander Friedmann (1888–1925), a Russian mathematician who worked chiefly as a meteorologist, published a paper in a German physics journal in which, following Einstein, he assumed a spherical space with a uniform distribution of matter and radiation inside it, but he allowed the radius R to depend on time. He found that R could either expand, shrink, or oscillate, depending how it started off, but what would someone living in that space see if it did such a thing? He doesn't say.

Most readers glanced at Friedmann's paper and forgot it. Nevertheless, at that time, there were signs that galaxies, at least, don't stand still. In Arizona an astronomer named Vesto Slipher measured galactic spectra and noticed that the colors of all the stars in a galaxy were usually a little different from those measured on Earth. If the galaxies were moving toward or away from Earth we would see such an effect, and he chose to interpret them that way. He found that the nearby Andromeda nebula is approaching us at some 300 kilometers per second but that the more distant ones are moving away. Other astronomers, looking at his numbers, noticed that the smallest and faintest nebulae move the fastest. Are these the most distant ones?

By 1929 Hubble had measured several galactic distances. It turned out, and further work has verified, that except for the closest galaxies the speeds of recession are quite accurately proportional to their distance from us. Does this mean that those other galaxies are trying to get away from us? No, for if the whole celestial arrangement is expanding uniformly, an observer on any galaxy will look out and see the other galaxies moving away just as we see them. We are not at the center of anything.

10.2 THE BIG BANG

Through the 1920s, the few astronomers who thought the universe as a whole might be in motion were seen as people who let dreams outrun the evidence. Of course the universe doesn't change. The sky doesn't change— what could possibly make it change? Slipher's departing galaxies could be a local phenomenon, and besides, a shift in the colors of a spectrum doesn't have to be a Doppler effect that indicates motion. There may be another explanation. Einstein's closed-in universe, wrapped in mathematics that few could fathom, provided an attractive solution to the problem of infinity, but in 1930 the British astronomer Arthur Eddington showed that such a universe would be as unstable as a pencil balanced on its point: the slightest change in the densities of matter and energy would set it either expanding or contracting, and of course, as stars radiate energy, those densities are always changing. This encouraged astronomers to look at a paper called "The Expanding Universe" that Eddington "discovered" in 1931 in an "obscure" Belgian journal and caused to be published in the *Monthly Notices of the Royal Astronomical Society*. Its author was Abbé Georges Lemaître (1894–1966). He had been an artillery officer in the war, then trained for the priesthood, studied at Louvain and Cambridge and finally at MIT, where he had learned about Hubble's measurements. Friedmann had thought of the changing universe in mathematical terms, but Lemaître saw it as a physical problem and asked himself: If the universe expands or contracts as Friedmann proposes, what should we see? For an expanding universe, the answer was exactly what Hubble and Slipher were seeing: a redshift of galactic spectra proportional to the galaxies' distance away from us. Slipher's measurements, instead of showing galaxies spreading out into previously empty space, showed space itself expanding and taking the galaxies with it. The message of Maxwell's electromagnetic theory was that the electromagnetic field is not just force that pushes and pulls; it changes obedient to its own law of motion. Fifty years later came Einstein's theory, in which gravity obeys a similar law, but with a new implication, for gravity is the shape of spacetime, and so spacetime geometry also moves and changes. This insight guides all the later speculations.

Lemaître's equations, derived from Einstein's, have a variety of solutions. He chose one in which the universe began infinitely long ago at a certain size, grew infinitely slowly at first, and then grew faster. But almost simultaneously with the appearance of this paper he put a two-paragraph letter into the journal *Nature* in which he wrote,

> I would . . . be inclined to think that the present state of quantum theory suggests a beginning of the world different from the present order of Nature. Thermodynamical principles from the point of view of quantum theory may be stated as follows: (1) Energy of constant total amount is distributed in discrete quanta. (2) The number of distinct quanta is ever increasing. If we go back in the course of time we must find fewer and fewer quanta, until we find all the energy of the universe packed in a few or even in a unique quantum. . . . If the world has begun with a single quantum, the notions of space and time would altogether fail to have any meaning at the beginning: they would only begin to have a sensible meaning when the original quantum had been divided into a sufficient number of quanta. If this suggestion is correct, the beginning of the world happened a little before the beginning of space and time.

The Abbé, responding perhaps to an intellectual need for a beginning in a single creative act, has just invented the Big Bang. His universe is not preplanned like those of Augustine and Gassendi. He continues: "The whole story of the world need not have been written down in the first quantum like a song on the disc of a phonograph. The whole matter of the world must have been present at the beginning, but the story it has to tell may be written step by step."

• • •

Here this book says good-bye to the past, and I will try to describe the situation as it is now. Starting in the 1960s, two streams of discovery have converged on the science of cosmology and transformed it. The first is new observations through new instruments. The second is new understanding of the elementary particles and their transformations. By now, cosmology and particle physics have become different branches of the subject known as quantum cosmology, each providing information, insight, and tough questions for the other. The whole field is in a state of rapid development, and from here to the end of the chapter every statement I make, as the physicist Niels Bohr used to say, should be taken as a question. Cosmology is getting complicated—though it still may be easier to understand a galaxy than a housefly—and by now it takes a book to say in accessible language what its workers think about the universe and why they think it. Here there are only a few pages for a quick sketch.

What needs to be explained? The universe's enormous size and profusion: over 100 billion galaxies visible from here, the farthest of them billions of light-years away. Galaxies are unevenly distributed in space. Many belong to clusters; the Coma cluster contains more than a thousand, but there is no sign of clusters of clusters. The overall distribution of galaxies and clusters is like a sponge, composed of walls and open spaces, but on the largest scale there are about as many galaxies in one direction as in another. Did it all start in one small region with a "big bang"? If so, when and how? Where did all that material come from, and how did it get organized into stars, galaxies, and clusters? These are questions to which reasonable answers are available today; that is not to claim that they are correct or even nearly correct—a single new discovery can upset the wheelbarrow—but they reinforce one another, and for the time being, most workers feel fairly comfortable with them.

First, when did the universe begin? This can be estimated from telescopic observations of the receding galaxies. Imagine the universe as a small sphere with the seeds that will later become galaxies scattered through it. Stand on one of them while the whole arrangement doubles in size. A nearby seed that started one foot away is now two feet away; it has moved one foot. A seed originally ten feet away is now twenty feet away; it has moved ten feet in the same time, so it must have moved ten times as fast. Thus speed is proportional to distance, as Hubble found, and if you can measure the speed and distance of just one galaxy, you can figure out how long ago it started. After a few corrections the resulting figure is something like 14 billion years.

That's about *when* the universe got started, but *how* did it get started? In 1980 Alan Guth, a postdoc at Stanford, proposed a model that shows what may have happened; he called it the *inflationary universe*. The main idea has survived through several versions, and what follows is a simplified account. The story may seem implausible, but it does not contradict any well-established physics, and when it ends, the world it pictures is remarkably like the one we know.

Alan Guth's construction project starts by imagining a new kind of field, similar to the fields of gravitation and electromagnetism but not the same; he called it *inflaton*. This field existed for a while, and there is no trace of it now. It had no dimensions of space or time; one cannot say if it was spread out or bunched together, or where or when anything happened to it, but in it occurred a fluctuation that produced a bubble at a very high temperature. The bubble started to expand and cool, and spacetime was born. Somehow, the four dimensions of spacetime sorted themselves out, and time can be measured from that moment on. Once the bubble was past this initial stage, the Einstein-Friedmann-Lemaître cosmological equations determined what happened next. Fields of the inflaton type (they are called *scalar fields*) have the bizarre property that they can reverse the usually attractive force of gravity so that it becomes a repulsion. Under this force, for a very brief interval between about 10^{-35} and about 10^{-33} second after the clock started, the bubble expanded exponentially, ending something like $10^{10^{12}}$ times as big as when it started.[1] The symbols denote a number that, if you bothered to write it out, would be a 1 with a trillion zeroes after it. The time interval may seem a little short for such a big change, but reflect: the universe defines its own timescale. 10^{-33} second is one hundred times as long as 10^{-35} second, the bubble's age when inflation started, so that by local standards it took quite a long time.

It would be nice to know how big the bubble is now, but there is as yet no way to tell because if the universe is only 14 billion years old we can't see to infinity. There is no sign of an end, but if the whole universe is contained inside a bubble we should expect to find that three-dimensional space is curved, as in Einstein's universe. The spatial curvature of the universe can be estimated from astronomical measurements. The answer, at the time of writing, is that there seems to be a very slight curvature, suggesting that we may perhaps live in a very large bubble.

Let's stop for a moment to consider a question that must occur to everyone: Are these scientists so naïve as to think they know how the universe was 10^{-35} second after it started? Actually they aren't. At the beginning of this account I said I would describe a simple model that accounts for a universe that resembles the one we see. The model's assumptions are based on principles some of which are well established while others aren't, and the numbers pertain to the model. If the model explains what

---

[1] 10^1 is 10, 10^2 is 100, etc., 10^0 is 1, 10^{-1} is $1/_{10}$, 10^{-2} is $1/_{100}$, etc.

is observed, we can imagine that if some creature had been there at the beginning, outside the universe (whatever that may mean) and looking in, it would have seen something like the chain of events I am describing. But that is an idle speculation. All we know with reasonable certainty is what we observe now. The theory of the Big Bang, though it is surely far from its final form, explains much that is observed and does not seriously conflict with any observations, and there are observed facts that support it simply and forcefully. One is as follows.

The theory states that energy released during and after inflation created the particles of matter that we now see scattered around the universe in the form of stars and masses of gas and dust. At present the universe contains many kinds of atoms, and in the 1940s the Russian-American physicist George Gamow began to wonder how freshly created neutrons and protons were persuaded to stick together and form the nuclei of heavier atoms. Protons repel one another because they are electrically charged; Gamow realized that only violent impacts would drive them close enough to make them stick, and therefore at the beginning the soup must have been very hot. To test this assumption he suggested the following argument. At that kind of temperature there would have been an immense blast of radiation. Originally space was opaque; then, about 380,000 years after the Big Bang, wandering charged particles coalesced into atoms and space gradually became as transparent as it is now. Since the radiation that existed then was not absorbed and there is nowhere for it to go, we should be able to see it today. Gamow's rough calculation, refined by his students Ralph Alpher and Robert Herman, led to the estimate that by now expansion has cooled this cosmic background radiation (CBR for short) to about 5 degrees Centigrade above absolute zero. Almost twenty years later, scientists at the Bell Laboratories investigating the source of unwanted noise in their radio antennas found that the noise comes from every direction in the sky, and later measurements showed that it has a temperature of 2.726 degrees Centigrade. Definitive proof is a rare thing in science but the existence of the CBR, plus other evidence to be mentioned in a moment, makes it hard to deny that there was something like a Big Bang, and that in the beginning it was very hot.

The temperature of the CBR has a texture that is uneven, with variations at the level of a few hundred-thousandths of a degree that correspond to unevenness in the density of the matter that produced it. Inflationary theory explains them. The shock of inflation must have produced tremors in the inflaton field when the little bubble was formed. As it expanded, these fluctuations froze and grew with it, and as the inflated universe became full of hot gas they were preserved as unevenness in its density, pressure, and temperature.[2] These show up in measurements as slight variations in the temperature of the CBR, and they have recently been studied very carefully.

These variations are not randomly distributed in size; there is a pattern in them. Suppose you beat on a drumhead. Depending on where you hit it, a pattern of standing waves is set up that produces the fundamental note with overtones that make the complex sound we hear. Pushing on a drumhead doesn't produce a sound, you have to hit it. The vibrating drumhead in two dimensions is analogous to the bubble's contents in three dimensions. Analogous to the sudden blow on the drumhead is the shock of inflation. Measurements reported in 2003 showed the fundamental vibration and four of the overtones that inflationary theory predicts. Theory also relates the texture of the background radiation to the age of the universe, and the resulting figure of 13.6 billion years, give or take 200 million, confirms and sharpens the estimate that came from study of galactic motions.

The idea of inflation is so outlandish that I had better repeat it: the clumps of matter that separated out and evolved into galaxies and clusters of galaxies began as submicroscopic fluctuations in an inflating universe.

By now at least the general idea of inflation has settled into the minds of most astronomers and it gives a vivid picture of how the Grand Contraption got started, even if it doesn't say why. The theory's weakest point is the inflaton field, which is pure invention. Is it necessary? There are people working on the quantum theory of gravity who think they see inflation as a quantum effect that needs no inflaton to explain it. Newton said, "More is vain when less will serve." I hope they are right.

• • •

The bang continues: the fields released in inflation change rapidly, their heavy particles disintegrating into lighter ones until space is full of neutrons, protons, electrons, photons, and very lightweight particles called neutrinos, all of which we shall meet again in section 11.2. Unattached neutrons are unstable, and in a few minutes the only ones left have stuck to protons, mostly to form helium nuclei (two neutrons plus two protons). A simple calculation shows that the resulting mix should be close to 75 percent hydrogen and 25 percent helium, and that is what is observed. The same process soon produced a scattering of other light elements, but when they collided they didn't go on to synthesize heavier

---

[2] Since most of the particles of this gas were electrically charged, it is correctly called a plasma, but I will use the more familiar word.

nuclei as Gamow had imagined. Those came much later, in small amounts and by a different route. Most matter in the universe is still the 75-25 mixture.

After about 400,000 years the universe had cooled down enough so that stable atoms of hydrogen and helium could form, and by one billion years gravity was at work. I have mentioned that the mottled temperature of the cosmic background radiation corresponds to fluctuations in the density of the matter that produced it. Now the increased gravity of the denser bits pulled in less dense matter producing huge and more or less isolated masses of gas that began to form galaxies and clusters of galaxies, and within them, at about the same time, smaller gas clouds began to condense into stars. As gravity pulled these smaller condensations together, they heated up. When the temperature at their cores reached about 10 million degrees Centigrade, nuclear reactions started; they began to shine. These nuclear reactions are analogous to the burning of a piece of wood: the wood is consumed, heat comes out, and ash is left. The difference is that when nuclei react, the energy emitted from each reaction is about a million times greater than from each atomic reaction in the burning wood. In a young-to-middle-aged star, the fuel is hydrogen and the ash is mostly helium. Conditions change when the hydrogen in the core is gone, for then the star heats up until the ash itself begins to burn and its ash, in turn, consists of heavier nuclei. The heaviest stars finally die in a great explosion that fills the surrounding space with atoms of heavy elements. They mix with surrounding hydrogen clouds from which new stars form. Look around you, look at your own hand. There is hydrogen in flesh, most of it primeval. The other elements that compose you and the things you see were cooked up for you about 5 billion years ago, in the death struggle of an exhausted star.

• • •

Planets seem to form at about the same time as their central sun. The mathematician Pierre-Simon de Laplace (1749–1827) proposed a scenario. First there is a huge and gaseous sun, very hot and slowly rotating. As it cools it shrinks. This makes it turn faster, and centrifugal force throws out matter, like mud from a wheel, that ultimately condenses into planets. The trouble with Laplace's theory was that it did not explain how the Sun, whose equatorial region rotates only once in twenty-five days, managed to throw off all that material. A century and a half after Laplace, astronomers discovered that the Sun emits a continuous blast of matter known as the solar wind. This wind carries with it more than its share of angular motion, and calculation shows that the Sun turns much more slowly now than it did when it was young. Perhaps Laplace was right.

Whatever happened gave the Sun, an ordinary star, eight planets. And other stars? At present there is no way of seeing a planet the size of the Earth, since a telescope trying to find a planet right next to a star goes blind. But when a very massive planet revolves around a modest star it makes the star's position wobble visibly. About a hundred big planets have been found in this way. An ideal instrument would be an optical device that blanks out the central star, allowing its planets to be seen and their atmospheres analyzed for signs of life. Such a device has been designed. It is called a *nulling interferometer* and should fly in a few years. Watch your newspaper.

Galaxies, stars, planets, floating in empty space. A few generations ago I would have been corrected: floating in ether. Now the ether is gone, but might these celestial bodies be floating in something else? What else is there to float in?

10.3 WHAT'S OUT THERE?

An inventory of the universe starts with stars, virtually all of them in one galaxy or another, and a study of their spectra shows that they are made of the same kind of matter as the Sun and planets. The matter is composed of neutrons, protons, and electrons and is called *baryonic matter* (from Greek *barys,* heavy). There is even more baryonic matter in the form of clouds of gas and dust that have not yet condensed into stars. Until 1922 no one doubted that all cosmic matter was in stars and clouds and therefore visible; then the Dutch astronomer Jan Oort noticed that stars in the Galaxy move in a way that seems to defy Newton's laws. Only if he assumed that the Galaxy contains a large amount of invisible matter could he make sense of what he saw. In Switzerland a decade later, Fritz Zwicky noticed that the galaxies of the Coma cluster move so fast that gravitational forces couldn't keep them together. The cluster would have dispersed long ago unless it also contains invisible heavy matter, lots of it. Apparently most of the mass of every galaxy and cluster of galaxies is of a kind that creates gravity but is invisible because it neither emits nor absorbs light. It took about forty years for this radical and unwelcome suggestion to sink in. Now experiments are under way to find particles of the stuff, but if they interact with the baryonic matter of astronomers' instruments only through gravity, it will be a hard job. They seem to be heavily concentrated in galaxies and clusters, suggesting that the particles are not moving fast enough to escape galactic gravity. It is therefore called *cold dark matter.*

The gravitational force of cold dark matter stabilizes a cluster of galaxies, but it seems there is another, more pervasive force that acts in the

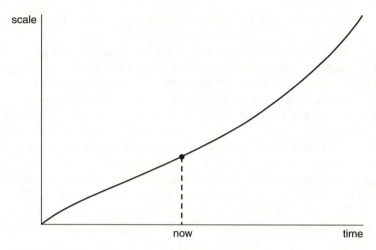

Figure 10.1   Starting about 4 billion years ago, the scale of the universe has steadily increased. At first the expansion was slowed by gravity but now it is accelerating. The future part of the curve is a plausible extrapolation.

other direction. Until the 1990s astronomers thought that the force of gravity must be slowing the expansion of the universe, but by now there is abundant evidence that for some time the expansion has been speeding up. There seems to be a force stronger than gravity that acts the other way. This force has a history. You will remember that in 1917 Einstein, trying to invent a model universe that would be finite, unbounded, and stationary, added to his cosmological equation a "cosmological constant" representing a general outward force to balance the inward force of gravity. Now it is back again, called *dark energy* or *quintessence* (the two are not quite synonymous), and unconnected with anything else that is known. Here, scaled so that they add up to 1, are numbers that show how the ingredients I have mentioned affect the universe's rate of expansion:

Baryonic matter	.04
cold dark matter	.29
dark energy	.67

The numbers are subject to change without notice, but they are striking when one considers that until a few years ago most people thought that except for electrons and other almost massless particles, baryonic matter was all there is. The astrophysical community has responded to this emergency with a blizzard of papers in which new fields are proposed, as well as new versions of spacetime geometry and new quantum effects. To sustain or demolish these notions new observational data are needed.

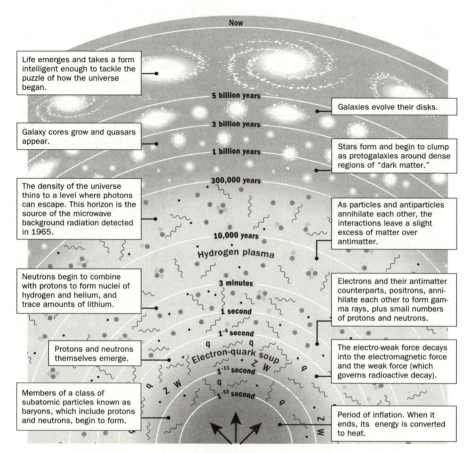

Now

Life emerges and takes a form intelligent enough to tackle the puzzle of how the universe began.

5 billion years

Galaxies evolve their disks.

3 billion years

Galaxy cores grow and quasars appear.

1 billion years

Stars form and begin to clump as protogalaxies around dense regions of "dark matter."

300,000 years

The density of the universe thins to a level where photons can escape. This horizon is the source of the microwave background radiation detected in 1965.

As particles and antiparticles annihilate each other, the interactions leave a slight excess of matter over antimatter.

10,000 years

Hydrogen plasma

Neutrons begin to combine with protons to form nuclei of hydrogen and helium, and trace amounts of lithium.

3 minutes

Electrons and their antimatter counterparts, positrons, annihilate each other to form gamma rays, plus small numbers of protons and neutrons.

1 second

1^{-5} second

q   q

Protons and neutrons themselves emerge.

Electron-quark soup

1^{-11} second

q

Z   W   W

The electro-weak force decays into the electromagnetic force and the weak force (which governs radioactive decay).

1^{-33} second

Members of a class of subatomic particles known as baryons, which include protons and neutrons, begin to form.

Period of inflation. When it ends, its energy is converted to heat.

Figure 10.2   The history of the universe in a single diagram. (Dave Herring/ 1999 © *The Christian Science Monitor.*)

The cosmic background radiation bears traces of the universe as it was at age 100,000 years. To learn more about its infancy, the only instruments contemplated at present are telescopes designed to detect gravitational radiation from the Big Bang and other cosmic explosions. Preliminary versions are functioning now but they have yet to detect anything. More sensitivity is needed. Watch your newspaper for results from the Laser Interference Gravitational Wave Observatory, LIGO for short.

The universe continues to expand, but this does not mean that its edges are being pushed back—nothing is known about edges. It is its scale, defined in terms of the distance between two free-floating objects far enough apart that their gravitational attraction is negligible, that increases, but faster now than before. Figure 10.1 shows how, after the inflationary jump, the scale changes with time in response to the three constituents. For past times, the theoretical curve is adjusted to fit the observations, and for the future it predicts what our remote descendants may expect. Figure 10.2 summarizes the history of the universe thus far.

Is this cosmological model correct? It's hard to say. Ether is gone but we still float in a mysterious ocean. Dark energy and dark matter are new arrivals. They seem to dominate the universe, yet all anyone knows about them is that one pushes and the other pulls. Until they are understood better, any astrophysical theory in which they play a large part will skate on thin ice.

So much for the history of the universe and some of the things in it, including our planet, compiled from the evidence of many different kinds of observations and a good dose of imagination. Perhaps it will be safe a few pages farther on to say something about the future, but first an important question. The universe is vast; the number of stars is amazing and many seem to have planets. One planet is inhabited. Only one?

ELEVEN

The View from Here

> Part of infinity seems to lie within the grasp of those who look
> across the sea.
>
> —Niels Bohr

THIS CHAPTER RETURNS TO a question that has arisen several times in
these pages: Are we alone in the universe? Anaxagoras and his friends
saw no reason why there shouldn't be other worlds. Thomas Wright estimated 60 million planets in our galaxy but didn't say how he did it, and
today astronomers have found more than a hundred in our immediate
galactic neighborhood. So that question is settled, but it opens two
others: Is there life anywhere else, and if so, is there life that thinks? Section 11.1 discusses what makes a planet suitable as a home for life and
describes recent attempts to detect any signals from outer space that
might be reaching our neighborhood. Section 11.2 discusses ways in
which our whole universe seems to welcome the possibility of life and
compares it with other less hospitable universes that might exist, arguing
that they may exist. Section 11.3 looks toward the future of the Earth
and predicts the approximate date when our remote descendants may
have to look around for a new home. At the end come some personal
reflections.

11.1 IS THERE ANYONE ELSE?

There are billions of stars in our galaxy, and many of those closest to us
have planets. Do they support any form of life? If a probe sent to Mars
finds anything that is or was alive (nothing more impressive than a bac-
terium is considered likely) it will encourage the hope of finding someone
farther away to talk to. Let's imagine extraterrestrial people as solid
rather than liquid or gaseous, made of a variety of chemical substances,
able to reproduce themselves, living in some kind of organized society,
able to pick up a screwdriver, and able to think. People like that need a
home. On a star there would be no place to sit down; a planet or a satellite of a planet seems the only choice. What would these people be made
of? I think it is safe to say that however much their chemistry may differ

from ours, the principles of chemistry will be the same wherever we look. The specialized tissues of terrestrial creatures are made of large molecules based on carbon, of which there would be plenty on almost any planet or satellite. It seems likely, though perhaps not absolutely necessary, that living tissue elsewhere would be carbon-based and would also contain that multpurpose solvent, liquid water. Water may be the most stringent requirement. There must be lots of it on a planet because it evaporates, and when the evaporated vapor is exposed to a star's ultraviolet radiation it dissociates into hydrogen and oxygen. The hydrogen flies away and so oxygen is gained, but water is continually lost. Too hot a planet would lose it all, as has happened on Venus; too cold would freeze it, as may have happened on Mars. There are micro-organisms on Earth that live at temperatures below freezing or above ordinary boiling temperature, but they have nothing to say to us. Without insisting that all inhabited planets must have a climate like the Earth's, it seems clear that in *our own* Solar System only the Earth can support creatures that run or swim or fly around and invent things and are not always looking for something to eat. The Earth carries a million living species, many of which need light and water and air. Estimates vary, but there is agreement that if the Earth's orbit around the Sun were 20 percent larger or smaller, life, intelligent or not, would have been less likely to develop.

Of other happy circumstances that make life easy for us I mention one that is less obvious but very important. Jupiter, enormously massive and only two planets away, is a magnet for wandering objects like comets. They enter the inner Solar System from a great distance and, if they crash into a planet, they do great damage. One that reached Earth ended the Cretaceous era by exterminating most living species, and without Jupiter to attract and absorb them we would be hit much more often.

But how many other planets are as lucky as ours, and how many intelligent civilizations are there? Estimates range from one (this one) to millions. Is there any basis for a guess? Advocates of a large number might quote Metrodorus of Chios: "It is strange for one stalk of wheat to stand in a great plain, and for one world to exist in boundless space." Advocates of a small number quote the Italian physicist Enrico Fermi who asked, "Then where is everybody?"

I think Fermi's question has a reasonable answer: they are in their homes, if they have homes. But the most naïve assumption we can make is that aliens' brains and minds are in any way like ours—that they are, for example, adventurous, curious, and on the whole well intentioned. What if their brains and bodies are electronically supplemented or replaced? What kind of human contact can we make with a Darth Vader? But let's pretend that they are something like us, that they are happy at home, not driven to desperation by stinking atmospheres and exhausted natural resources, and just want to make new friends. The closest stars (including, possibly, our Sun) are a few light-years away from them. If they could make the whole trip at 99 percent of the speed of light, the Einstein effect would cut the time as experienced by the passengers to a few months, but that is not how it is, for in order not to harm them the craft would have to be brought gradually up to speed and just as gradually down again. Travel time would be more like years than months. Further, even assuming the most efficient possible propulsion, the amount of fuel required to get them going and slow them down again would raise the weight of the craft to thousands of tons, and if the travelers plan to return home it would have to be bigger than that. For star hopping, the only reasonable alternative is to travel more slowly, say at 1 percent of the speed of light, and not worry about time. This would require technology that could shut down a living creature for five hundred or a thousand years and then wake it up, ask it how it feels, give it a back-rub, and serve it coffee. Or they could build a search robot that turns itself on and off. In either case, those at home would not hear from the spacecraft for a thousand years. Barring space warps or other distance-shrinking technology unknown to us, it does not seem that casual space tourism, by us or anybody else, has much of a future. But someone might at least phone.

In 1959 Giuseppe Cocconi and Philip Morrison, then at Cornell, published a paper in *Nature* that outlined a systematic search for radio signals from outer space. They estimated that an advanced civilization on a nearby planet (no extrasolar planets had yet been discovered) could quite easily beam a radio signal that we could detect above the background noise of distant astrophysical processes. The first searches were made in 1960 by Frank Drake and his colleagues at the 85-foot radio telescope in Green Bank, West Virginia. No significant signals showed up, but news of the project (called Ozma) got around, and since then public (and, more important, private) enthusiasm has grown. NASA was briefly involved, but the sarcasm of a senator from Nevada chopped off the funds and since then a succession of automated searches have been mounted by the SETI (Search for Extraterrestrial Intelligence) Institute and another called SERENDIP, both centered in California, as well as groups at Harvard and Ohio State and in France, Italy, Australia, and no doubt elsewhere. Funds come from private contributions and the computer industry. Detectors run day and night, and much of the signal processing is done without charge by computers all across this country that are wired up to work while their owners sleep. So far the phone hasn't rung. True believers labor on, but SETI's ambitious and technically sophisticated Project Phoenix gave up and stopped listening in March 2004.

Surely, it would be a great moment for our philosophical view of the cosmos if Earth-based antennas ever capture strange and beautiful music from outer space, uninterrupted by commercials. Imagination blossoms: we could form a sort of galactic club, or join one that already exists, to learn and compare the thoughts and arts of enlightened civilizations everywhere. But for a moment let history be our guide. When Europeans began to spread out across our own globe, they found music in many places—Africa, India, the Andean highlands—but no one invited the musicians into his club; what happened to them was very different. A club of the weak and the strong is no club at all, and the idea that inhabitants of a few planets attached to some nearby stars might all, by coincidence, be at a cultural level corresponding precisely to the latter part of the final dot in that Book of the World evaporates on consideration. Let us listen for other planets' music, but is it really wise, as some have suggested, to spray our own message into space? Since the consequences could affect everyone, perhaps everyone should have a chance to register an opinion.

11.2 THE BEST OF ALL POSSIBLE WORLDS?

Time and luck and hard work may some day reveal that we have neighbors. That would be enough for most people and more than enough for some, but inflation theory leads to a further question: Is our universe alone or is it one unit of what Britain's astronomer royal Martin Rees calls a multiverse? At first glance we may want to file this question along with What happens when an irresistible force meets an immovable object?, but a look at the facts reveals an angle that might lead to an opinion. The angle is the following.

According to inflation theory, our universe began as a quantum hiccup in some realm of existence (or nonexistence) of which we have little idea. But is there any reason to think it only hiccuped once? Inflation theory seems to imply that there may be many closed universes forever separated from our own, perhaps very many, and that conditions in them might be very different. According to the theory, at the moment inflation ends and new forms of matter start to pour out into expanding space, a phase transition takes place. A phase transition is a change in the physical properties of a substance even though the substance doesn't change. When a liquid freezes or boils, for example, it is the same stuff but it acquires properties that it didn't have before. When water freezes, it always produces the same thing, ice, but in particle theory there are more variables, and at least a few numbers relating to the masses of various particles in each universe and the strengths of the weak, strong, and electromagnetic interactions involving them could be determined not by any causal mechanism but randomly during that first rush of matter. It seems that in our universe, the values of these numbers were almost miraculously adjusted so that life could develop and evolve. There are two leading explanations. One is Intelligent Design. Its weakness is that even if it is correct, the scarcity of scientifically testable evidence of a Designer walls us off from knowing what happened. The other explanation is the multiverse, together with the remark that if some of the numbers necessary to life really occur at random, and if there are enough universes, there ought to be, and there is, at least one in which life could develop. This argument too has obvious weaknesses, but at least it can be strengthened or destroyed as more is learned about particles and fields.

Now for some examples of numbers that seem to have been nicely tuned. Excuse me if I get a bit technical, but the book is almost over and this is the first time. Neutrons and protons are very similar particles except that protons are a little lighter than neutrons and carry an electric charge that makes two protons—but not, for example, a proton and a neutron—repel each other. There is also a force that holds these particles together if they get close enough. This force is barely strong enough to bind a neutron (n) and a proton (p) into a particle called a deuteron (np). Stars produce light and heat because of nuclear reactions that continuously occur inside them. In the core of a star like the Sun, most of the relevant particles are protons, and in one of the reactions two protons interact to produce a positive electron (e^+), a neutrino (ν), and a deuteron:

$$p + p \longrightarrow e^+ + \nu + np.$$

Interactions involving neutrinos rarely occur. A proton can bang against other protons in the Sun's core for a billion years before this ever happens to it. When it does, energy is released, and this reaction supplies most of the energy that has kept the Sun steadily hot and luminous ever since it first formed. The reaction also produces deuterons. Those in the Sun are of no use to anybody, but in earlier generations of stars they were the first crucial step in the formation of the heavy atoms that make the solid Earth and everything around us.

The force that barely binds a neutron to a proton is almost but not quite strong enough to bind two protons into a particle called a di-proton (pp), which does not and cannot exist. If it did, two protons could react this way:

$$p + p \longrightarrow \gamma + pp,$$

where γ is a gamma ray, an energetic photon. This reaction would take place billions of times faster than the other one. A new star, forming in a cloud of hydrogen, would explode as soon as it got hot enough, and the night sky would be dark except for an occasional flash of light. No stars, no planets, no life. If the nuclear force were only a few percent weaker, there would be no deuterons; if a little stronger there would be di-protons. In either case we would not be here to talk about it.

There are other examples of what seems like marvelous luck that collaborate to make possible a world with us in it. Neutrons bound into a stable nucleus are stable, but an isolated neutron turns into a proton after a few minutes by shooting off a couple of other particles. It can do this because it is about a tenth of a percent heavier than a proton. The neutron disappears, and one more proton is born to increase the global supply. If protons were only a little heavier than neutrons, the reaction would go the other way. There would be no free hydrogen in the world, no stars, no planets, only neutrons and some wandering lighter particles. These two examples, and there are others, suggest that we live in a universe delicately tuned to our requirements. For those not satisfied with an explanation based on Providence or luck, the multiverse offers another.

Arguments of this kind are called *anthropic*. They are based on the unarguable fact that as we look out into our own universe we must expect to find all the conditions necessary for our existence, and on the conviction that these conditions deserve an attempt at a physical explanation. It is still possible that a new and better theory will show that some or all of the lucky numbers are determined necessarily, by a universal physical law. To the extent that this happens, the anthropic argument is weakened.

There are, of course, other happy coincidences that do not need a multiverse to explain them. If the Solar System had started with no planet in the zone of habitability that contains the Earth's orbit, there would probably be no intelligent life in it and very likely no life at all. But that was a chance event, pure luck, that has nothing to do with the nature of the universe. If not the Sun, then some other star surely has a lucky planet.

There is an argument that opposes the multiverse. In the fourteenth century, an English Franciscan named William of Ockham proposed a principle designed to clarify and simplify tangled theological discussions. It is known as *Ockham's razor,* or more formally as the *principle of parsimony,* and it reads: Don't introduce any more entities than necessary into an argument. The word "multiverse" refers to the possibility that inflation happened so many times that one universe has a lucky combination of physical qualities that allow us to exist. Introduce a wilderness of uninhabitable universes in order to solve the anthropic puzzle? William might not have been pleased, or he might just have laughed.

11.3 WILL IT EVER END?

After several pages about the beginning of this universe and perhaps others, it seems only fair, before closing the book, to talk again about endings. The scenarios of section 5.4 assume a final catastrophe after which the heavens will be rolled up like a scroll and tossed into storage. For much of the twentieth century, cosmologists thought things would end pretty much that way. It seemed that the force of gravity would finally conquer the impetus of the Big Bang, expansion would turn into contraction, and all would end by falling into a gigantic black hole and ceasing to exist. Now the indications are that dark energy will win and the universe will expand forever. Stars will die and after a while no more will be born. Whether their cold remains will evaporate depends on whether protons and bound neutrons are stable or whether they will finally decay into lighter particles; that is not known. Black holes will slowly evaporate and finally disappear in a flash of light, and after something like 10^{100} years, whatever was ever going to happen will have happened. At the moment, scientists are being entertained by another possibility. Nobody knows what the recently discovered dark energy is, and its properties are anybody's guess. One possibility is that the expansive force slowly becomes stronger with time; if so, dramatic effects will ensue as it breaks up various systems now held together by stronger forces. After about 35 billion years the Milky Way will disband, and then the Solar System. The Sun will go out and after that things will happen quickly. The Earth will come apart, its molecules and atoms (held together by electric forces) will break up, and a few seconds later. . . . But notions like these are born and published in a haze of speculation.

It is more interesting to focus on the Earth's nearer future. Let's divide it into short, medium, and long. In second 9.4, I described a 900-page Book of the World in which mankind is mentioned on the last page and modern industrial society takes up only part of the dot that ends the volume. I can't guess how long the present changes in the Earth's biosphere will continue into volume 2. That depends on poorly understood causes of global warming and on the resilience of the Earth and its passengers. Up till now the survival of all species has depended on their toughness (think of bacteria, which appeared somewhere around page 200 of the Book), and their ability to deal with predators and accidents of nature. Humans seem to be their own predator, producing environmental changes faster than any natural agent except a wandering asteroid. If one looks at history from the Earth's point of view, our species appears as something like a sudden viral infection; how long the Earth will support complex life forms depends on our ability to mutate into a more benign form. Will we? Anyone's guess is as good as mine, but the Earth will still be there, so let's go on to the medium future, say the next hundred pages of volume 2.

Concerning the medium future, we know an inescapable fact: the Sun burns thousands of tons of hydrogen every second. The supply is finite; therefore the light will finally go out. But stars do not go out like a lamp that runs out of oil. In the Sun, the hydrogen that actually burns sits in the small hot core, and as hydrogen fuel is replaced by helium ash the Sun keeps shining by slowly raising its temperature. Instead of cooling down the Earth is getting warmer, and after 500 million years, about a hundred pages into volume 2, its mean atmospheric temperature will reach 60° Centigrade, which is 140° F. That is pretty warm, but at least up to 400 million years—eighty pages—unless something else happens, whatever descends from us might be able to handle it.

What else could happen? In the past there have been mass extinctions of species. A few of them were so serious that many or most species disappeared; others were less complete. Paleontologists have good evidence that these extinctions have come periodically. It is not a simple periodicity, but the intervals between extinctions vary around 31 million years. One explanation is that they happen when the Solar System, orbiting the center of the galaxy, crosses and recrosses a layer of asteroids in the galactic plane. Another is the discovery by way of computer simulations that at long intervals the orbits of Jupiter and Saturn become unstable, with the possibility of propelling asteroids from their usual belt between Mars and Jupiter into Earth's orbit. When even one object of kilometer size comes down to Earth, the dust raised by its impact and the fires started by its heat can produce years of darkness that cut the food chain and start an extinction. In the past, recovery from a severe one has taken millions of years, and the life forms after it have been very different from those before. The most recent and therefore best documented ones occurred about 11, 37, 66, 92, and 113 million years ago; you can see a pattern of repetition. It doesn't seem likely that Polyphemus throws exactly one huge rock at us each time we sail past him; more likely he throws several. The next scheduled encounter is some time off, but if the Earth remains habitable for 400 million years we can expect about thirteen passages through the danger zone.

There will also be random encounters with material of our own Solar System and the swarm of comets that surrounds it. These can come at any time but are perhaps not as dangerous as they seem, for orbiting the Sun they are lit up and can be spotted and their orbits calculated years before they are due to come near the Earth. If calculation shows that one of them is going to hit, a steady gentle shove by some form of spacecraft when it is still far away can cause it to miss. If the asteroid is not too large, a force of as little as half a pound, applied for a few months a decade or so before the calculated impact, would be enough. The technology does not exist now but it could come soon.

After 500 million years, oceans will be evaporating and the air will be heavy with moisture. This may be good for some future forms of plant life but they will have to survive with less carbon dioxide, for much of it will have been absorbed by what is left of the sea. The remaining species, including whatever creatures may descend from us, will have adapted to these conditions. By the time water is gone, who knows what will still be alive; possibly life will end as it began, with a few hardy bacteria.

All of this is what will happen if nothing is done. Long before then, if the Earth still has living creatures and technology, they will have at least three choices. They may move the Earth to a new orbit farther from the Sun but not too close to Mars (they may have to move Mars to make this possible). They may erect a screen of some kind that orbits with the Earth to control the amount of sunlight it receives. Or they may find a planet to colonize that has the right temperature and plenty of water and oxygen and no one who strenuously objects to being colonized. It would be good if this planet circled a star smaller than the Sun, for these are more economical of energy and can burn steadily for billions of years. There are plenty of candidate stars, but probably fewer suitable planets.

• • •

About 7 billion years (1,400 pages) from now, the hydrogen in the Sun's core will be gone and the steady production of energy will end. Combustion will move outside the core, the Sun will cool and redden and expand, and in the next half-billion years it will spread out as far as the orbit of Mercury. As the core produces less energy, gravity will begin to contract it again and it will heat until its helium ash begins to burn. Another half-billion years and it will swell again, into a giant like Betelgeuse, the red star in Orion. At its surface, so far from the center, gravity is weak, and radiation from its hot core will drive off much of the remaining matter. Nuclear reactions will slow and finally stop, and what is left will settle down as a white dwarf. Astronomers know these little points of light very well, for there are plenty of stars near the Sun that have already followed the same course. It will be 10 or 12 miles across, made mostly of carbon and oxygen ash from burning helium. The material is unimaginably dense—a little matchbox full of it would weigh a ton. The Sun might still have some planets left, first scorched and now frozen, but just as likely, in that long time some other star will have wandered close enough to scatter them into eternal night.

The story of the Grand Contraption has a beginning and a middle, but when shall we say it ends? When life ends on Earth? When it ends in the universe? In the theater a tragedy is apt to end when somebody dies. Elsewhere life goes on, of course, but it is time for the audience to go home.

11.4 REFLECTIONS

The Sun will get warmer, but if nothing worse happens sentient beings will be able to live on this blessed planet for about 400 million years before it gets too hot. Species or families of species ordinarily last for a million years or so; then they seem to get out of date and disappear. Among astronomy's billions of years, 400 million don't sound so long—less than a tenth of the Earth's present age—but reflect: 400 million years ago was the age of the trilobites. What will happen to reasoning creatures during the coming thousands of millennia? I don't know. Blaise Pascal wrote, "The silence of those infinite spaces makes me afraid." It isn't the infinite spaces that bother me, for telescopes show us strange and beautiful sights that enrich our thinking and invite us to learn more; it's the empty time that lies ahead. We explore space and learn about the past, but as to the future we are children lost in the dark, encouraging us to make absurd bets on the future as children might, and take absurd chances.

I think of events that would destroy the delicate system of arrangements that make civilized life possible. There have been mass extinctions. We are a long way from the next scheduled one but some have happened out of schedule, and at present we are in the middle of a quiet mass extinction that is one of the most extensive so far. Species are disappearing so quickly that at least a third of them, and possibly twice that, will be gone in the foreseeable future. Some will die out because we kill them, some because we are destroying their homes, and others because there is no way for them to live in the world we are making. How this extinction will affect civilization, beyond impoverishing the texture of our lives, is hard to guess. The only firm prediction I have read is an increase in the populations of mice, rats, and cockroaches. A recent survey by the World Conservation Monitoring Center, a United Nations agency, estimates that 6.3 percent of the Earth's land surface is under some kind of protection covering its forest and wildlife. What this actually means depends on the willingness and ability of nations to enforce regulations and varies widely from country to country. Environmentalists tend to value places like tropical rain forests where there is great diversity of species, but most of these are in poor countries where population grows quickly. In Nigeria, for example, it is expected to double in the next forty years.

Nuclear extinction, unlike the other two, aims specifically at civilization. The present supply of nuclear weapons is enough to turn life, perhaps for generations, into a struggle for survival. Civilization requires a little more than the minimum of food and leisure. It allows the arts and sciences to thrive and gives most people a chance to enjoy their products. Russell Hoban's brilliant novel *Riddley Walker* imagines life in southern England perhaps a century after a nuclear catastrophe. The structures that support civilized society are gone and only some memory of their names is left: the Pry Mincer and the Ardship of Cambery are comic characters who wander around with nothing useful to do, while huge dogs threaten everything that breathes. It doesn't have to be dogs. New bacterial or viral agents may be created that the human immune system does not recognize or resist, and now that the human genome has been mapped and made public, this becomes easier. It takes money, time, and heavy equipment to make an atomic bomb, but a few malignant (or, possibly, careless) persons may soon, or even now, hold the lives of millions in their hands.

At present the Earth is the subject of a number of large-scale experiments; I will mention only two. The largest is the one that will show what happens when the transparency of a planet's atmosphere to infrared light is suddenly decreased. Every year, humans consume fossil fuels amounting to about 6.5 billion tons of carbon, of which half remains in the atmosphere as carbon dioxide, adding to what was already there. The odd thing is that the experiment has no plan and no safeguards, and having got started it cannot be stopped. Carbon dioxide, as well as a few other gases, contributes to global warming, and even if its production comes under control tomorrow, what is already in the atmosphere will remain for a century or more. Under its influence, the seas will continue to warm and the planet's glaciers and coastal ice will continue to melt. The question is, How far will this go? During the twentieth century, the sea level rose about six inches and during this one it will probably rise more than that, largely because water expands as it warms. The icecaps of Greenland and Antarctica are still frozen, but they are changing, and it is idle to assume that no melting will occur. Some icecaps are two miles thick, and if the worst happens and they disappear, every coastal city in the world will have either to build dikes or be abandoned, and some whole countries will be in trouble.

Global warming affects agriculture, and it may have other significant effects. For example, it helps to thin the layer of ozone in the atmosphere that protects people underneath it from the Sun's ultraviolet radiation. The Antarctic hole grew to 27 million square kilometers in the autumn of 2003. Luckily, few people live under it, but the northern layer is also getting thinner as conditions over the North Pole start to resemble those over Antarctica. A northern hole of that size would expose Scandinavia, the British Isles, much of Russia, and most of Canada to dangerous levels of radiation. So the carbon dioxide experiment entails a radiation experiment as well.

Another experiment concerns water and the overpumping of reserves. The largest aquifer in this country is the Ogallala, which supplies eight states from South Dakota to Texas with water for irrigation and general consumption. It was stocked with water at the end of the last ice age and will not be replenished before the next one. Pumping has lowered its level by different amounts, as much as 100 feet in parts of Texas. If pumping continues at this rate the aquifer will run dry, first in one region, then another. This is the area from which most of our wheat comes, and in fact half the world's wheat is nourished from aquifers that are being pumped dry. Billions of people depend heavily on wheat in their traditional food supply. What will people eat when wheat and other irrigated crops become scarce commodities?

•••

The dot in figure 9.6 is something like 350 years wide. Does anyone know what the conditions of life on Earth will be even at the end of another fifty? There is no point in discussing here the dangers that face this planet and its passengers from environmental degradation and simple malevolence, but it behooves everyone to know that dangers are there and some may be imminent. Several are soberly discussed in a recent book, *Our Final Hour,* by Britain's Astronomer Royal, Sir Martin Rees, and I recommend that you read it.

The world will change, as it must. Impersonal laws of nature will determine the levels of temperature and humidity in which the Earth's surviving creatures will live in the coming century (I do not even think of a millennium), and conditions will be changing long after we stop, if we ever do, treating the loving and fertile Earth as a doormat on which to wipe our feet.

Plato identifies the world's creation as an attempt to unite soul with matter in search of the Good, and Aristotle judges—I quote it for the last time—that nature always strikes after the better. Nature perhaps, but not man. It is easy to blame money and politics for the destruction of the environment, and we should, but responsibility is wider than that. In the "developing" world, where life is risky and death comes soon, you can't blame people for putting survival first, but in the "first world" it is we the consumers whose modest desire for a bit of personal space, easy transportation, and a pleasant place to live makes us spread into areas where we crowd out birds and animals and soak up irreplaceable resources. We do this even though we know that Earth is vulnerable. Who can help us to do better? In this country, citizens' groups collect money and press legislators to save what can be saved, and in Europe Green parties are growing. Some great religions see holiness in the poor and the sick and minister to them with love and hard work, but how many teach that Earth too is holy? At least one does.

Ecumenical Patriarch Bartholomew I, the spiritual leader of the world's 250 million Orthodox Christians, sees the preservation of the Earth as a duty imposed by religion. In 1995 he declared that "crime against the natural world," including actions that extinguish species, change the climate, destroy forests, or pollute air, water, or land, is a sin. His chief theologian, Metropolitan John of Pergamon, has gone farther, declaring that enlightened management is not enough; as "priests of creation" we must adopt the care of nature as a sacred duty. "The human being is almost by its very constitution the link between creation and God. We are part of nature." In June 2003, from the deck of a cruise ship carrying two hundred scientists, political leaders, and journalists, the Patriarch and his fellow clerics blessed the Black Sea, which has been specially troubled by human actions.

The Okanagon Indians live in the American Northwest, and one of them recalls the Earth Mother, "The Earth was formerly a human being and lives still, but transformed so we cannot see the person she was. But she still has the parts of a person—legs, arms, flesh, and bone. Her flesh is the soil; her hair is the trees and other plants. Her bones are the rocks, and her breath is the wind. She lies, her limbs and body extended, and on her body we live."

In early spring the Earth Mother is pregnant. She gives birth, and in that season she is treated gently; some Pueblo Indians even take the shoes off their horses. When agents from the Bureau of Indian Affairs offered to supply tractors and steel plows, the reaction was what one might imagine. Barre Toelken writes: "I asked a Hopi whom I met in that country, 'Do you mean to say, then, that if I kick the ground with my foot, it will botch everything up, so that nothing will grow?' He said, 'Well I don't know whether that would happen or not, but it would really show what kind of person you are.'" The Earth is the body of an old woman? Not true at all, obviously a myth. Truth isn't the point. All over the Earth, people leap up to defend myths that are just as unlikely to be literally true. The point is love. A wise man, crossing the Danube with me one day on the old Chain Bridge in Budapest, said, "We must love the universe even though the universe does not love us." Perhaps each of us must love it in his or her own way, and that, at any rate, explains why I wrote this book.

(The numbers refer to section numbers in the book.)

CHAPTER 1. VOICES FROM THE SANDS

1. Epigraph: Simpson 1972, 305, edited.

1.1 Blacker and Loewe 1975; Oxford Bible 1992; Pritchard 1969.
The creation story begins the book of Genesis. God's design for the world: Psalm 104. Elihu chastises Job: Job 37:18. Ezekiel's vision: Ezekiel 1. "And about three thousand angels": Apocalypse of Paul, ch. 23, at Elliott 1993, 630. The sapphire pavement: Genesis 24:10. "No man shall see me": Genesis 33:20. "Why did he separate them?": Kasher 1953, 35. The City of God: Psalm 46:4, Isaiah 40:22. "One and the same fate": Ecclesiastes 9:2. "Many of those": Daniel 12:2. The *Mishnah:* quoted in Blacker and Loewe 1975, 72. It is generally assumed that "before" and "after" refer to time, though they may also refer to what lies outside the bubble.

1.2 Anon. 1992 (*Gilgamesh*); Blacker and Loewe 1975; Dalley 1989; Heidel 1949, 1951; Hesiod 1914; Horowitz 1998; Kramer 1981; Pritchard 1969; Redford 2001.
"After heaven had been moved": Kramer, 81. *Enuma elish:* Heidel. Tiamat and her progeny: Dalley, 228ff. Oannes: Verbrugghe and Wickersham 1996, 44. "First Chaos": Hesiod, *Theogony,* ll. 115, 147ff. "Sky and earth": Diodorus 1933, I.7. Rangi and Papa: Grey 1961, ch. 1. The Babylonian world: Horowitz, mostly chs. 1 and 11. Strange Babylonian sky: Kramer, 76. Map in the British Museum: Horowitz, ch. 2. "Whose roots reach a hundred leagues": Ibid., 245. "One day Death comes": Dalley, 107f, edited. I recommend David Ferry's excellent translation of *Gilgamesh,* Anon. 1992. Shield of Achilles: *Iliad,* bk. 18; see also Hesiod's description of the shield of Heracles: Hesiod 243. "He made on it": *Iliad* XVIII, l.606. "To misty Tartarus": Hesiod, *Theogony,* 46f; see also the *Iliad,* beginning of bk. 8. A god named Nun: Blacker and Loewe, 25. Geb and Nut: Ibid., 31.

1.3 Blacker and Loewe 1975; Hesiod 1914; Kongtrul and Tayé 1995; Williamson 1984.
Traditional Hindu cosmology: Blacker and Loewe, ch. 5. This Buddhist version: Kongtrul and Tayé, 108ff. The Chumash: Williamson; Grant 1965; Hudson et al. 1981; Hudson and Underhay 1978; Krupp 1984. Ceremonies to turn the sun northward: Williamson, ch. 5; Stephen 1936.

1.4 Heidel 1951; Oxford 1992; Pritchard 1969; Ryan and Pitman 1998.

Sumerian flood story: Pritchard, 42, translation edited. Gilgamesh flood story: Tablet XI, in Pritchard or Dalley 1989. Enlil couldn't sleep: Pritchard, 104. Biblical account: Genesis 6–8. Flood story told in Greece: Apollodorus 1975, I.46–48. Ovid's version: Bk. 1 of the *Metamorphoses*. "Immense lakes hidden underground": Seneca 1971, III.27–30. "Another catastrophe": Ryan and Pitman.

1.5 Nonnos 1940; Plato 1961 or Cornford 1937.

Phaethon's ride: Plato's version is in *Timaeus* 22c–e; the story without its astronomical coda is in book 2 of Ovid's *Metamorphoses*. There is an interesting book, de Santillana and von Dechend 1969, which finds traces of the twisted axle as well as the precession of the equinoxes in the myths of a wide range of cultures. The evidence is impressive, although the conclusions occasionally seem to outrun it. The poles are in the sky: Aristotle, *On the Heavens* 392a1.

CHAPTER 2. MANAGING THE WORLD

2. Epigraph: Epicurus 1926, 91, much edited.

2.1 Charlesworth 1983; Wise et al. 1996.

"Primitive man": Dickinson 1896, I.2. No other God: Exodus 20:3, rather flatly rendered in Oxford Bible. "They forsook the Lord": Judges 2. "When men began to increase": The recent Jewish translation, Anon. 1985, is clearer than the Oxford Bible and is used here. Augustine 1957, v.4, XV.23, interprets the text in an entirely different way. Nephilim in Canaan: Numbers 13:33. Only three Nephilim: Judges 1:20. Survival of Nephilim: Joshua 11:21 (where they are called Anakim). Demons in Babylonia: Jastrow 1911. Old lament: Jastrow 1911, 311, edited. Angel as manager: Samuel 24:13. Angel as executioner: Isaiah 37:36. Shouted for joy: Job 38:7. *Book of Enoch*: Charlesworth. *Book of Giants*: Wise et al., 246–50. Painful plastic surgery: 1 Maccabees 1:11; see also 2 Maccabees 4.

2.2 Hesiod 1914; The *Iliad*; Harrison 1903 or 1955; Russell 1981.

"Zeus and the other *daimones*": *Iliad* I.222. "Kindly, delivering from harm": Hesiod, *Works and Days*, l. 123. "In the history of human thought": Durkheim 1995, 36; where the translation says *profane* I have written *secular*. "There are seven": remodeled from various translations from the Akkadian, e.g., Foster 1996, v.2, 833. "*Keres* by the thousands": *Iliad* XII.326; see also Harrison, ch. 5. "All the air": Plutarch 1927, v.2, 176; the translation is from Murray 1935, 34. Excellent books on Satan: Russell 1977, 1981, 1984. Book of Zechariah: Zech. 3:1. His title and function: Oxford Bible, 511, n. 6.

2.3 Manetho 1940; Schaff and Wace 1890–1900.

"For one thing": Marcus Aurelius 1964, XII.28. "The Pythagoreans marvelled": Aristotle 1984, v.2, 2442. "When she wanes": Basil's *Hexaëmeron*, homily IX, in Schaff and Wace, v.8, 88. "It is Saturn's quality": Ptolemy *Tetrabiblos* I.4 in Manetho. Taurus was the Bull of Heaven: O'Neil 1986, ch. 12. "When the Pleiads": Hesiod, *Works and Days* ll. 383f.

2.4 Charlesworth 1983; Leichty 1970; Neugebauer 1957.

"When a yellow dog": Lindsay 1971, 2. "If a *suppu* sheep": Leichty, nos. 18.8', 20.3', 1.12. "God Enlil": Swerdlow 1998, 7. "Slowly perfected": Neugebauer 1975, v.1, 474ff. "Let your astrologers": Based on Oxford Bible 47:13, but edited so as to be closer to the Vulgate, which was written down when people still took these things seriously. Abraham sits up: Jubilees 12:16 = Charlesworth 1983, v.2, 81. "A special prison house": Charlesworth, v.1, 23. Pliny on Berossus's predictions: Pliny 1938, VII, 123. A Greek-speaking Chaldean named Berossus: Verbrugghe and Wickersham 1996. Theophrastus, the scholar: Proclus 1968, III.151. Strabo visited Babylon: Strabo 1917, XVI.1.6.

2.5 Plutarch, *Life of Marcellus*; Thucydides, *History of the Peloponnesian War*, bk. 7.

"When they had put all things": Plutarch 1631, 555; the whole chapter is saturated with omens. "They were, as the saying goes": the last words of Thucydides' *History*. Caligula's bridge: Cassius Dio 1924, LIX.17.1. Meet without smiling: Cicero, *On the Nature of the Gods* I.26. The world embodied the future: See Seneca, *Natural Questions* II.46 and his book *On Providence*. Augustine on astrology: Augustine 1957, V.1–5.

2.6 Betz 1986; Gager 1992; Jordan et al. 1999.

The Egyptian charm: from the London-Leiden papyrus, Griffith and Thompson 1904, 111. This is a very late production, probably dating from the early third century CE, but many of the spells it contains seem to be much older. God told Isaiah: Isaiah 8:1. Blood of Hephaistos: Faraone and Obbink 1991, 159. "Take a little shaving": Ibid., 105, edited. A man desires the death: Griffith and Thompson 1904, 145. "How was it that you": Dickie 2001, 51. "Lady Dimeter": Gager, 166, slightly edited; Jordan et al., 125; more *defixiones* in Betz and in Jordan et al. Tacitus reports: Tacitus 1925, *Histories* II.69. "As for the messengers": Reiner 1995, 98. Plato mentions: *Gorgias* 513a. "Suppose I buy": Aristophanes *Clouds* 749. Plutarch's essay: Plutarch 1927–69, XIII. King Sisebut: Isidore 1960, 330. See also Tupet 1976, ch. 7. Pliny's view of medicine: Beagon 1992, ch. 6. Eagerly absorbed by the Greeks: Pliny 1938, 30.2.

CHAPTER 3. GUESSWORK

3. Epigraph: Aristotle, *Eudemian Ethics*, 1216a11, quoted in Gershenson and Greenberg, 1. The passages from Aristotle quoted in this section will all be as translated by them. Anaxagoras was responding to a cliché of Greek tragedy (Sophocles, *Oedipus at Colonus* l. 1224.): The happiest fate would be never to have been born.

3.1 Gershenson and Greenberg (G&G) 1964; Kirk et al. 1983.

"The moon is made of earth": Kirk et al., 381. "Here lies Anaxagoras": Aelianus 1997, VIII.19. Plato mentions the rolls: *Apology*, 26d. Almost 70 references: G&G have collected 707 references to Anaxagoras in writings before the seventh century.

3.2 Herodotus 1921–24 or 1928; Kirk et al. 1983.

Carian language completely different: Herodotus I.142. Homer, writing in the eighth century at the start of the era of colonization, mentions "Carians of barbarous speech" in book 2 of the *Iliad*. "There would be a great harvest of olives": Aristotle, *Politics* 1259a9; for another version of the story, see Michel de Montaigne, Essay 24, "On Pedantry." Thales is said to have taught: Aristotle, *On the Heavens* 294a28. Most of what we are told about Thales's ideas comes through Aristotle, but Seneca, *Natural Questions* II.14 mentions the floating Earth and calls it an antiquated notion. "So they say nothing comes to be": Aristotle, *Metaphysics* 983b6, slightly edited. "All things are full of gods": Aristotle, *On the Soul* 405a19; 411b8. History of Miletus: Greaves 2002.

3.3 Freeman 1978; Kirk et al. 1983.
Anaximander's cosmos: Kirk et al., 134. "He says that it is neither": Simplicius of Cilicia, c. 530 CE: Kirk et al., 107. "We know nothing in reality: Freeman, Democritus fr. 117. "The source from which existing things": Freeman, Anaximander fr. 19, edited. "Anaximander said": Kirk et al., 141, edited. "He says that in the beginning": Ibid. "Man was originally": Ibid.

3.4 K. S. Guthrie 1987; Hopper 1938; Kirk et al. 1983.
"Don't poke the fire," etc.: Guthrie, 131. "One is that": Euclid 1925, v.2, beginning of bk. 7. "Then comes two": See Nicomachus 1926. History is built out of numbers: Hopper. Samuel Pepys learns the multiplication table: Pepys 2000, v.3, 131ff. "All other things seemed": Aristotle, *Metaphysics* 985b33. "Numbers occupy space": Ibid., 1080b17, 1083b12. "The whole universe is arranged": Sextus Empiricus, *Adversus mathematicos,* quoted at Kirk et al., 233, edited. The ratio 9/4 in the Parthenon: Dinsmore 1950, 161.

3.5 Diogenes Laertius 1925; Freeman 1978; Kirk et al. 1983.
"When you have listened," etc.: Freeman, 24–34. Related to the first two is Seneca's equally obscure remark that "eternity consists of opposites." "What I understood": Diogenes Laertius, II.22. "Set his own state in order": Plutarch 1927, v.14, "Reply to Colotes," 1126.32. "And remaining the same": Freeman, Parmenides fr. 8, edited. Short and confusing summary by Aëtius: W.K.C. Guthrie 1965, v.2, 61; Kirk et al., 258. For a brave effort to make sense of the Parmenidean cosmos, see Morrison 1955. "I met him when I was quite young": Plato, *Theatetus* 183e.

3.6 Freeman 1978; Gershenson and Greenberg (G&G) 1964; Kirk et al. (KRS) 1983.
"An immortal god": Empedocles fr. 112, from KRS. "From these elements": Fr. 107, Freeman, 63. "It is impossible": Fr. 12, KRS, 292, edited. "Here sprang up": Fr. 57, ibid., 303. "Many creatures were born": Fr. 61, Ibid., 304. The fittest: ibid., no. 380, 304. "A plant does not breathe": G&G, 79, from Aristotle, *On Plants* 816b26. "All things contain": G&G, 289. "The most insubstantial": Ibid., 285. "It is strange": Diels 1952, v.2, 231. "Our senses are weak": Frs. 21 and 21a, Freeman, 86, edited. Doesn't believe a word of it: Aristotle, *On Generation and Corruption* 325a1. "Sweet exists by convention": First half from Freeman, Democritus, fr. 9; second from KRS, 412. "The existence of nothing": Freeman, Democritus fr. 126. "To maintain that . . . is divisible": Aristotle, *On*

Generation and Corruption 325a9. Aristotle's theory of minimal particles: *Physics* 187b. "We know nothing": Freeman, Democritus fr. 117. Schiller enlarges these words in a graceful couplet in his *Sonnets to Confucius:*

> Nur die Fülle führt zur Klarheit,
> Und im Abgrund wohnt die Wahrheit.

(Only fullness leads to clarity / and truth dwells in the depths.) "Many bodies": KRS, 417. "Nothing happens": Freeman, Leucippus fr. 2.

3.7 Bailey 1926, 1979; Diogenes Laertius 1925; Lucretius 1968.
"A blessed and eternal being": Diogenes Laertius v.2, 663. In this boundless space: Epicurus, Letter to Pythocles quoted in Diogenes Laertius X.83–116. A swerve is possible: Lucretius II, ll. 218, 275ff. The interpretation I have given is not the only one possible. Epicurus says clearly: Letter to Pythocles cited above. Vitruvius on atoms: Vitruvius 1931, II.2.

CHAPTER 4. EARTH AND HEAVEN

4. Epigraph: *Metaphysics* 1026a16, edited.

4.1 Aristotle 1984; Granet 1958; Plato 1961.
"The sun will not transgress": Freeman 1978, Heraclitus, fr. 94. "The twelve-spoked wheel": Flaherty 1981, 77. "Everything happens": Freeman, Leucippus, fr. 2; Plato and Aristotle both reject: Plato, *Republic* 525; Aristotle, *Metaphysics* 1001b26. "What Pythagoras": Proclus 1989 (bilingual ed.), 6. "The Tao that can be told:" adapted from Lao Tsu 1972 and Waley 1942. *Tao-te:* Granet, 250. *Tao-te* rules the universe: Needham 1954, 2: 543, 578–83; Ronan 1981, ch. 16. "She takes her stand," and following quotations: Proverbs 8. "The breath of the power of God": Wisdom 7:25. Wisdom is God's wife: Philo 1929, v.2, 39. The creative part changed: Paul, 1 Corinthians 1:24; Origen 1989, bk. 1, pars. 115, 243, 246, and 249 (Origen is very diffuse). "She brought them through": Wisdom 10:18. "To the gods she is dear": Pritchard 1969, 428 (square brackets indicate restoration of missing symbols). "What remains, then": Cicero 1933, *On the Nature of the Gods* II.16.44f.

4.2 Neugebauer 1957 = 1962, 1975; O'Neil 1986.
Chinese astronomers of the Shang dynasty: Ronan 1981, v.2, 134, 183.

4.3 Cornford 1937.
"Socrates occupied himself": *Metaphysics* 1078b17. "The Good is not": *Republic* 509b. "Desiring that all things": *Timaeus* 30a,b, 31. "Eternal but moving" and following quotes start ibid., 37d. "They were the first to broach": Herodotus 1928 (Rawlinson trans.) II.23. "Space, which is everlasting": *Timaeus* 52b. "We must imagine": Ibid., 56c. "There still remained": Ibid., 55c. "The living universe itself": Aristotle, *On the Soul* 404b20, edited. "Thus if the Ideas": Syrianus, *Commentarius in Metaphysica* 159.5–160.3 quoted in Aristotle 1984, 2391.

4.4 Aristotle 1984; Dreyer 1953.

"The first principles": *Metaphysics* 982b2. "All things that are in motion": *Physics* 241b34. "Of no great size": *On the Heavens* 297b32ff. "The jostling of parts": Ibid., 297a19. The fifth element, Ether: Ibid., bk. I. "The shape of the heaven": Ibid., 286b10. Eudoxus's planetary spheres: Dreyer 1953, chs. 4, 5. 55 concentric material spheres: *Metaphysics* XII.8. The spheres and planets are made of ether: *On the Heavens,* II. Ether gives heat, life, and motion: *Meteorology* 340b9. "Something beyond the bodies": *On the Heavens* 269b13. Ether is divine: Ibid., 270b8. "Our forefathers": *Metaphysics* 1074b. "We think of the stars as mere bodies": *On the Heavens* 282a19. "Soul is the actualization": *On the Soul* 412a27. "If there is any way of acting": Ibid., 403a10. The part of the soul that is pure reason: *Metaphysics* 1070a26. These are natural motions: *Physics* VIII.4. "Everything that is in motion": Ibid., 241b34. Two-part trajectories: Santbech 1561. "There must necessarily be something eternal": Ibid., 258b10. "Produces motion by being loved": *Metaphysics* XII.7, *Physics* VIII.6. "There is neither place": *On the Heavens* 279a17. Sun and moon act to produce motion: *Physics* VIII.6. "We must assign causality": *Meteorology* 339a30. "God and nature": *On the Heavens* 284b. "The good and the beautiful": *Metaphysics* 1013a21. "In all things": *On Generation and Corruption* 336b27. "The more the universe seems comprehensible": Weinberg 1977, 154. "For thought, foresight": Cicero, *Tusculan Disputations* I.x, 22, quoted in Aristotle 1984, 2397.

CHAPTER 5. BEGINNINGS AND ENDINGS

5. Epigraph: The beginning of John Dryden's *Ode for Saint Cecilia's Day.*

5.1 Aristotle 1984.

"What is time?": Augustine, *Confessions* XI.14. "A moving image": *Timaeus* 37d. Time, says Aristotle: *Physics* IV. Aristotle argues: *On the Heavens* I.10. "The universal cause": *Meditations* VII.19. "Time is but the stream": *Walden,* end of chapter 2 = Thoreau 1971, 98. "The Pythagoreans . . . construct": *Metaphysics* 1080b19. "The Pythagoreans held": *Physics* 213b22, revised from Aristotle 1984 with the help of Sachs 1995, 108. "Space, which is everlasting": *Timaeus* 52b. "The place of a thing": Ibid., IV.4. "Space is not the limiting surface": Duhem 1913–59, v.1, 317.

5.2 Aristotle 1984; Kirk et al. 1983; Maimonides 1963.

"He explains the creation": Kirk et al. 129, no. 119, much edited. "A kind of sphere of flame": Ibid., 131, no. 121. "From the Boundless": Ibid., 107. "Thy almighty hand": Wisdom 11.17. "The air was lighted up": Schaff and Wace 1895, v. 8, Homily IX, 2.7ff. The Old Latin version: Fanciers of Latin may wish to compare this text with the Vulgate of Saint Jerome, which was coming out at about the time Augustine wrote but was not immediately welcomed and accepted. The Old Latin of the first three chapters of Genesis is in Augustine 1982, v.2, 325. "Before [God] formed": Augustine 1912, XIII.8. "It is one thing": Augustine 1968, 143. For this purpose, certain points: Maimonides, 346 = II.66a; see also 7 = I.5a. (The first number refers to the quoted translation; the second to the standard Arabic text, Munk 1856–66.) At least eighteen ethereal spheres:

Ibid., 185 = I.99b. Maimonides argues in another place: Ibid., 259f = II.15b. The spheres are angels: Ibid., 266 = II.19a.

5.3 Aristotle 1984; Kirk et al. 1983.
Anaximander talked of cycles: Kirk et al., 108. Pythagoras went further: Guthrie 1987, 126. Heraclitus believed in exact repetition: Aristotle, *On the Heavens* 279b15. Plato takes up the story: *Critias* 108ff; *Laws* 676–80; see also Aristotle *Metaphysics* 1074b; fr. 25. Greatest year: Aristotle, 2395. Ptolemy on exact alignments: *Tetrabiblos* in Manetho 1940, I.2. Yugas, Kalpas, etc.: put together from many passages in Biruni 1910. A fourth-century commentator: Kasher 1973, v.1, 73. "It is both impious and absurd": Origen 1966, 238f. Successive worlds do not repeat: Duhem 1913–59, v. 2, 449. "There shall be a new heaven": Isaiah 66:22. "What is it that hath been?": Ecclesiastes 1:9. Jerome on Ecclesiastes: Origen 1966, 239, fn. 1. Augustine's ready answer: Augustine 1957, XII.14. "Of old has thou laid": Psalm 102:25; also Isaiah 51:6 and Paul's letter to the Hebrews 1.11.

5.4 Cohn 1993.
"The universe itself": Sallustius 1966, 13. "Fate will flatten mountains": Seneca 1923, v.3, "Ad Marciam de consolatione," XXVI.6. "The Lord's anger": Isaiah 34:2. "Age after age": Ibid., 34:10. "The first earth had vanished": Revelation 21. "One-year-old children": Charlesworth 1983–85, v. 2, 576. "A great river of blazing fire": Ibid., v.1, 350, edited. "Springs of wine": Ibid., v.1, 353. Not to have been born: Sophocles, *Oedipus at Colonnus* l. 1224, but it was a cliché at the time. Quotations from the Koran are from the Penguin edition, Anon. 1997. The ancient tradition (Rapture as early as the third century): B. E. Daley, S.J., "Apocalypticism in early Christian theology," in McGinn et al. 1999, v.1, 140–78. Nineteenth-century apocalypticism: P. Boyer, "The growth of fundamentalist apocalyptic in the United States," Ibid., v.2, 3–47.

CHAPTER 6. PHILOSOPHY CONTINUED

6. Epigraph: Aristotle, *Generation of Animals* 760b29.

6.1 Dreyer 1906 or 1953; Heath 1913, 1897 or 1953; Plutarch 1927–69; Ptolemy 1984, 1996.
Important books: Russo 1996. Hellenistic science and mathematics: Russo 1996. Cleanthes the Stoic: Plutarch, On the Face in the Moon, v.12, 922f. Plutarch says in another place: Platonic questions, Ibid., v.13.1, 1006. "Now the first cause": Ibid., 85. Ptolemy mounts them on spheres: Ptolemy 1984, 46. 34 spheres: Neugebauer 1975, v.2, 926.

6.2 Nicomachus 1926, Ptolemy 1980.
"We must assign causality": Aristotle *Meteorology* 339a30. "The weakness and unpredictability": Ptolemy I.1. "A certain power": Ibid., I.2 (Manetho, much edited.) Things of our world: Ibid., I.2. In another place: Ibid., XXIX.15. "Comets generally foretell": Ibid., II.13 (comets have occasionally been seen to break up as they approach the Sun). "The so-called beams": Ibid., II.9. "The qualities of

number": Plato, *Republic* 525b. "Nature is everywhere": Aristotle, *Physics* 252a13. The pattern of the universe: Nicomachus, 189. "The superabundant": Ibid., 207.

6.3 Dijksterhuis 1961 or 1986; Flint 1991; Pharr 1952; Thorndike 1923–58, vols. 1–3.

"Gave our communities": Plato, *Laws* 713c. Suetonius says: Suetonius 1998, v.1, 196. "A class of men": Tacitus 1925–37, Histories I.22. Constantine's edicts: Pharr, 237. "I judge that divination": Pharr, 238. *Daimones* have a necessary function: "On the disappearance of Oracles," in Plutarch 1927–69, v.5, 13. Augustine on twins leading different lives: *City of God* V.1–7. A rich and cultivated man: *Confessions* VII.6. *The Divination of Demons*: Deferrari 1955; see also *City of God* VIII.15. Thomas Aquinas on demons: *Summa theologica* (Aquinas 1945, v.1), questions 63 and 109. The merchant from Hamburg: Kieckhefer 1998, 232.

6.4 Brehaut 1912; Brown 1971; Grosseteste 1942, 1982; Isidore 1911, 1960; Lindberg 1988; McEvoy 1982; Plato 1961; Thompson and Johnson 1937; Thorndike 1923.

Cassiodorus's *Introduction to Divine and Human Readings*: Cassiodorus 1946. "The number, the strength": Gibbon, *Decline and Fall,* XXVI. "At the hour of midnight": Ibid., XXXI. *Etymologies*: English abridgment in Brehaut; Latin in Isidore 1911. "*Mundus*": Brehaut III.29.1. Babel: Ibid., IX.1.1. *De rerum natura*: Isidore 1960. "The world is formed this way": Ibid., 206. "As to the equatorial zone": Ibid., 211. "The sphere of heaven": Bits from Brehaut III.32.1, 40.1, 52.1. "The world is formed": Isidore 1960, 206. "As to the equatorial zone": Ibid., 211. "The sphere of heaven": Put together from *Etymologies* III.32.1, 40.1, 52.1.

6.5 Brehaut 1912; Dronke 1988; Gregory 1955; Isidore 1960; Pullman 1998.

"Philosophers call": Brehaut XIII.2, edited. "The *Kalam* is an entity": *Encyclopedia of Islam,* v.4, 470. Thierry of Chartres: Gregory, 182. "The smallest and simplest: Gregory, 204. "An element is what is found": Dronke, 312. "True elements are things": Dronke, 316. "We condemn and excommunicate": Lasswitz 1890, 73, fn. 2; also Gregory, 210. "This disease of curiosity": Augustine, *Confessions* X.35. Buridan's theory of *impetus*: Clagett 1979, 534. "Since the Bible": Ibid., 536.

6.6 Brehaut 1912; Coopland 1952; Hansen 1985; Hopkins 1981; Lovejoy 1936 or 1960; Thorndike 1923–58, vols. 2, 3.

Timaeus proposes: Cornford 1937, 40. "Nature proceeds": Aristotle, *History of Animals* 588b4. "Since Mind emanates": Macrobius 1952, 1.14. For the chain of being in Islam, see Nasr 1964, 70. A golden chain: *Iliad* 8:19. "A universe in which": Lovejoy, 77. Augustine on Noah's ark: Augustine 1957, XV.26. "Pythagoras said": K.S. Guthrie 1987, 139. "Man is a little World": Freeman 1978, 99, fr. 34. "The idea appeals to me": Seneca 1971, 233. "Its true perfection": Munk 1859, 364f, quoted in Nasr 1964, 259, fn. 77. Bodily fluids known as humors: Hippocrates 1978; Riesman 1936, ch. 29. "The world is actually": Isidore 1960, IX.1. Table 6.1: Siegel 1968, ch. 4. Table 6.2: Riesman 1936, 101.

Cities and their stars: Ptolemy's *Tetrabiblos,* bk. II, in Manetho 1940; App. 1 in Abu Ma'sar 2000. "The sky": Isidore XII.1. "The earth is not": Hopkins II.11. "[If a man]": Ibid., II.11, 117. "We surmise": Ibid., II.12, 120.

INTERLUDE. THE WORLD MAP

I. Epigraph: Parry 1968, 87.

I.1 Bunbury 1883; Kimble 1938.
"Flows back on itself": *Iliad* XVIII, l.399; Hesiod 1914, *Theogony* l. 776. "When Odysseus went": *Odyssey* XI. "He sleeps peacefully": Mimnermus, Anon. 1929, 8, trans. C. Park. "The pygmy men": *Iliad* III, l.5. "The flat extremities of the earth": Tacitus 1914, 12.

I.2 Cary and Warmington 1963; Herodotus 1921–24; Kimble 1938; Pliny 1938; Strabo 1917–32.
"There are trees": Herodotus III.106. "The Chinese": Pliny VI.20. The historian Ammianus Marcellinus, fourth century CE, has little to add to the account written by Herodotus eight centuries earlier; see Ammianus 1935, XXIII.6.

I.3 Cosmas 1897; Lactantius 1964.
"To investigate": Lactantius III.3. "Is there anyone so stupid": Lactantius III.24. Augustine on the round earth: *City of God* XVI. 9. Cosmas supplies drawings: Bk. XI.

I.4 Herodotus 1921–24; Mandeville 1900; Pliny 1938; Rogers 1962; Solinus 1587 (the only English translation); Yule 1913–15.
"In this desert": Herodotus III.102. "Winged horses": Pliny VIII.30. "There are some people" and the following marvels: Ibid., VII.2. Desdemona's eyes: *Othello* I.3. "People remarkable for having only one eye": Pliny VII.1. Gold-digging ants: Solinus XLII. "A beast called Alce": Ibid., XXXI, language modernized. "Many wondrous things": Ibid., XXXIX. Augustine on marvels: *City of God* XVI.9. For the narratives of Carpini, Ruysbroeck, and Pordenone, see Mandeville, 213ff. For Ruysbroeck's, see also Rubruck 1990. For other explorations, Yule. "Wild men are certainly reported": Mandeville, 223. "Certain monsters": Mandeville, 233. "In that country be folk": Mandeville, 105, fig. 106b. More from Mandeville: snails, 129; geese, 132; headless folk, 134; 72 provinces, 180. "Prester John, most powerful lord": Rogers 1962, 127. "With the favor of divine clemency": Rogers 1962, 139. As to how the amassadors fared in Abyssinia, see the account written by the priest who accompanied the mission as its chaplain, Álvares 1961. "Giving many blows to the people": Parry 1968, 86.

I.5 Estensen 1998; Jane 1930–33; Hildebrand 1924; Morison 1942; Pigafetta et al. 1874.
"Eastward in Eden": Genesis 2:8. Columbus: Morison. "Sent two men inland": Jayne, v.1, 5. Fernão de Magalhães: Hildebrand. "The Clowds near the horizon": Beaglehole 1974, 365. "Ropes like wires": Ibid., 362.

CHAPTER 7. TOWARD A NEW ASTRONOMY

7. Epigraph: Galileo 1968, v.6, 232.

7.1 Armitage 1947; Copernicus 1972–85; Dreyer 1953; Koestler 1959; Kuhn 1957.
Koestler's book is a novelist's carefully researched account of Copernicus, Kepler, and Galileo. I suggest that it be read along with the review by G. de Santillana and S. Drake in *Isis* 50, 255 (1959). *Commentariolus:* Copernicus, v.3. "The people gave ear": These words are reported in various forms; see, for example, Kuhn 191. *On the Revolutions of the Celestial Orbs:* Reproduction of the manuscript: Copernicus, v.1; English translation: Ibid., v.2. "We hold it as a fact": Ibid., 16. Ptolemy's claim that a rotating earth: Ptolemy 1984, 44f. "Are deeply offended": Copernicus, v.2, xx. 34 spheres: Neugebauer 1975 v.2, 926. "Only the earth": Davies 1947, no. 51. "And new philosophie": Donne 1941, 202.

7.2 Caspar 1993; Dick 1982; Galileo 1968 (EN); Kepler 1937–75 (GW), 1992. EN stands for the *Edizione nazionale,* GW is Kepler's *Gesammelte Werke.*
"I wasted a great deal of time": Kepler 1981, 63. "The Earth is the circle": Kepler 1981, 69. *Epitome of Copernican Astronomy:* Kepler GW, VII. "The bare and solitary power": Kepler 1992, 407. For a careful discussion of Kepler's "physical astronomy," see Stephenson 1987. Ancient optical theory: Park 1997. Kepler's third law: Kepler 1997, 411.

7.3 S. Drake 1957; Galileo 1968 (EN); Kepler 1965 (GW); Langford 1971; Reston 1994.
Siderial Messenger, also known as *Starry Messenger:* Drake, 23–58. "Might almost be called": *De republica* VI.18 in Cicero 1928. Galileo's letter to the astrologer: EN XI, 112. *Conversation with Galileo's Sidereal Messenger:* Kepler 1965. "Our moon exists for us": Kepler 1965, 42. "False and damnable," from a letter on sunspots written in 1612: Drake, 137. Galileo's trial: Langford. "Once I measured heaven":

> Mensus eram coelos, nunc terrae metior umbras.
> Mens coelestis erat, corporis umbra jacet.
> > From Caspar 1959, 359.

7.4 Descartes 1964; Gaukroger 1995; translations mentioned below.
No person should either hold or express: Partington 1961–70, v.2, 459. "I think": from *Discourse on the Method,* in Descartes 1997, 92. "Having remarked": Ibid.; *Principles of Philosophy:* The modern translation used here is Descartes 1983. "But what do we know": Descartes 1964, v.5, 168. "He was wrong": Voltaire 1931, 14th letter.

7.5 Alexander 1956; Densmore 1995; Dobbs 1975; Inwood 2003; Newton 1952, 1959, 1999; Westfall 1980.
Biography: Westfall on Newton, Inwood on Hooke. "The vortices of the sun and planets": Newton 1959, 368. "The matter of the heavens is fluid": Koller-

strom 1999, 332. The laws of motion: Newton 1999, 416f. Derivation of the inverse-square law of gravity: Newton 1999, 462. The law of gravity: Ibid., 802. "A supernatural thing": Alexander, 43. "I have not as yet": Newton 1999, 943. "He is omnipresent": Ibid., 941. "By his will": Newton 1952 (*Opticks*), 403. "What hinders the fixed Stars": Ibid., query 28. "Until this system": Ibid., query 31. "Nay, the machine of God's making:" Alexander, 11. "The case is quite different": Ibid., 14. Newton on the cause of gravity: Newton 1959, v.3, 338. "This most elegant system": Newton 1999, 940. "A god without dominion": Ibid., 942, first appears in the second edition, 1713. "What though in solemn silence": Quiller-Couch 1939, no. 444.

CHAPTER 8. WHAT IS THE WORLD MADE OF?

8. Epigraph: Newton 1934, 398: Rules for Philosophizing, no. 1.

8.1 Boyle 1965; Descartes 1983, 1997; Joy 1987; Partington 1961–70; Pullman 1998.
Unheard-of questions: Mersenne 1634 or 1985. "It may be supposed": Gassendi 1972, 400, much cut and edited; see also Sambursky 1974, 253f.

8.2 Boas 1958; Cobb and Goldwhite 1995; Partington 1961–70; Principe 1998; Roberts 1994; Taylor 1949.
Jabir ibn-Hayyan: Taylor 1949, 78–85. Paracelsus: Paracelsus 1951.

8.3 Densmore 1995; Dobbs 1975; Newton 1952, 1959, 1999; Westfall 1980.
"Observe the products": Newton 1959, v.1, 9. Newton was deeply into alchemy: Dobbs; Westfall, ch. 8. "If Gold could once": Newton 1958, 258. "All matter duly formed": B.J.T. Dobbs, "Newton's *Commentary on the Emerald Tablet*," in Merkel and Debus 1982. (The *Emerald Tablet,* a document precious to alchemists, is one of the so-called Hermetic texts, written 100 to 300 CE and attributed to a thoroughly mythical character called Hermes Trismegistus [Thrice-Greatest]. For text and commentary, see Taylor 1949, 89.) "Though it be mentally": Boyle 1672, v.3, 27ff, from a book *The Origin of Forms and Qualities According to the Corpuscular Hypothesis,* written in 1666.

8.4 Brush 1976; Partington 1961; Perrin 1910. Pullman 1998.
"A uniform and blind attraction": From Maupertuis 1984, quoted in Pullman, 147. "Enslave the mind": Wordsworth, *The Prelude,* XIII. 141. Dalton on oxides of nitrogen: Partington, v.3, 792. Young measures a water molecule: Young 1855, v.1, 454–83. Ostwald's text: Ostwald 1890; biography by E. H. Hiebert and H.G. Körner in DSB v.15, suppl., 1978. Einstein biography: There are many; I recommend Frank 1947. A second paper by Einstein: Einstein 1987, English trans., v.2, 123–34. Perrin's observations and conclusions: Perrin.

8.5 Buchwald 1989; Kuhn 1978; Newton 1730.
"The superior glory": Aristotle, *On the Heavens,* 269b. "Light is emitted": from the last paragraph of Newton's *Principia.* "On the theory of light and colors":

Young 1855, 140–69. Maxwell's long paper: Maxwell 1865. "The agreement of the results": Ibid., 580. "We know not": Young 1807, 610. *The Unseen Universe:* Anon. 1875. "Not grossly material": Ibid., 161. "A large portion": Ibid., 158. "Maxwell's theory is": Hertz 1893, 21. "When a light ray": Einstein 1987, v.2, 87. De Broglie now proposed: de Broglie 1925.

CHAPTER 9. THE UNIVERSE MEASURED

9. Epigraph: Wisdom 11:20.

9.1 Herschel 1785; Pannekoek 1961; Van Helden 1935; Wright 1971.
Hipparchus measures the distance to the moon: Van Helden, 11. The solar parallax: Ibid., chs. 12, 13. Wright proposes a model: Wright, 63f. "It is here": Ibid., 170. "An intelligent principle": Ibid., 168. "Whitish clouds": Galileo 1957, 49. Kant's ideas about nebulae and the solar system: Kant 1900. Herschel biography: DSB. Figure 9.3: Herschel. A scale for the picture: Ibid. "Probably all the stars": Ibid. Bessel biography: DSB.

9.2 Cutler 2003; Haber 1959; Hutton 1970; Roger 1997a; Steno 1968.
Annals of the Old Testament: in Ussher 1658. "I do not see": da Vinci 1938, v.1, 351. *The Prodromus:* Steno. "Above all": quoted by Roger, 101, from Buffon 1744, v.1, 203. Earth's rate of cooling: Roger 409, from Buffon Suppl. I (1774), 157–58. Hutton's *Theory of the Earth:* Hutton. "If the succession of worlds": Hutton, 128. "Certainly no conclusion": Deluc 1809. Between 20 and 400 million years ago: Thomson 1882–1911, v.2, 295–311; result on 300. "Within a finite period": Thomson 1882, v.1, 511. "It would, I think": Thomson 1889–94, v.1, "On the Sun's Heat," 369–422, quote from 390.

9.3 Browne 1995, 2002; C. Darwin 1839, 1859–1966, 1871, 1958, 1977; Fichman 2004; Irvine 1955; Lyell 1830; Raby 2001; Roger 1997b; Secord 2000.
C. Darwin biography: Browne. A. R. Wallace biography: Raby, Fichman. Maupertuis: DSB; Roger; Maupertuis 1980, pt. I, chs. 17f. *Zoonomia:* E. Darwin 1803, 395ff. *Vestiges:* Chambers 1994. "We have seen powerful evidence": Chambers 1994, 153f. "His organization": Ibid., 199. "Whole races became extinct": Miller 1852, 229. *On the Origin of Species:* Darwin 1859–1966 and later editions. *Principles of Geology:* Lyell 1830 and later editions. "Never was there a dogma": Lyell, ch. 13. "There are no less": Darwin 1839, ch. 17. *Essay on Population:* Malthus 1976 is a good edition. "Suppose that in a certain spot": Darwin 1977, vv.2, 5. Birds do better: Grant and Grant 1989, 280. Darwin's and Wallace's simultaneous papers: Darwin 1977, 3–18. "Pleaded pathetically": Irvine 1955, 107. "And it is since that time": Darwin 1958, 93. "There can hardly be": *Descent of Man,* ch. 6 (Darwin 1915, 156). Becquerel: see DSB. The new catastrophism: Ryder et al. 1996.

CHAPTER 10. THE EXPLODING UNIVERSE

10. Epigraph: From Paul Valéry's poem *L'Ebauche d'un serpent* (Valéry 1944, 167) in which a snake, *the* snake, speaks: Sun, Sun. . . . You dazzling mistake! . . .

You keep hearts from knowing that the universe is only a defect in the purity of Nonexistence!

10.1 Einstein 1920, 1987; Kogut 2002; Lorentz et al. 1923.

"On the electrodynamics": Lorentz et al., 37, or Einstein 1987 (English translation) v.2, 140. Announcement of the general relativity theory: Einstein 1915; expanded version: Lorentz et al., 109; Einstein 1987 (English trans.) v.6, 146. "Cosmological considerations": Ibid. 421; Lorentz et al., 177. The radius *R*: Einstein 1987, v.6, 552, fn. 16. Friedmann's calculation: Friedmann 1922. Slipher's observations: Slipher 1914, 1915.

10.2 Barrow 1994, 2003; Greene 2003, 2004; Guth 1997; Rees 1997, 2001; Silk 1994.

Eddington's proof: Eddington 1930. "The expanding universe": Lemaître 1931a. "I would [] be inclined": Lemaître 1931b. Inflationary theory: Barrow, Rees, Silk, and, especially, Guth. Inflation answers several technical questions: See, for example, Silk, ch. 4. Gamow's deduction: Gamow 1946, 1948. Alpher and Herman: Alpher and Herman 1948. Scientists at the Bell Laboratories: Penzias and Wilson 1965. For more on CBR, see Lemonick 2003. Laplace's nebular hypothesis: Laplace 1984, 548. Immanuel Kant had a similar idea in 1755 (Kant 1900), but it was little noted.

10.3 Kirshner 2002; Rees 1997, 2001; Rubin, V. 1997; Silk 1994.

Dark matter: Rubin. Dark energy: Kirshner. Table of the principal contents of the universe: Bennett et al. 2003.

Chapter 11. The View from Here

11. Epigraph: quoted in Heisenberg 1971, 60. Thomas Wright's estimate: Wright 1750, 76.

11.1 Drake and Sobel 1992; Goldsmith and Owen 1980; Koerner and LeVay 2000; Shapiro 1999.

Discovery of planets around nearby stars: Mayor and Frei 2003. Estimates of the habitable zone around a star: Kasting et al. 1993. Metrodorus, fourth century BCE: Freeman 1978. The practical impossibility of space travel: Pierce 1959. The first serious SETI proposal: Cocconi and Morrison 1959. More on SETI: McConnell 2001; Ekers et al. 2002.

11.2 Barrow and Tipler 1986; Davies 1982; Rees 1997, 2001.

A force that is barely strong enough: Dyson 1971 = Dyson 1992, ch. 12. More on the anthropic argument: Barrow and Tipler; Davies; Rees 1997, 2001. Several other coincidences: Davies.

11.3 Adams and Laughlin 1997; Dyson 1979; Ryder et al. 1996.

Scientists are being entertained: Caldwell et al. 2003. Evidence of periodic mass extinctions: M. R. Rampino and B. M. Haggerty, "Impact Crises and Mass Extinctions: A Working Hypothesis," in Ryder et al. 11-30; see also M. R.

Rampino, "Role of the Galaxy in Periodic Impacts and Mass Extinctions on the Earth," in Koeberl and MacLeod 2002, 667–78. Computer simulations: Chown 2004. A force of half a pound: Schweickart et al. 2003. For more speculations on the universe's ultimate future, see Ellis 2002.

   11.4 McDonald and Jehl 2003; Wilson 2002.
   "The silence": B. Pascal, *Pensées* III.206. A recent survey: from www.wcmc .org.uk/protected_areas/data/summstat, courtesy of Kai N. Lee. Ozone hole etc.: www.theozonehole.com Ogallala aquifer: D. Jehl in McDonald and Jehl, xv. Half the world's wheat: L. R. Brown, ibid. 79. Facts, figures, and proposals for retarding environmental degradation: Wilson. Patriarch Bartholomew: *Christian Science Monitor*, 24 July 2003, 17. The Hopi Earth Mother: B. Tolkien in Capps 1976, 14.

Dictionary of Scientific Biography (DSB). 1970–90, 18 vols. Ed. C. C. Gillispie. New York: Scribner.

The Encyclopedia of Islam. 1960–. 11 vols. Ed. H.A.R. Gibbs et al. Leiden: Brill.

Abu Ma'sar. 2000. *On Historical Astrology.* 2 vols. Trans. K. Yamamoto and C. Burnett. Leiden: Brill.

Adams, F. C., and G. Laughlin. 1997. "A Dying Universe: The Long-term Fate and Evolution of Astrophysical Objects." *Reviews of Modern Physics* 69, 337–72.

———. 1999. *The Five Ages of the Universe.* New York: Free Press.

Aelianus, C. 1997. *An English Translation of Claudius Aelianus'* Varia Historia. Trans. D. O. Johnson. Lampeter, Wales: Edwin Mellen.

Alexander, H. G. 1956. *The Leibniz-Clarke Correspondence.* Manchester, U.K.: University of Manchester Press.

Alpher, R. A., and R. Herman. 1948. "Evolution of the Universe." *Nature* 162, 774–75.

Álvares, F. 1961. *The Prester John of the Indies.* 2 vols. Trans. Lord Stanley of Alderly, rev. and ed. C. F. Beckingham and G.B.W. Huntingford. Cambridge, U.K.: Hakluyt Society.

Ammianus Marcellinus. 1935–39. *Ammianus Marcellinus.* Trans. J. C. Rolfe. London: Heinemann.

Anon. [Chambers, R.] 1844. *Vestiges of the Natural History of Creation.* London: Churchill. Many later editions, most recently 1994, ed. J. A. Secord. Chicago: University of Chicago Press.

Anon. [Balfour Stewart with the help of Peter Guthrie Tait] 1875. *The Unseen Universe.* New York: Macmillan.

Anon. 1929. *Selections from the Elegiac, Iambic, and Lyric Poets.* Cambridge, Mass.: Harvard University Press.

———. 1985. *Tanakh: A New Translation of the Holy Scriptures.* Philadelphia: Jewish Publication Society. 2nd ed., 1999.

———. 1992. *Gilgamesh.* Trans. D. Ferry. New York: Farrar Straus Giroux.

———. 1997. *The Koran.* 5th rev. ed. Trans. N. J. Dawood. London: Penguin.

Apollodorus. 1975. *The Library of Greek Mythology.* Ed. and trans. K. Aldrich. Lawrence, Kans.: Coronado Press.

Apuleius (Apulée). 1973. *Opuscules philosophiques.* Ed. and trans. E. Beaujeu. Paris: Les Belles Lettres.

Aquinas, Saint Thomas. 1945. *Basic Writings of Saint Thomas Aquinas,* 2 vols. Ed. A. C. Pegis. New York: Random House.

Aristotle. 1928. *Meteorologica.* Trans. H.D.P. Lee. Cambridge, Mass.: Harvard University Press.

————. 1984. *The Complete Works.* 2 vols. Ed. J. Barnes. Princeton, N.J.: Princeton University Press.

Armitage, A. 1947. *Sun, Stand Thou Still.* New York: Schuman.

Augustine, Saint. 1912. *St. Augustine's Confessions.* 2 vols. Trans. W. Watt. London: Heinemann.

————. 1957–72. *The City of God against the Pagans.* 7 vols. Trans. G. E. McCracken et al. London: Heinemann.

————. 1968. *De Trinitate libri XV.* 2 vols. Ed. W. J. Mountain. Turnhout, Belgium: Brepols.

————. 1982. *The Literal Meaning of Genesis.* 2 vols. Trans. J. H. Taylor. New York: Newman Press.

Bailey, C. 1926. *Epicurus: The Extant Remains.* Oxford: Clarendon Press. Repr. Hyperion, Westport, 1979.

————. 1928. *The Greek Atomists and Epicurus.* Oxford: Clarendon Press.

Ball, C. J. [1899]. *Light from the East.* London: Eyre and Spottiswoode.

Barrow, J. D. 1994. *The Origin of the Universe.* New York: Basic Books.

————. 2000. *The Book of Nothing.* New York: Pantheon.

————. 2003. *The Texture of the Cosmos.* New York: Knopf.

Barrow, J., and F. Tipler. 1986. *The Anthropic Cosmological Principle.* Oxford: Clarendon Press.

Beagelhole, L. C. 1974. *The Life of Captain James Cook.* Stanford, Calif.: Stanford University Press.

Beagon, M. 1992. *Roman Nature.* New York: Oxford University Press.

Bennett, C. L., et al. 2003. "First-Year *Wilkinson Microwave Anisotropy Probe (WMAP)* Observations: Preliminary Maps and Basic Results. *Astrophysical Journal Supplement Series* 148, 1–28.

Betz, H. D., ed. 1986. *The Greek Magical Papyri in Translation.* Chicago: University of Chicago Press.

Biruni, M. 1910. *Alberuni's India.* 2 vols. Trans. and ed. E. C. Sachau. London: Kegan Paul, Trench, Truebner.

Blacker, C., and M. Loewe, eds. 1975. *Ancient Cosmologies.* London: Allen and Unwin.

Boas, M. 1958. *Robert Boyle and Seventeenth-Century Chemistry.* Cambridge, Mass.: Cambridge University Press.

Boyle, R. 1661. *The Sceptical Chymist.* London: Crooke. Repr. Dawson's, London, 1965.

————. 1672. *The Works.* 2nd ed., 6 vols. London: Rivington et al. Repr. Olms, Hildesheim, Germany, 1965–66.

Brehaut, E. 1912. *An Encyclopedist of the Dark Ages: Isidore of Seville.* New York: Columbia University Press.

Broglie, L. de. 1925. *Recherches sur la théorie des quanta. Annales de physique* 3, 22–128.

Brown, P. 1971. *The World of Late Antiquity.* London: Thames and Hudson. Repr. Harcourt Brace Jovanovich, New York, 1974.

Browne, J. 1995. *Charles Darwin: Voyaging.* New York: Knopf.

————. 2002. *Charles Darwin: The Power of Place.* New York: Knopf.

Brush, S. G. 1976. *The Kind of Motion We Call Heat*. Amsterdam: North-Holland. Repr. 1986, 1996.

Buchwald, J. Z. 1989. *The Rise of the Wave Theory of Light*. Chicago: University of Chicago Press.

Buckland, W. 1820. *Vindiciae Geologicae, or the Connexion of Geology with Religion Explained*. Oxford: University Press.

Buffon, G. 1749–67. *Histoire naturelle*. 44 vols. Paris: Imprimérie Royale.

———. 1962. *Les Epoques de la nature*. Ed. J. Roger. Paris: Editions du Muséum.

Bunbury, E. H. 1883. *A History of Ancient Geography*. 2 vols. London: Murray.

Caldwell, R., M. Kamiankowski, and N. N. Weinberg. 2003. "Phantom Energy: Dark Energy with $w < -1$ Causes a Cosmic Doomsday." *Physical Review Letters* 91, 071301.

Capps, W. H., ed. 1976. *Seeing with a Native Eye*. New York: Harper and Row.

Cary, M., and E. H. Warmington. 1929. *The Ancient Explorers*. London: Methuen; 2nd ed. Pelican, London, 1963.

Caspar, M. 1959. *Kepler*. Trans. C. D. Hellmann. London: Abelard-Schuman. Repr. Dover, New York, 1993.

Cassiodorus. 1946. *An Introduction to Divine and Human Readings*. Trans. L. W. Jones. New York: Columbia University Press.

[Chambers, R.] 1844. *Vestiges of the Natural History of Creation*. London: Churchill. Many later editions, most recently 1994, ed. J. A. Secord. Chicago: University of Chicago Press.

Charlesworth, J. H., ed. 1983–85. *The Old Testament Pseudephigraphia*. 2 vols. Garden City, N.Y.: Doubleday.

Chown, M. 2004. "Chaotic Heavens." *New Scientist,* 28 Feb. 2004, 32–35.

Cicero. 1928. *De re publica & de legibus*. Ed. and trans. C. W. Keyes. London: Heinemann.

———. 1933. *De natura deorum & Academica*. Trans. H. Rackham. London: Heinemann.

Clagett, M. 1979. *The Science of Mechanics in the Middle Ages*. 2nd ed. Madison: University of Wisconsin Press.

Cobb, C., and H. Goldwhite. 1995. *Creation's Fire*. New York: Plenum.

Cocconi, G., and P. Morrison. 1959. "Searching for Interstellar Communications." *Nature* 184, 844–46.

Cohn, N. 1993. *Cosmos, Chaos, and the World to Come*. New Haven: Yale University Press.

Coopland, G. W. 1952. *Nicole Oresme and the Astrologers*. Cambridge, Mass.: Harvard University Press.

Copernicus, N. 1972–85. *Complete Works*. 3 vols. Trans. E. Rosen. London: Macmillan (vols. 1, 3); Baltimore: Johns Hopkins University Press (vol. 2).

Cornford, F. M. 1937. *Plato's Cosmology*. London: Routledge and Kegan Paul. Repr. Library of Liberal Arts, Bobbs-Merrill, Indianapolis, Ind.

Cosmas (Indicopleustes). 1897. *The Christian Topography*. Ed. J. W. McCrindle. London: Hakluyt Society.

Cutler, A. 2003. *The Seashell on the Mountaintop*. New York: Dutton.

Dalley, S., trans. 1989. *Myths from Mesopotamia*. Oxford: Oxford University Press.

Darwin, C. 1839. *The Voyage of the Beagle*. New York: Harper. Repr. Harper, 1959.

———. 1859. *On the Origin of Species by Means of Natural Selection*. London: Murray. Repr. Harvard University Press, Cambridge, Mass., 1966.

———. 1871. *The Descent of Man and Selection in Relation to Sex*. London: Murray.

———. 1915. *The Descent of Man and Selection in Relation to Sex*. 2nd ed. New York: Appleton.

———. 1958. *The Autobiography of Charles Darwin*. Ed. N. Barlow. London: Collins.

———. 1977. *The Collected Papers of Charles Darwin*. 2 vols. Ed. P. H. Barrett. Chicago: University of Chicago Press.

Darwin, E. 1794. *Zoonomia*. 2 vols. London: J. Johnson. Repr. AMS Press, New York, 1974. 2nd American ed., 1803, 2 vols. Boston: Thomas and Andrews.

Davies, Sir John. 1947. *Orchestra*. Ed. E.M.W. Tillyard. London: Chatto and Windus.

Davies, P.C.W. 1982. *The Accidental Universe*. Cambridge: Cambridge University Press.

da Vinci, Leonardo. 1938. *The Notebooks of Leonardo da Vinci*. 2 vols. Ed. and trans. E. MacCurdy. New York: Reynal and Hitchcock.

Deferrari, R. J., ed. 1955. *Saint Augustine, Treatises on Marriages and Other Subjects. The Fathers of the Church*, vol. 27. Washington, D.C.: Catholic University of America Press.

Deluc, J. A. 1809. *An Elementary Treatise on Geology*. Trans. H. De La Fite. London: Rivington.

Densmore, D. 1995. *Newton's Principia: The Central Argument*. With translations by W. H. Donahue. Santa Fe, N.M.: Green Lion Press.

Descartes, R. 1647. *Les principes de la philosophie*. Trans. [C. Picot]. Paris: Le Gras.

———. 1664. *L'homme de René Descartes*. Paris: Le Gras.

———. 1925. *The Geometry of René Descartes*. Trans. D. E. Smith and M. L. Latham. Chicago: Open Court. Repr. Dover, New York, 1954.

———. 1964–74. *Oeuvres*. 11 vols. Ed. C. Adam and P. Tannery. Paris: Vrin.

———. 1983. *Principles of Philosophy*. Trans. V. R. Miller and R. P. Miller. Dordrecht: D. Reidel.

———. 1977. *Key Philosophical Writings*. Ed. E. Chávez-Arvizo, trans. E. S. Haldane and G.R.T. Ross. Ware, U.K.: Wordsworth Eds.

Dick, S. J. 1982. *Plurality of Worlds*. Cambridge: Cambridge University Press.

Dickie, M. W. 2001. *Magic and Magicians in the Greco-Roman World*. London: Routledge.

Dickinson, G. L. 1896. *The Greek View of Life*. London: Methuen.

Diderot, D., ed. 1751–72. *Encyclopédie, ou dictionnaire raisonné des arts, des métiers, et des letters*. 28 vols. Paris: Briasson et al. Repr. in small format, 5 vols., Pergamon, Elmsford, N.Y., 1966.

Diels, H. 1952. *Die Fragmente der Vorsokratiker.* 9th ed., 3 vols. Ed. W. Kranz. Berlin: Weidmann.

Digges, T. 1556. *A Prognostication everlasting of ryte goode effect.* London: Thomas Gemini.

Dijksterhuis, E. J. 1961. *The Mechanization of the World Picture.* Trans. C. Dikshoorn. Oxford: Oxford University Press. Repr. Princeton University Press, 1986.

Dinsmore, W. B. 1950. *The Architecture of Ancient Greece.* 3rd ed. London: Batsford.

Diodorus of Sicily. 1933–67. *The Library of History.* 12 vols. Trans. C. H. Oldfather et al. London: Heinemann.

Diogenes Laertius. 1925. *Lives of Eminent Philosophers.* 2 vols. Trans. R. D. Hicks. London: Heinemann.

Dobbs, B.J.T. 1975. *The Foundations of Newton's Alchemy.* Cambridge: Cambridge University Press.

Donne, J. 1941. *Complete Poetry and Selected Prose.* Ed. J. Hayward. London: Nonesuch Press.

Drake, F., and D. Sobel. 1992. *Is Anyone Out There?* New York: Delacorte Press.

Drake, S. 1957. *Discoveries and Opinions of Galileo.* Trans. S. Drake. New York: Doubleday Anchor Books.

Dreyer, J.L.E. 1953. *A History of the Planetary Systems from Thales to Kepler.* Cambridge: Cambridge University Press. Repr. as *A History of Astronomy from Thales to Kepler,* Dover, New York, 1953.

Dronke, P. 1988. *A History of Twelfth-Century Western Philosophy.* Cambridge: Cambridge University Press.

Duhem, P. 1913–59. *Le Système du monde.* 10 vols. Paris: Hermann.

Durkheim, E. 1995. *The Elementary Forms of Religious Life.* Trans. K. E. Fields. New York: Free Press.

Dyson, F. J. 1971. "Energy in the Universe." *Scientific American* 225(9), 50–59.

———. 1979. "Time without End: Physics and Biology in an Open Universe." *Reviews of Modern Physics* 51 (3), 447–60.

———. 1992. *From Eros to Gaia.* New York: Pantheon. Repr. Penguin, London.

Eddington, A. S. 1930. "On the Instability of Einstein's Spherical World." *Monthly Notices, Royal Astronomical Society* 90, 668–78.

Einstein, A. 1915. "Die Feldgleichungen der Gravitation." *Sitzungsberichte der Preussischen Akademie der Wissenschaften* (1915), 844–47.

———. 1920. *Relativity: The Special and the General Theory.* Trans. R. W. Lawson. London: Methuen. Many reprint editions.

———. 1987–. *The Collected Papers of Albert Einstein.* 9 vols. Ed. J. Stachel et al. Princeton, N.J.: Princeton University Press. References are to the accompanying English translations.

Ekers, R. D., et al., eds. 2002. *SETI 2020.* Mountain View, Calif.: SETI Press.

Elliott, J. K., ed. 1993. *The Apocryphal New Testament.* Oxford: Clarendon Press.

Ellis, G.F.R., ed. 2002. *The Far-Future Universe.* Philadelphia: Templeton Foundation Press.

Epicurus. 1926. *Epicurus*. Ed. and trans. C. Bailey. Oxford: Clarendon Press. Repr. Hyperion, Westport, Conn., 1979.

Estensen, M. 1998. *Discovery*. Sydney: Allen and Unwin. Repr. St. Martin's Press, New York, 1999.

Euclid. 1925. *The Thirteen Books of the Elements*. 3 vols. Ed. and trans. T. L. Heath. Cambridge: Cambridge University Press. Repr. Dover, New York, 1956.

Faraone, C. A., and Obbink, D., eds. 1991. *Magika Hiera*. New York: Oxford University Press.

Faulkner, R. O., trans. 1969. *The Ancient Egyptian Pyramid Texts*. Oxford: Clarendon.

Fichman, M. 2004. *An Elusive Victorian*. Chicago: University of Chicago Press.

Flaherty, W. D. 1981. *The Rig Veda*. London: Penguin.

Flint, V.I.J. 1991. *The Rise of Magic in Early Medieval Europe*. Princeton, N.J.: Princeton University Press.

Foster, B. R. 1996. *Before the Muses*. 2 vols. Bethesda, Md.: CDL Press.

Frank, P. 1947. *Einstein: His Life and Times*. Trans. G. Rosen, ed. S. Kusaka. New York: Knopf. More complete German ed.: *Albert Einstein, sein Leben und seine Zeit*. Braunschweig, Germany: Vieweg, 1979.

Freeman, K. 1978. *Ancilla to the Presocratic Philosophers*. Cambridge, Mass.: Harvard University Press.

Friedmann, A. 1922. "Über die Krummung des Raumes." *Zeitschrift für Physik* 10, 377–386.

Gager, J. G. 1992. *Curse Tablets and Binding Spells from the Ancient World*. New York: Oxford University Press.

Galileo, G. 1914. *Dialogues Concerning Two New Sciences*. Trans. H. Crew and A. de Salvio. New York: Macmillan. Repr. Dover, New York, 1953.

———. 1968. *Opere* (EN). 20 vols. Ed. A. Favaro. Firenze: Barbara.

Gamow, G. 1946. "Expanding Universe and the Origin of the Elements." *Physical Review* 70, 572–73.

———. 1948. "The Evolution of the Universe." *Nature* 162, 680–82.

Gassendi, P. 1972. *The Selected Works of Pierre Gassendi*. Ed. and trans. C. B. Brush. New York: Johnson.

Gaukroger, S. 1995. *Descartes, an Intellectual Biography*. Oxford: Clarendon.

Gershenson, D. E., and D. A. Greenberg. 1964. *Anaxagoras and the Birth of Physics*. New York: Blaisdell.

Gesner, K. 1604. *Historia Animalium*. Frankfurt a.M.: Cambieri.

Gingerich, O. 1992. *The Great Copernicus Chase and Other Adventures in Astronomical History*. Cambridge, Mass.: Sky Publishing.

Goldsmith, D., and T. Owen. 1980. *The Search for Life in the Universe*. Menlo Park, Calif.: Benjamin Cummings.

Granet, M. 1958. *Chinese Civilization*. Trans. K. E. Innes and M. R. Braisford. New York: Meridian Books.

Grant, B. R., and Grant, P. R. 1989. *Evolutionary Dynamics of a Natural Population*. Chicago: University of Chicago Press.

Grant, C. 1965. *The Rock Paintings of the Chamash*. Berkeley: University of California Press.

Graves, R., and R. Patai. 1964. *Hebrew Myths: The Book of Genesis.* Garden City, N.J.: Doubleday.

Greaves, A. M. 2002. *Miletos, a History.* London: Routledge.

Greene, B. 2003. *The Elegant Universe.* New York: Norton.

———. 2004. *The Fabric of the Cosmos.* New York: Knopf.

Gregory, T. 1955. *Anima Mundi.* Firenze: Sansoni.

Grey, G. 1961. *Polynesian Mythology.* 2nd ed. Auckland, N.Z.: Whitcombe and Tombs.

Griffith, F. Ll., and Thompson, H., eds. 1904. *The Demotic Magical Papyrus of London and Leiden.* London: Grevel. Repr. as *The Leyden Papyrus,* Dover, New York, 1974.

Grosseteste, R. 1942. *On Light.* Trans. C. Riedl. Milwaukee: Marquette University Press.

———. 1982. *Hexaëmeron.* Ed. R. C. Dales and S. Gieben. London: Oxford University Press.

Guth, A. H. 1997. *The Inflationary Universe.* Reading, Mass.: Addison-Wesley. Repr. Addison-Wesley-Longman, 2000.

Guthrie, K. S., ed. and trans. 1987. *The Pythagorean Sourcebook and Library.* Grand Rapids, Mich.: Phanes Press.

Guthrie, W.K.C. 1962–81. *A History of Greek Philosophy.* 6 vols. Cambridge: Cambridge University Press.

Haber, F. 1959. *The Age of the World: Moses to Darwin.* Baltimore: Johns Hopkins University Press.

Hansen, B. 1985. *Nicole Oresme and the Marvels of Nature.* Toronto: Pontifical Institute of Medieval Studies.

Harrison, J. 1903. *Prolegomena to the Study of Greek Religion.* Later eds. 1908, 1922. Cambridge: Cambridge University Press. Third ed. repr. Meridian, New York, 1955.

Heath, T. H., ed. and trans. 1897. *The Works of Archimedes.* Cambridge: Cambridge University Press. Repr. together with *The Method of Archimedes,* Dover, 1953.

———. 1913. *Aristarchus of Samos, the Ancient Copernicus.* Oxford: Clarendon. Repr. Dover, New York, 1981.

Heidel, A. 1949. *The Gilgamesh Epic and Old Testament Parallels.* 2nd ed. Chicago: University of Chicago Press.

———. 1951. *The Babylonian Genesis.* Chicago: University of Chicago Press.

Heisenberg, W. 1971. *Physics and Beyond.* Trans. A. J. Pomerans. New York: Harper and Row.

Herodotus. 1858. *The History of Herodotus.* 4 vols. Trans. Sir G. Rawlinson. London: Murray. Later edition ed., M. Komroff, Dial, New York, 1928. Repr. Tudor, New York, 1943.

———. 1921–24. *History.* 4 vols. Trans. A. D. Godley. London: Heinemann. (This is the standard Greek-English edition, but the Rawlinson translation is better reading.)

Herschel, W. 1785. "On the Construction of the Heavens." *Philosophical Transactions of the Royal Society* 75, 213–65.

Hertz, H. 1893. *Electric Waves.* Trans. D. E. Jones. London: Macmillan.

Hesiod. 1914. *The Homeric Hymns and Homerica.* Trans. H. G. Evelyn-White. Cambridge, Mass.: Harvard University Press. Many reprints.

———. 1973. *Hesiod and Theogonis.* Trans. D. Wender. London: Penguin.

Hildebrand, A. S. 1924. *Magellan.* New York: Harcourt Brace.

Hippocrates. 1978. *Hippocratic Writings.* Ed. G.E.R. Lloyd, trans. J. Chadwick et al. London: Penguin.

Hoban, R. 1980. *Riddley Walker.* New York: Summit Books.

Hopkins, J. 1981. *Nicholas of Cusa on Learned Ignorance.* Minneapolis: Banning.

Hopper, V. F. 1938. *Medieval Number Symbolism.* New York: Columbia University Press. Repr. Dover, New York, 2000.

Horowitz, W. 1998. *Mesopotamian Cosmic Geography.* Winona Lake, Ind.: Eisenbrauns.

Hudson, T., et al., 1981. *The Eye of the Flute.* 2nd ed. Santa Barbara, Calif.: Santa Barbara Museum of Natural History.

Hudson, T., and E. Underhay. 1978. *Crystals in the Sky.* Los Altos, Calif.: Ballena Press.

Hutton, J. 1788. "Theory of the Earth." *Transactions of the Royal Society of Edinburgh* 1, part 2, 209–304.

———. 1970. *James Hutton's System of the Earth, 1785; Theory of the Earth, 1788, Observations on Granite, 1794; together with Playfair's* Biography of Hutton. Darien, Conn.: Hafner.

Inwood, S. 2003. *The Forgotten Genius.* San Francisco: MacAdam/Cage.

Irvine, W. 1955. *Apes, Angels, and Victorians.* New York: McGraw-Hill. Repr. Meridian, New York, 1959.

Isidore, Saint. c. 1473. *Etymologiae.* Strassburg: Mentelin.

———. 1911. *Etymologiarum sive originum.* 2 vols. Ed. W. M. Lindsay. Oxford: Clarendon. Repr. Clarendon, 1971, 1985.

———. 1960. *Traité de la nature.* Ed. J. Fontaine. Bordeaux: Feret.

Jane, C., trans. and ed. 1930–33. *Select Documents Illustrating the Four Voyages of Columbus.* 2 vols. London: Hakluyt Society.

Jastrow, M. 1911. *Religious Belief in Babylonia and Assyria.* New York: Putnam.

Jordan, D. R., H. Montgomery, and E. Thomassen, eds. 1999. *The World of Ancient Magic.* Bergen: Norwegian Institute at Athens.

Joy, L. S. 1987. *Gassendi the Atomist.* Cambridge: Cambridge University Press.

Kant, I. 1900: *Universal Natural History and Theory of the Heavens.* Trans. W. Hastie. Glasgow: Maclehose. Repr. Johnson, New York, 1990.

Kasher, M. M., ed. 1953–79. *Encyclopedia of Biblical Interpretation.* Ed. and trans. H. Freedman. 9 vols. New York: American Biblical Encyclopedia Society.

Kasting, J. F., D. P. Whitmire, and R. T. Reynolds. 1993. "Habitable Zones around Main Sequence Stars." *Icarus* 101, 108–28.

Kepler, J. 1937–75. *Gesammelte Werke* (GW), 22 vols. Ed. M. Caspar et al. Munich: Beck.

———. 1965. *Kepler's Conversation with Galileo's Sidereal Messenger.* Trans. E. Rosen. New York: Johnson.

———. 1981. *Mysterium Cosmographicum: The Secret of the Universe.* Trans. A. M. Duncan. New York: Abaris Books.

———. 1992. *New Astronomy*. Trans. W. H. Donahue. Cambridge: Cambridge University Press.

———. 1997. *The Harmony of the World*. Trans. E. J. Aiton, A. M. Duncan, and J. V. Field. Philadelphia: Americal Philosophical Society.

Kieckhefer, R. 1998. *Forbidden Rites*. University Park: Pennsylvania State University Press.

Kimble, G.H.T. 1938. *Geography in the Middle Ages*. London: Methuen.

Kirk, G. S., J. E. Raven, and M. Schofield. 1983. *The Presocratic Philosophers*. 2nd ed. Cambridge: Cambridge University Press.

Kirshner, R. P. 2002. *The Extravagant Universe*. Princeton, N.J.: Princeton University Press.

Koeberl, C., and K. G. MacLeod, eds. 2002. *Catastrophic Events and Mass Extinctions: Impacts and Beyond*. Special paper 356. Boulder, Colo.: Geological Society of America.

Koerner, D., and S. LeVay. 2000. *Here be Dragons*. Oxford: Oxford University Press.

Koestler, A. 1959. *The Sleepwalkers*. New York: Macmillan. Repr. Grosset and Dunlap, New York, 1963.

Kogut, J. B. 2002. *Introduction to Relativity*. San Diego: Harcourt/Academic.

Kollerstrom, N. 1999. "The Path of Halley's Comet, and Newton's Late Apprehension of the Law of Gravity." *Annals of Science* 56, 331–57.

Kongtrul, J. and Tayé, L. 1995. *Myriad Worlds*. Ithaca, N.Y.: Snow Lion.

Kramer, S. N. 1981. *History Begins at Sumer*. 3rd ed. Philadelphia: University of Pennsylvania Press.

Krupp, E., ed. 1984. *Archaeoastronomy and the Roots of Science*. Boulder, Colo.: Westview Press.

Kuhn, T. S. 1957. *The Copernican Revolution*. Cambridge, Mass.: Harvard University Press. Repr. Random House, New York, 1959.

———. 1978. *Black-Body Theory and the Quantum Discontinuity, 1894–1912*. Oxford: Clarendon.

Lactantius. 1964. *The Divine Institutes, Books I–VII*. Trans. M. F. McDonald. Washington, DC: Catholic University of America Press.

Langford, J. J. 1971. *Galileo, Science, and the Church*. Ann Arbor: University of Michigan Press.

Lao Tsu. 1972. *Tao Te Ching*. Trans. G.-F. Feng and J. English. New York: Vintage (Random House).

Laplace, P. S. 1984. *Exposition du système du monde*. Paris: Fayard.

Lasswitz, K. 1890. *Geschichte der Atomistik*. 2 vols. Hamburg: Voss.

Lavoisier, A. L. 1789. *Traité élémentaire de chimie*. 2 vols. Paris: Cuchet.

———. 1796. *Elements of Chemistry*. 3rd ed. Trans. R. Kerr. London: Creech.

Layard, A. H. 1856. *Discoveries among the Ruins of Nineveh and Babylon*. New York: Harper.

Leichty, E. 1970. *The Omen Series Summa Izbu*. Locust Valley, N.J.: Augustin.

Lemaître, G. 1931a. "The Expanding Universe." *Monthly Notices, Royal Astronomical Society* 21, 491–503.

———. 1931b. "The Beginning of the World from the Point of View of Quantum Theory." *Nature* 127, 706.

Lemonick, M. D. 2003. *Echo of the Big Bang*. Princeton, N.J.: Princeton University Press.

Lindberg, D. 1988. "The Genesis of Kepler's Theory of Light: Light Metaphysics from Plotinus to Kepler." *Osiris*, 2nd ser., *2*, 5–42.

Lindsay, J. 1971. *Origins of Astrology*. London: Frederick Muller.

Lorentz, H. A., A. Einstein, H. Minkowski, and H. Weyl. 1923. *The Principle of Relativity*. Ed. A. Sommerfeld. Trans. W. Perrett and G. H. Jeffery. London: Methuen. Repr. Dover, New York, 1952.

Lovejoy, A. O. 1936. *The Great Chain of Being*. Cambridge, Mass.: Harvard University Press. Repr. Harper, New York 1960.

Lucretius. 1968. *The Way Thing Are*. Trans. R. Humphries. Bloomington: Indiana University Press.

———. 1992. *De rerum natura*. Trans. W.H.D. Rouse, rev. M. F. Smith. Cambridge, Mass.: Harvard University Press.

Lyell, C. 1830–33. *The Principles of Geology*, 3 vols. London: Murray.

Macrobius. 1952. *Commentary on the Dream of Scipio*. Trans. W. H. Stahl. New York: Columbia University Press.

Maimonides, M. 1963. *The Guide of the Perplexed*. Trans. S. Pines. Chicago: University of Chicago Press.

Malthus, T. 1976. *An Essay on the Principles of Population*. Ed. P. Appleman. New York: Norton.

Manetho. 1940. *Manetho*. Trans. W. G. Waddell. Contains Ptolemy's *Tetrabiblos*. Ed. and trans. F. E. Robbins. London: Heinemann.

Mandeville, J. 1900. *The Travels of Sir John Mandeville*. Ed. A. W. Pollard. London: Macmillan. Repr. Dover, New York, 1964.

Marcus Aurelius. 1964. *Meditations*. Trans. M. Staniforth. Harmondsworth: Penguin.

Maricq, A. 1952. "Tablette de défixion de Beyrouth." In *Byzantion*. Brussels: Fondation Byzantine et Néo-grecque.

Masi, M. 1983. *Boethian Number Theory*. Amsterdam: Rodopi.

Mather, K. F. and S. L. Mason 1939. *A Source Book in Geology*. New York: McGraw-Hill.

Maupertuis, P.L.M. de. 1745. *Vénus physique*. Paris: [N.p.] Repr. Aubier Montaigne, Paris, 1980.

———. 1750, 1751. *Essay de cosmologie*. Eds. in Amsterdam, Berlin, and Leiden. Repr. Vrin, Paris, 1984.

Maxwell, J. C. 1890. *The Scientific Papers of James Clerk Maxwell*. 2 vols. Ed. W. D. Niven. Cambridge: Cambridge University Press.

———. 1865. "A Dynamical Theory of the Electromagnetic Field." In Maxwell 1890, vol. 1, 526–97.

Mayor, A. 2000. *The First Fossil Hunters*. Princeton, N.J.: Princeton University Press.

Mayor, M., and Frei, P-Y. 2003. *New Worlds in the Cosmos*. Cambridge: Cambridge University Press.

McConnell, B. 2001. *Beyond Contact*. Beijing: O'Reilly.

McDonald, B. and D. Jehl 2003. *Whose Water Is It?* Washington, D.C.: National Geographic.

McEvoy, J. J. 1982. *The Philosophy of Robert Grosseteste*. New York: Oxford University Press.

McGinn, B. et al., eds. 1999. *The Encyclopedia of Apocalypticism*. 3 vols. New York: Continuum.

Merkel, I. and Debus, A. G., eds. 1988. *Hermeticism and the Renaissance*. Washington, D.C.: Folger Shakespeare Library.

Mersenne, M. 1634. *Questions inouyes*. Paris: Villery. Repr. Fayard, Paris, 1985.

Miller, H. 1852. *The Old Red Sandstone*. From 4th English ed. Boston: Gould and Lincoln.

Mohammed. 1956. *The Koran*. Trans. N. J. Dawood. London: Penguin.

Morison, S. E. 1942. *Admiral of the Ocean Sea*. Boston: Little, Brown.

Morrison, J. S. 1955. "Parmenides and Er." *Journal of Hellenic Studies* 75: 59–68.

Münster, S. 1551. *Rudimenta mathematica*. Basel: Petrus.

Munk, S. 1856–66. *Le Guide des égarées*. 3 vols. Paris: A. Franck.

———. 1859. *Mélanges de philosophie juive et arabe*. Paris: A. Franck.

Murray, G. 1935. *Five Stages of Greek Religion*. London: Watts.

Nasr, S. H. 1964. *An Introduction to Islamic Cosmological Doctrines*. Cambridge: Belknap Press (Harvard).

Needham, Joseph, et al. 1954–[2000]. *Science and Civilization in China*. [7] vols. Cambridge: Cambridge University Press.

Neugebauer, O. 1957. *The Exact Sciences in Antiquity*. 2nd ed. Providence: Brown University Press. Repr. Harper Torchbooks, New York, 1962.

———. 1975. *A History of Ancient Mathematical Astronomy*, 3 vols. New York: Springer-Verlag.

Newton, I. 1730. *Opticks*. 4th ed. London: Innys. Repr. Dover, New York, 1952.

———. 1934. *Sir Isaac Newton's Mathematical Principles of Natural Philosophy and His System of the World*. Ed. and trans. F. Cajori. Berkeley: University of California Press. Repr. Dover, New York.

———. 1958. *Isaac Newton's Papers and Letters on Natural Philosophy*. Ed. I. B. Cohen. Cambridge, Mass.: Harvard University Press.

———. 1959–77. *The Correspondence of Isaac Newton*. 7 vols. Ed. H. W. Turnbull et al. Cambridge: Cambridge University Press.

———. 1999. *The Principia*. Trans. I. B. Cohen and A. Whitman. Berkeley: University of California Press.

Nicomachus. 1926. *Introduction to Arithmetic*. Trans. M. L. d'Ooge. New York: Macmillan.

Nonnos. 1940. *Dionysiaca*. Trans. H. J. Rouse, 3 vols. London: Heinemann.

Nordenskiöld, A. E. 1973. *Facsimile-Atlas*. New York: Dover.

O'Neil, W. M. 1986. *Early Astronomy*. Sydney: Sydney University Press.

Origen. 1966. *On First Principles*. Trans. G. W. Butterworth. New York: Harper and Rowe. Repr. 1973.

———. 1989. *Commentary on the Gospel According to John, Books 1–10*. Trans. R. E. Heine. (Fathers of the Church, vol. 80.) Washington, D.C.: Catholic University of America Press.

Ostwald, W. 1890. *Outlines of General Chemistry*. Trans. J. Walker. London: Macmillan.

Ovid. 1955. *Metamorphoses.* Trans. R. Humphries. Bloomington: Indiana University Press.

Oxford Bible. 1992. *The Oxford Study Bible.* Ed. M. J. Suggs, K. D. Sakenfeld, and J. R. Mueller. New York: Oxford University Press.

Pannekoek, A. 1961. *History of Astronomy.* London: Allen and Unwin.

Paracelsus. 1951. *Selected Writings.* Ed. J. Jacobi, trans. N. Guterman. Princeton, N.J.: Princeton University Press. Repr. 1988.

Park, D. 1988. *The How and the Why.* Princeton, N.J.: Princeton University Press.

———. 1990. *Classical Dynamics and its Quantum Analogues.* 2nd ed. New York: Springer-Verlag.

———. 1997. *The Fire within the Eye.* Princeton, N.J.: Princeton University Press.

Parry, J. H. 1968. *The European Reconnaissance.* New York: Harper and Row.

Partington, J. R. 1961–70. *A History of Chemistry.* 4 vols. London: Macmillan.

Penzias, A. and R. W. Wilson. 1965. "A Measurement of Excess Antenna Temperature at 4080 Mc/s." *Astrophysical Journal* 142, 419.

Pepys, S. 1971. *The Diary of Samuel Pepys,* 11 vols. Ed. R. Latham and W. Matthews. London: Bell and Hyman. Repr. HarperCollins, London, 2000.

Perrin, J. 1910. *Brownian Movement and Molecular Reality.* Trans. F. Soddy. London: Taylor and Francis.

Pharr, C., ed. and trans. 1952. *The Theodosian Code.* Princeton, N.J.: Princeton University Press.

Philo of Alexandria. 1929–62: *Philo.* 10 plus 2 supplementary vols. Trans. F. H. Colson and G. H. Whitaker. London: Heinemann.

Pierce, J. R. 1959. "Relativity and Space Travel." *Proceedings, Institute of Radio Engineers* 47, 1053–61.

Pigafetta, A., et al. 1874. *The First Voyage around the World by Magellan.* Trans. Lord Stanley of Alderley. London: Hakluyt Society.

Plato. 1961. *The Collected Dialogues.* Ed. E. Hamilton and H. Cairns. Princeton, N.J.: Princeton University Press.

Pliny (C. Plinius Secundus). 1938–63. *Natural History.* 10 vols. Trans. H. Rackham and W.H.S. Jones. London: Heinemann.

Plutarch. 1631. *The Lives of the Noble Greeks and Romains.* Trans. T. North. London: G. Miller.

———. 1914. *Plutarch's Lives.* 11 vols. Trans. B. Perrin. London: Heinemann.

———. 1927–69. *Plutarch's Moralia.* 16 vols. Trans. F. C. Babbitt et al. London: Heinemann.

Principe, L. M. 1998. *The Aspiring Adept.* Princeton, N.J.: Princeton University Press.

Pritchard, J. B. 1969. *Ancient Near Eastern Texts Relating to the Old Testament.* 3rd ed. Princeton, N.J.: Princeton University Press.

Proclus Diadochus. 1968. *Commentaire sur le Timée.* 5 vols. Trans. A. J. Festugière. Paris: Vrin.

———. (Proclo). 1989. *Lezioni sul "Cratilo" di Platone.* Trans. and notes. F. Romano. Catania, Italy: Università.

Ptolemy (Claudius Ptolemaeus). 1980. *Tetrabiblos.* Ed. and trans. F. E. Robbins. Cambridge, Mass.: Harvard University Press.

———. 1984. *Ptolemy's Almagest.* Trans. G. J. Toomer. New York: Springer-Verlag.

Pullman, B. 1998. *The Atom in the History of Human Thought.* New York: Oxford University Press.

Quiller-Couch, A., ed. 1939. *The Oxford Book of English Verse.* New ed. Oxford: Clarendon.

Raby, P. 2001. *Alfred Russel Wallace, a Life.* Princeton, N.J.: Princeton University Press.

Redford, D. B. 2001. *The Oxford Encyclopedia of Ancient Egypt.* 3 vols. Oxford: Oxford University Press.

Rees, M. 1997. *Before the Beginning.* Reading, Mass.: Addison-Wesley.

———. 2001. *Our Cosmic Habitat.* Princeton, N.J.: Princeton University Press.

———. 2003. *Our Final Hour.* New York: Basic Books.

Regiomontanus, J. 1496. *Epytoma Joannis De monteregio in almagestum ptolemei.* Venice: Hamman.

Reiner, E. 1995. *Astral Magic in Babylonia.* Transactions of the American Philosophical Society 85, pt. 4.

Reston, J., Jr. 1994. *Galileo: A Life.* New York: HarperCollins.

Riesman, D. 1936. *The Story of Medicine in the Middle Ages.* New York: Harper.

Roberts, G. 1994. *The Mirror of Alchemy.* Toronto: University of Toronto Press.

Roger, J. 1997a. *Buffon.* Trans. S. L. Bonnefoi. Ithaca, N.Y.: Cornell University Press.

———. 1997b. *The Life Sciences in Eighteenth-Century French Thought.* Ed. K. R. Benson, trans. R. Ellrich. Stanford, Calif.: Stanford University Press.

Rogers, F. M. 1962. *The Search for Eastern Christians.* Minneapolis: University of Minnesota Press.

Ronan, C. A. 1981. *The Shorter Science and Civilisation in China.* 2 vols. Cambridge: Cambridge University Press.

Rubin, V. 1997. *Bright Galaxies, Dark Matters.* Woodbury, N.Y.: American Institute of Physics.

Rubruck, W. 1990. *The Mission of Friar William Rubruck.* Trans. P. Jackson. London: Hakluyt Society.

Russell, J. 1977. *The Devil.* Ithaca, N.Y.: Cornell University Press.

———. 1981. *Satan.* Ithaca, N.Y.: Cornell University Press.

———. 1984. *Lucifer.* Ithaca, N.Y.: Cornell University Press.

Russo, L. 1996. *La rivoluzione dimenticata.* Milan: Feltrinelli.

Ryan, W., and W. Pitman. 1998. *Noah's Flood.* New York: Simon and Schuster.

Ryder, G., D. Fastovsky, and S. Gartner, eds. 1996. *The Cretaceous-Tertiary Event and Other Catastrophes in Earth History,* Special Paper 307. Boulder, Colo.: Geological Society of America.

Sachs, J. 1995. *Aristotle's Physics.* New Brunswick, N.J.: Rutgers University Press.

Sallustius. 1966. *Sallustius Concerning the Gods and the Universe.* Ed. and trans. A. D. Nock. Hildesheim, Germany: Olds.

Sambursky, S. 1974. *Physical Thought from the Presocratics to the Quantum Physicists.* New York: Pica Press.

Santbech, D. 1561. *Problematum astronomicorum et geometricorum septem.* Basel: Petrus and Perna.

Santillana, G. de. 1955. *The Crime of Galileo*. Chicago: University of Chicago Press.

Santillana, G. de, and H. von Dechend. 1969. *Hamlet's Mill*. Boston: Gambit.

Schaff, P., and H. Wace. 1895. *A Select History of Nicene and Post-Nicene Fathers of the Christian Church*. 2nd ser., 14 vols. New York: Christian Literature Company.

Schedel, H. 1493. *Liber chronicarum*. Nürnberg, Germany: Koberger.

Schweickart, R. L., et al. 2003. "The Asteroid Tugboat." *Scientific American*, November, 54–61.

Secord, J. A. 2000. *Victorian Sensation*. Chicago: University of Chicago Press.

Seneca. 1923. *Dialogues*. 4 vols. Trans. R. Waltz. Paris: Belles Lettres.

———. 1971. *Naturales quaestiones*. 2 vols. Trans. T. H. Corcoran. London: Heinemann.

Shapiro, R. 1999. *Planetary Dreams*. New York: Wiley.

Siegel, R. E. 1968. *Galen's System of Physiology and Medicine*. Basel: Karger.

Silk, J. 1994. *A Short History of the Universe*. New York: Scientific American Library.

Simpson, W. K., ed. 1972. *The Literature of Ancient Egypt*. New Haven, Conn.: Yale University Press.

Slipher, V. M. 1914, 1915. "Spectroscopic Observation of Nebulae." *Popular Astronomy* 22, 146; 23, 21–24.

Solinus, J. 1587. *The Excellent and Pleasant Work of Julius Solinus, Polyhistor*. Trans. A. Golding. London: Hacket. Repr. Scholars' Facsimiles and Reprints, Gainesville, Fla., 1955.

Steno, N. 1968. *The Prodromus of Nicolaus Steno's Dissertation Concerning a Solid Body Enclosed by Process of Nature in a Solid*. Ed. and trans. J. G. Winter. New York: Hafner.

Stephenson, B. 1987. *Kepler's Physical Astronomy*. Princeton, N.J.: Princeton University Press. Repr. 1994.

Stephen, A. 1936. *Hopi Journal of Alexander M. Stephen*. Ed. E. C. Parsons, 2 vols. New York: Columbia University Press.

Stewart, Balfour. See Anon., 1875.

Stolcius, D. 1624. *Viridarium chymicum*. Frankfurt am Main, Germany.

Strabo. 1917–32. *The Geography of Strabo*. 8 vols. Trans. H. L. Jones. London: Heinemann.

Suetonius, C. 1998. *Suetonius*. 2 vols. Trans. J. C. Rolfe. Cambridge, Mass.: Harvard University Press.

Swerdlow, N. M. 1998. *The Babylonian Theory of the Planets*. Princeton, N.J.: Princeton University Press.

———. 1999. *Ancient Astronomy and Celestial Divination*. Cambridge, Mass.: MIT Press.

Tacitus, C. 1925–37. *The Histories and the Annals*. 3 vols. Trans. C. H. Moore and J. Jackson. London: Heinemann.

Taylor, F. S. 1949. *The Alchemists*. New York: Schuman.

Thomson, J. W., and E. N. Johnson 1937. *Introduction to Medieval Europe*. New York: Norton.

Thomson, W. 1882–1911. *Mathematical and Physical Papers*. 6 vols. London: R. Clay and Cambridge University Press.

————. 1889–94. *Popular Lectures and Addresses.* 3 vols. London: Macmillan.

Thoreau, H. D. 1971. *Walden.* Ed. J. L. Shanley. Princeton, N.J.: Princeton University Press. Repr. Princeton University Press, 1989.

Thorndike, L. 1923–58. *A History of Magic and Experimental Science.* 8 vols. New York: Macmillan.

Thorne, Kip. 1994. *Black Holes and Time Warps.* New York: Norton.

Tupet, A.-M. 1976. *De la magie dans la poésie latine.* Paris: Les Belles Lettres.

Ussher, J. 1658. *The Annals of the World.* London: Crook and Bedell.

Valéry, P. 1944. *Poésies.* Paris: Gallimard.

Van Helden, A. 1985. *Measuring the Universe.* Chicago: University of Chicago Press.

Verbrugghe, G. P., and Wickersham, J. M. 1996. *Berosus and Manetho, Introduced and Translated.* Ann Arbor: University of Michigan Press.

Vitruvius (Marcus Vitruvius Pollio). 1931. *On Architecture.* Trans. F. Granger. London: Heinemann.

Voltaire. 1931. *Lettres sur les anglais.* Ed. A. Wilson-Green. Cambridge: Cambridge University Press.

Waley, A. 1942. *The Way and the Power.* Boston: Houghton-Mifflin.

Weinberg, S. 1977. *The First Three Minutes.* New York: Basic Books.

Westfall, W. 1980. *Never at Rest.* Cambridge: Cambridge University Press.

Williamson, R. A. 1984. *Living the Sky.* Boston: Houghton Mifflin.

Wilson, E. O. 2002. *The Future of Life.* New York: Knopf.

Wise, M., M. Abegg, Jr., and E. Cook. 1996. *The Dead Sea Scrolls.* San Francisco: Harper.

Wright, T. 1750. *An Original Theory or New Hypothesis of the Universe.* London: Chapelle. Repr. Elsevier, New York, 1971.

Young, T. 1807. *A Course of Lectures on Natural Philosophy and the Mechanical Arts.* 2 vols. London: Johnson. Repr. Johnson, New York, 1971.

————. 1855. *Miscellaneous Works of the Late Thomas Young.* 3 vols. Ed. G. Peacock and J. Leitch. London: Murray. Repr. Johnson, New York, 1972.

Yule, H., ed. and trans. 1866. *Cathay and the Way Thither,* 2 vols. Second ed., ed. H. Cordier, 1913–15, 4 vols. London: Hakluyt Society.

Academy, 88, 96, 136
Achilles, fights the Scamander, 31; his
    shield, 13
Addison, Joseph, quoted, 223
Albuquerque, Alfonso de, 179
Alexander, 165
Alvarez, Luis, explains an extinction,
    267
Ananke, 85
Anaxagoras, 55; on nature of matter, 74;
    on plurality of worlds, 283
Anaximander, 60ff
angels, 28, 116
anthropic argument, 288
Apocalypse of Paul, 3
Apsu and Tiamat, 6
Aquinas, Saint Thomas, on evil, 155
Aristarchus, 126f
Aristotle, 95–106; on Anaxagoras, 54; on
    atoms, 77; on four kinds of cause, 105;
    On the Heavens, 99; on the ladder of
    nature, 154; on motion, 103; on nature's
    tendency to improve, 106, 228, 294; on
    occult properties of numbers, 136; on
    Pythagoreans, 36, 66; on science, 83; on
    the soul, 102; on substance and
    qualities, 227; on Thales, 58f; on
    theory and observation, 126; on
    vacuum, 110
arts and sciences, 98
astrology, 42; in medicine, 158; in
    Ptolemy's *Tetrabiblos*, 133
astronomical unit, 246
Athens: conquered by Macedon, 96;
    defeated at Syracuse, 45
atoms: in Epicurean philosophy, 80;
    invented by Leucippus, 76; popularized
    by Democritus, 77; popularized by
    Lucretius, 81; rediscovered, 232–39
augury, 46
Augustine, Saint, 113, 298; on astrology,
    139; on Creation, 114; on demons, 139;
    on divination, 47
Averroës, on religion and philosophy, 190
Azaz'el, fallen angel, 29

Baade, Walter, astronomer, 270
Babylonian cosmos, 10; mathematics, 39,
    62, 118; time measurements, 87; world-
    map, 10
Balboa, Vasco de, explorer, 185
Bartholomew I, Patriarch, on the
    sacredness of the environment, 295
Basil of Caesarea, on lunar influences, 36
Becquerel, Henri, discovers radioactivity,
    266
Berossus, 43
Bessel, Friedrich Wilhelm, 250; measures
    distance to a star, 250
Big Bang, 272
Boethius, author and translator, 136, 143,
    174
Bohr, Niels, physicist, 274; quoted, 283
Book of the World, 267; continued, 289
Boyle, Robert, chemist, 232
Brahe, Tycho, astronomer, 200
Brahmagupta, mathematician, 118
Broglie, Louis de, on matter as a field,
    243
Buckland, William, reinterprets Genesis,
    256
Buddhist cosmos, 16
Buffon, Comte de, naturalist, 253
Buridan, Jean, philosopher, 60n, 152

Caligula, emperor, 46
Carpini, Giovanni del, explorer, 176
Carthage, 164
Cassiodorus, encyclopedist, 143
catastrophe, universal, 117
causality, classified by Aristotle, 105
Cavendish, Henry, analyzes water, 234
CBR (cosmic background radiation), 276;
    texture, 277
Cepheid variables, 269f
chain of being, 154
Chambers, Robert, on evolution, 258
Chumash Indians, their cosmos, 17
Cicero: on Aristotle's causes, 106; on astral
    influences, 139; on divination, 47; on the
    number seven, 208

Classicists and Romantics in science, 234, 238
Clazomenae, 55
Cocconi, Giuseppe, physicist, 285
Columbus, Christopher, 181ff
comets: may strike the Earth, 290; of 1577, 196; of 1604, 202; of 1682, 218
Conches, William of, 151, 225, 227
Constantine, edict against divination, 138
constellations, 35
Cook, Captain James, explorer, 188
Copernicus, Nicolaus, 191–97; *Commentariolus*, 192; *On the Revolutions . . .* , 193
Cosmas Indicopleustes, sailor, 169
cosmology: cyclic, 102, 116, 119; popular, 130
Creation: according to Anaximander, 111; to Augustine, 114; to Basil of Caesarea, 112; to Buffon, 253; to Chambers, 259; to Genesis, 1; to Laplace, 278; in Egypt, 14
Cusanus, Nicholas, *On Learned Ignorance*, 159

daimon. See demon
Dalton, John, on atoms, 236
Dante, 30
Darwin, Erasmus, on origin of life, 258
Darwin, Charles, 260–65; *Descent of Man*, 265; observations in the Galápagos, 261; *Origin of Species*, 264; *Voyage of the Beagle*, 261
Davies, Sir John, MP, 196
days of the week, 137
defixio, 50
Deluc, André, on biblical chronology, 254
Democritus, 76
demon, 31ff, 136ff
Descartes, René, 210–16; *Discourse on Method*, 210; *Principles of Philosophy*, 212
design, 84, 153ff, 287
Diaz, Bartolomeu, explorer, 180
Dickinson, Goldsworthy Lowes, critic, 26
Diogenes Laertius, historian, 77, 226
Diu, Battle of, 180, 186
divination, 137
Donne, John, quoted, 197

Drake, Frank, astronomer, 285
Dryden, John, quoted, 107, 124
Durkheim, Emile, quoted, 32

Earth: age, according to Aristotle, 109; to Buckland, 256; to Buffon, 253; to Hutton, 254; to Kelvin, 255; to Patterson, 166; to scripture, 251; floats on water, 58; size, 99, 163, 245
Earth Mother, 295
ecliptic, plane, 23
Eddington, Arthur, astronomer, 271
Einstein, Albert, 238; "Cosmological Considerations . . . ," 271; "Electrodynamics . . . ," 270; on the existence of atoms, 239; on the nature of light, 242
elements, 73, 93; their creation, 276
ellipse, 201
Emerald Tablet, 232, 307
Empedocles, 73
energy, dark, 280
end of the world, 120; according to American fundamentalists, 124; to *IV Ezra*, 122; to Isaiah, 121; to the Koran, 123; to modern astronomy, 291; to Revelations, 121; to Seneca, 20, 120; to Sibylline Oracles, 122
Enkidu, 12, 38
Enlil, 19, 27
Enuma Anu Enlil, 40
Epicurus, quoted, 26, 78ff, 225
epistēmē, 98
equatorial, plane, 23
Eratosthenes, 163, 245
eschatology, 30
ether, 99, 102, 112, 116, 133f, 195, 239
Eudoxus, 100
Euripides, quoted, 9
evolution: according to Anaximander, 62; to Darwin, 261ff; to Empedocles, 74; puzzles, 265
extinctions, 267, 290
extraterrestrial beings, 284; intelligence, search for (SETI), 285, a problematic enterprise, 286

Fermi, Enrico, quoted, 284
fields, 240, 242f
firmament, 2

flood: evidence, 21; legend, Akkadian, 12, 18; Greek, 19; biblical, 19, 29
Friedmann, Alexander, cosmological theory, 271, 273

Gaia, 8
Galaxy, 248
Galilei, Galileo: quoted, 190, 204–209; discovers extragalactic nebulae, 248; makes a telescope, 205; observes the Moon and discovers Jupiter's satellites, 205; *Sidereal Messenger*, 205
Gama, Vasco da, explorer, 162, 180
Gamow, George, on creation of elements and CBR, 276
Gassendi, Pierre, physicist, 226f
Geb and Nut, 14
Genghis Khan, 178
Germanicus, soldier, 50
Gilgamesh, 12, 18, 29, 38
God: in Newton's *Principia*, 221f; one among many, 27
gods and goddesses, 27
Golden Horde, 174
gravity, 218, 220
Great Year, Greatest Year, 117
Guth, Alan, inflationary universe, 274

Henry the Navigator, 180
Heracleides, 100, 126
Heraclitus, 69ff, 83
Herodotus, 166, 172
Herschel, William, 249; discovers Uranus, 249; maps the Galaxy, 250
Herschel, Caroline, astronomer, 249
Hesiod, 8, 162
Homer, 137, 162f, 165
Hooke, Robert, physicist, 218
House of Wisdom, 97
Hubble, Edwin, astronomer, 270
humors, 156
Hutton, James, geologist, 254

Idea, 90; of the Good, 90
igigi, 10, 27
impetus, 152
inflation, theory, 275, 286
Ionia, 57
Ishtar, 12
Isidore, Saint, encyclopedist, 145; on atoms, 148; *Etymologies*, 145; on geography, 146, 174; on man as microcosm, 156; *On the Nature of Things*, 148

Jesus, quoted, 230
Jupiter, planet, as a magnet, 284
Justinian, 89

Kalam, 149
kalpa, 118
Kant, Immanuel, 249
katádesmos. See *defixio*
Kelvin, Lord, 254
Kepler, Johannes, 197–209; *Astronomia Nova*, 201, 203; *Conversation with Galileo's . . .*, 208; epitaph, 209, 306; *Harmony of the World*, 204; laws of planetary motion, 201, 204, 223; *More Certain Foundations . . .*, 200; *Secret of the Universe*, 198, 204, 208
keres, 33

Lactantius, 168
Lampsacus, 55
Laplace, Pierre-Simon, astronomer, 278
Lavoisier, Antoine, lists elements, 235f
law, natural, 190
Leavitt, Henrietta, astronomer, 269
Leeuwenhoek, Anton, microscopist, 256
Leibniz, Gottfried von, 220
Leiden papyrus, 48
Lemaître, Abbé Georges, 272
Leucippus, atomist, 73
life on other worlds, 160, 209, 213, 283ff
LIGO, 281
logos, 69f; 84
Luther, Martin, 192
Lyell, Charles, geologist, 260

Ma'at, 84
Macrobius, 154
Magellan, Ferdinand, explorer, 186
magic: Egyptian, 48; German, 140
Maimonides, Moses, 115f, 149
Malthus, Thomas, *Essay on Population*, 261
Mandeville, John, fabulist, 178
Marcus Aurelius, Emperor, quoted, 36, 108
Marduk, 7, 27
Mars, orbit, 200f

matter: baryonic, 279; cold dark, 279
Maupertuis, Pierre-Louis de, 233;
    contributions to physics, 258n; on
    heredity, 256
Maxwell, James Clark, on light, 240
mechanization of the world, 216
Mediterranean, length, 167, 169, 170 (fig.
    I.1)
Mersenne, Marin, mathematician, 226
Metrodorus of Chios, quoted, 284
midrash, 114
Miletus, 57, 75
Miller, Hugh, on Creation, 260
Milton, John, 25
mishna, 5
Moerbeke, William of, translator, 97
molecula, 227
monotheism, 28
monsoon, 165
Moon: brought down by magic, 51; its
    distance, 245; eclipsed at Syracuse, 44,
    observed by Galileo, 205
Morrison, Philip, physicist, 285
Moses, 29
motion: defined by Aristotle, 103; natural
    and violent, 103, 192
mover, unmoved, 104; Prime, 105
multiverse, 286
Mutakallimun, 149

Nephilim, 28f
Nestorius, bishop, 176
Newton, Isaac, 217–33; heretical views,
    218; on light, 239; *Principia
    Mathematica*, 219; on scientific
    explanation, 225; speculation on atoms,
    233; work in alchemy and chemistry,
    231f
Nicias, Greek general, 44
Nicomachus, mathematician, 135
number, 62, 85; properties, 64, 135
Nürnberg Chronicle, 131

Oannes, 8
Oceanus, 11, 14, 162
Ockham, William of, philosopher,
    288
Odoric de Pordenone, explorer, 176
omens, 40, 44
oracles, Sibylline, 122, 137
Origen, 119

Osiander, Andreas, 195
Ostwald, Wilhelm, chemist, 238
Ouranos, 8,
Ovid, 20, 22
ozone layer, 293

Pascal, Blaise, quoted, 292
Pandora, 33
papyrus rolls, 56
Paracelsus, 229
parallax, 246
Parmenides, 70f, 89
Parthenon, proportions of, 67
Pepys, Samuel, 65
perfect numbers, 64, 135
Perrin, Jean, physicist, 239
Phaethon, 22
pharmakis, 49
Philoponus, John, philosopher, 110
Philosopher's Stone, 229f
Plato, 88–95; "On the Good," 94, 294;
    theory of Ideas, 90; theory of matter, 93;
    Timaeus, 90–95
Pliny (the Elder), 43, 51; on medicine, 52,
    137n, 166, 173
pneuma, 109
Polo, Marco, 177
Prester John, 176–80
Priestly, Joseph, analyzes air, 234
Proclus, on Pythagoras, 85
Ptolemaeus, Claudius (Ptolemy), 128;
    Almagest, 128; *Geography*, 159, 167;
    on planetary influences, 37; his spheres,
    130, 193, 195, 201; *Tetrabiblos*, 132
Pythagoras, 63ff, 85; on man as
    microcosm, 156; on music theory, 66
Pytheas, 165

quintessence, 229, 280

Rangi and Papa, 9
rapture, 124
Rees, Martin, astronomer, 286, 294
relativity: early (special) theory, 270;
    general theory, 271
Rheinhold, Erasmus, astronomer, 196
Rheticus, Georg, assistant to Copernicus,
    193
Richer, Jean, astronomer, 246
Roesbroeck, Willem, explorer, 176
Rome, sacked by Visigoths, 144

Rta, 84
Rudolph II, emperor, 200; dies, 209

Satan, 30, 34, 138
Schedel, Dr. Hartmann, encyclopedist, 131
Schele, Carl, analyzes air, 233
Schiller, Johann von, quoted, 301
Seneca, 20, 120; on eternity, 300
Shapley, Harlow, astronomer, 270
Sheol, 5
shield: of Achilles, 13; of Heracles, 297
silk, 165f
Solinus, fabulist, 173
slavery, in Athens, 79
Slipher, Vesto, astronomer, 271
Smith, George, epigraphist, 18
Solomon, King, quoted, 245
Solon, 22
Socrates, 89; quoted, 70
solids, Platonic, 94
Sophocles, quoted, 78, 299
souls: evolution of, 259; transmigration of, 92
Southern Continent, 188
space: according to Aristotle, 110; to Pythagoreans, 109
spacetime, 271; curved, 271
space travel, 285
Steno, Nicholaus, on formation of fossils, 252
Stewart, Balfour and Peter Tait, *The Unseen Universe*, 141
Stoicism, 45
Strabo, quoted, 44, 165
Sun: distance, 246; extinction, 291; source of its light, 287
Syracuse, besieged, 44

Tacitus, 137, 164
Tait, Peter. *See* Stewart, Balfour
Tao, 85
technē, 98

telescope, invented, 205
Thales, 57ff
Theophrastus, quoted, 43
Thoreau, Henry David, quoted, 108
Thucydices, quoted, 44
Timaeus, 90–95
time: defined by Aristotle, 108; in *Timaeus*, 108

universe: age, 274; cyclic, 78; expanding, 271; inflationary, 274; scale, 270f
up and down, 164, 168
Utnapishtim, 12

vacuum: exists, 77; does not exist, 110
Valéry, Paul, quoted, 269
Vinci, Leonardo da, on fossils, 252
Visigoths, 144f
Voltaire, on Descartes, 216
vortices, Cartesian, 213; demolished by Newton, 219

Wallace, Alfred Russel, naturalist, 263
warming, global, 289, 293
water, world supply, 294
Weinberg, Steven, quoted, 106
Wien, Wilhelm, on light, 242
Wisdom, 86
Wolfgang, Saint, church, 67
Wordsworth, William, quoted, 234
Wright, Thomas, model of the Galaxy, 248, 283

Xenocrates, 136

Young, Dr. Thomas: on molecular size, 237; on spiritual beings, 241
yuga, 118

Zeno of Citium, 45
Zeus, 27
zodiacal man, 156